T0353475

Fractional Evolution
Equations and Inclusions

Fractional Evolution Equations and Inclusions: Analysis and Control

Yong Zhou

Xiangtan University, P. R. China

AMSTERDAM • BOSTON • HEIDELBERG • LONDON
NEW YORK • OXFORD • PARIS • SAN DIEGO
SAN FRANCISCO • SINGAPORE • SYDNEY • TOKYO
Academic Press is an imprint of Elsevier

Academic Press is an imprint of Elsevier
32 Jamestown Road, London NW1 7BY, UK
525 B Street, Suite 1800, San Diego, CA 92101-4495, USA
225 Wyman Street, Waltham, MA 02451, USA
The Boulevard, Langford Lane, Kidlington, Oxford OX5 1GB, UK

Notices
Knowledge and best practice in this field are constantly changing. As new research and experience
broaden our understanding, changes in research methods, professional practices, or medical treatment
may become necessary.

Practitioners and researchers must always rely on their own experience and knowledge in evaluating
and using any information, methods, compounds, or experiments described herein. In using such
information or methods they should be mindful of their own safety and the safety of others, including
parties for whom they have a professional responsibility.

To the fullest extent of the law, neither the Publisher nor the authors, contributors, or editors, assume
any liability for any injury and/or damage to persons or property as a matter of products liability,
negligence or otherwise, or from any use or operation of any methods, products, instructions, or ideas
contained in the material herein.

ISBN: 978-0-12-804277-9

British Library Cataloguing in Publication Data
A catalogue record for this book is available from the British Library

Library of Congress Cataloging-in-Publication Data
A catalog record for this book is available from the Library of Congress

For information on all Academic Press publications
visit our website at http://store.elsevier.com/

Working together
to grow libraries in
developing countries

www.elsevier.com • www.bookaid.org

TABLE OF CONTENTS

PREFACE

Fractional evolution equations provide a unifying framework in order to investigate well-posedness of complex systems of various types describing the time evolution of concrete systems (such as time-fractional diffusion equations). Fractional evolution inclusions are a kind of important differential inclusions describing the processes behaving in a much more complex way on time, which appear as a generalization of fractional evolution equations through the application of multivalued analysis. Comparing with the classical integer order case, the researches on theory of fractional order evolution equations and inclusions are only on their initial stage of development.

A strong motivation for investigating this class of equations and inclusions comes mainly from two compelling reasons: differential models with the fractional derivative providing an excellent instrument for the description of memory and hereditary properties have recently been proved valuable tools in the modeling of many physical phenomena. The fractional order models of real systems are always more adequate than the classical integer order models, since the description of some systems is more accurate when the fractional derivative is used. The advantages of fractional derivative becomes evident in modeling mechanical and electrical properties of real materials, description of rheological properties of rocks, and various other fields. Such models are interesting not only for engineers and physicists but also for pure mathematicians. Another of the reasons is that a lot of phenomena investigated in hybrid systems with dry friction, processes of controlled heat transfer, obstacle problems, and others can be described with the help of various differential inclusions, both linear and nonlinear. The theory of differential inclusions is highly developed and constitutes an important branch of nonlinear analysis.

This monograph is devoted to a rapidly developing area of the research for fractional evolution equations and inclusions and their applications to control theory. The monograph is divided into five chapters.

In order to make the book self-contained, we devote the first chapter to a description of general information on fractional calculus, semigroups, space of functions, weak topology, multivalued maps, and stochastic process.

In the second chapter, we first study Cauchy problems for fractional evolution equations with almost sectorial operators. The suitable mild solutions of fractional Cauchy problems with Riemann-Liouville derivative and Caputo derivative are introduced. We give existence results of mild solutions in the cases that the almost sectorial operator is compact and noncompact, respectively. In the second part of this chapter, we discuss the existence and uniqueness of the bounded solutions on real axis for fractional evolution equations with the Liouville fractional derivative of order $\alpha \in (0, 1)$ with the lower limit $-\infty$. Applying Fourier transform we give the reasonable definitions of mild solutions. Then some sufficient conditions are established for the existence and uniqueness of periodic solutions, S-asymptotically periodic solutions, and other types of bounded solutions.

The third chapter deals with fractional evolution inclusions with Hille-Yosida operators. A definition of integral solutions for fractional differential inclusions is given. Further, the topological structure of solution sets is investigated. It is shown that the solution set is nonempty, compact, and, moreover, an R_δ-set. We apply these results to some control problems.

In the fourth chapter, we investigate control problems for systems governed by fractional evolution equations. First, we study the optimal control for fractional finite time delay evolution systems in infinite dimensional spaces. Next, we investigate optimal feedback controls of fractional evolution systems via a compact semigroup in Banach spaces. The existence of feasible pairs is proved. An existence result of optimal control pairs for the Lagrange problem is also presented. In addition, we investigate a class of Sobolev type fractional evolution systems in a separable Banach space. The controllability and approximate controllability results are established. Last but not least, we discuss the topological structure of solution sets for the control problems of fractional delay evolution equations. The information about the structure is employed to show the invariance of a reachability set of the control problem under nonlinear perturbations.

The fifth chapter is devoted mainly to the investigation of fractional stochastic evolution inclusions in Hilbert spaces. First, we give the existence results by means of weak topology. Then the topological structure of solution sets is investigated.

The materials in this monograph are based on the research work carried out by myself and collaborators during the past four years. In particular, we are interested in the existence theory and topological structure of solution sets for fractional evolution inclusions and control systems. Here we are able to present some new results and this represents the main difference between our work and other monographs completely or partially devoted to the theory of fractional differential equations (e.g., Miller and Ross (1993), Podlubny (1999), Kilbas, Srivastava and Trujillo (2006), Lakshmikantham, Leela and Devi (2009), Diethelm (2010), Tarasov (2010) and Zhou (2014)). The monograph provides the necessary background material required to go further into the subject and explore the rich research literature. Each chapter concludes with a section devoted to notes and bibliographical remarks and all abstract results are illustrated by examples.

I would like to thank Professors B. Ahmad, O.P. Agrawal, D. Baleanu, M. Benchohra, M. Fečkan, F. Liu, J.A.T. Machado, V.E. Tarasov, and J.J. Trujillo for their support. Special thanks to collaborators, Professors J.R. Wang, and R.N. Wang, for their support and help. I also wish to express my appreciation to my colleagues, Professors W. Jiang, H.R. Sun, X.F. Zhou, J. Mu, and my graduate students, L. Zhang, L. Peng, and H.B. Gu, for their help. Finally, I thank the editorial assistance of Elsevier Limited.

I acknowledge with gratitude the support of National Natural Science Foundation of China (10971173) and the Specialized Research Fund for the Doctoral Program of Higher Education (20114301110001).

Yong Zhou
Xiangtan University, P. R. China

CHAPTER 1

Preliminaries

Abstract In this chapter, we introduce some basic facts on fractional calculus, semigroups, space of functions, weak topology, multivalued analysis, and stochastic process which are needed throughout this monograph.

Keywords Caputo fractional derivatives, Riemann-Liouville fractional derivatives, Mittag-Leffler function, Wright function, Asymptotically periodic functions, C_0-semigroup, Almost sectorial operators, Weak topology, Multivalued maps, Measure of noncompactness, R_δ-set, Random variables, Brownian motion, Stochastic integral and differential.

1.1 BASIC FACTS AND NOTATION

As usual $\mathbb{N}^+$ denotes the set of positive integer numbers and $\mathbb{N}_0$ the set of nonnegative integer numbers. $\mathbb{R}$ denotes the real line, $\mathbb{R}_+$ denotes the set of nonnegative reals, and $\mathbb{R}^+$ denotes the set of positive reals. Let $\mathbb{C}$ be the set of complex numbers.

We recall that a vector space X equipped with a norm $|\cdot|$ is called a normed vector space. A subset Ω of a normed space X is said to be bounded if there exists a number K such that $|x| \leq K$ for all $x \in \Omega$. A subset Ω of a normed vector space X is called convex if for any $x, y \in \Omega$, $ax + (1-a)y \in \Omega$ for all $a \in [0, 1]$.

A sequence $\{x_n\}$ in a normed vector space X is said to converge to the vector x in X if and only if the sequence $|x_n - x|$ converges to zero as $n \to \infty$. A sequence $\{x_n\}$ in a normed vector space X is called a Cauchy sequence if for every $\varepsilon > 0$ there exists an $N = N(\varepsilon)$ such that for all $n, m \geq N(\varepsilon)$, $|x_n - x_m| < \varepsilon$. Clearly a convergent sequence is also a Cauchy sequence, but the converse may not be true. A space X where every Cauchy sequence of elements of X converges to an element of X is called a complete space. A complete normed vector space is said to be a Banach space.

Let Ω be a subset of a Banach space X. A point $x \in X$ is said to be a limit point of Ω if there exists a sequence of vectors in Ω which converges to

Fractional Evolution Equations and Inclusions: Analysis and Control. http://dx.doi.org/10.1016/B978-0-12-804277-9.50001-8

x. We say a subset Ω is closed if Ω contains all of its limit points. The union of Ω and its limit points is called the closure of Ω and will be denoted by $\overline{\Omega}$. Let X, Y be normed vector spaces and Ω be a subset of X. An operator $\mathscr{T} : \Omega \to Y$ is continuous at a point $x \in \Omega$ if and only if for any $\varepsilon > 0$ there is a $\delta > 0$ such that $|\mathscr{T}x - \mathscr{T}y| < \varepsilon$ for all $y \in \Omega$ such that $|x - y| < \delta$. Further, $\mathscr{T}$ is continuous on Ω, or simply continuous, if it is continuous at all points of Ω.

Let $J = [a, b]$ $(-\infty < a < b < \infty)$ be a finite interval of $\mathbb{R}$. We assume that X is a Banach space with the norm $|\cdot|$. Denote $C(J, X)$ by the Banach space of all continuous functions from J into X with the norm

$$\|x\| = \sup_{t \in J} |x(t)|,$$

where $x \in C(J, X)$. $C^n(J, X)$ $(n \in \mathbb{N}_0)$ denotes the set of mappings having n times continuously differentiable on J. Let $AC(J, X)$ be the space of functions which are absolutely continuous on J and $AC^n(J, X)$ $(n \in \mathbb{N}_0)$ be the space of functions f such that $f \in C^{n-1}(J, X)$ and $f^{(n-1)} \in AC(J, X)$. In particular, $AC^1(J, X) = AC(J, X)$.

Let $1 \le p \le \infty$. We denote by $L^p(J, X)$ the set of those Lebesgue measurable functions $f : J \to X$ for which $\|f\|_{L^p J} < \infty$, where

$$\|f\|_{L^p J} = \begin{cases} \left(\int_J |f(t)|^p dt \right)^{\frac{1}{p}}, & 1 \le p < \infty, \\ \operatorname*{ess\,sup}_{t \in J} |f(t)|, & p = \infty. \end{cases}$$

In particular, $L^1(J, X)$ is the Banach space of measurable functions $f : J \to X$ with the norm

$$\|f\|_{LJ} = \int_J |f(t)| dt,$$

and $L^\infty(J, X)$ is the Banach space of measurable functions $f : J \to X$ which are bounded, equipped with the norm

$$\|f\|_{L^\infty J} = \inf\{c > 0 : |f(t)| \le c, \text{ a.e. } t \in J\}.$$

Lemma 1.1. *(Hölder inequality) Assume that $p, q \ge 1$ and $\frac{1}{p} + \frac{1}{q} = 1$. If $f \in L^p(J, \mathbb{R}), g \in L^q(J, \mathbb{R})$, then $fg \in L^1(J, \mathbb{R})$ and*

$$\|fg\|_{LJ} \le \|f\|_{L^p J} \|g\|_{L^q J}.$$

We say that a subset Ω of a Banach space X is compact if every sequence of vectors in Ω contains a subsequence which converges to a vector in Ω. We say that Ω is relatively compact if every sequence of vectors in Ω contains a subsequence which converges to a vector in X, i.e., Ω is relatively compact if $\overline{\Omega}$ is compact.

Lemma 1.2. *(Arzela-Ascoli's theorem) If a family* $F = \{f(t)\}$ *in* $C(J,X)$ *is uniformly bounded and equicontinuous on* J, *and for any* $t^* \in J$, $\{f(t^*)\}$ *is relatively compact, then* F *has a uniformly convergent subsequence* $\{f_n(t)\}_{n=1}^{\infty}$.

Remark 1.1.

(i) *If a family* $F = \{f(t)\}$ *in* $C(J, \mathbb{R})$ *is uniformly bounded and equicontinuous on* J, *then* F *has a uniformly convergent subsequence* $\{f_n(t)\}_{n=1}^{\infty}$.
(ii) *Arzela-Ascoli's theorem is the key to the following result: A subset* F *in* $C(J, \mathbb{R})$ *is relatively compact if and only if it is uniformly bounded and equicontinuous on* J.

Theorem 1.1. *(Lebesgue's dominated convergence theorem) Let* E *be a measurable set and let* $\{f_n\}$ *be a sequence of measurable functions such that* $\lim_{n \to \infty} f_n(x) = f(x)$ *a.e. in* E, *and for every* $n \in \mathbb{N}^+$, $|f_n(x)| \leq g(x)$ *a.e. in* E, *where* g *is integrable on* E. *Then*

$$\lim_{n \to \infty} \int_E f_n(x)dx = \int_E f(x)dx.$$

Theorem 1.2. *(Bochner's theorem) A measurable function* $f : (a,b) \to X$ *is Bochner integrable if* $|f|$ *is Lebesgue integrable.*

Finally, we introduce some general concepts which will be used in the following chapters.

An inner product on a complex vector space X is a mapping $(\cdot, \cdot) : X \times X \to \mathbb{C}$ such that for all $x, y, z \in X$ and all $\lambda \in \mathbb{C}$:

(i) $(x, y) = \overline{(y, x)}$;
(ii) $(\lambda x, y) = \lambda(x, y)$;
(iii) $(x + y, z) = (x, z) + (y, z)$;
(iv) $(x, x) > 0$, when $x > 0$.

An inner product space is a pair $(X, (\cdot, \cdot))$, where X is a complex vector space and $(\cdot, \cdot)$ is an inner product on X. A Hilbert space is an inner product space which is a complete metric space with respect to the metric induced by its inner product.

Let X be a Banach space. By a cone $K \subset X$, we understand a closed convex subset K such that $\lambda K \subset K$ for all $\lambda \geq 0$ and $K \cap (-K) = \{0\}$. We define a partial ordering $\leq$ with respect to K by $x \leq y$ if and only if $y - x \in K$. Then

(i) K is called positive if the element $x \in K$ is positive;
(ii) K is regeneration if $K - K = X$, and total if $\overline{K - K} = X$;
(iii) K is called normal if $\inf\{|x + y| : x, y \in K \cap \partial B_1(0)\} > 0$.

1.2 FRACTIONAL INTEGRALS AND DERIVATIVES

A number of definitions for fractional derivative have emerged over the years, we refer the reader to Diethelm [81], Hilfer [119], Kilbas et al. [134], Lakshmikantham et al. [144], Miller and Ross [184], Podlubny [202], and Tarasov [221]. In this book, we restrict our attention to the use of the Riemann-Liouville and Caputo fractional derivatives. In this section, we introduce some basic definitions and properties of fractional integrals and fractional derivatives which are used further in this book. The material in this section is taken from Kilbas et al. [134].

The gamma function $\Gamma(z)$ is defined by

$$\Gamma(z) = \int_0^\infty t^{z-1} e^{-t} dt, \quad Re(z) > 0,$$

where $t^{z-1} = e^{(z-1)\log(t)}$. This integral is convergent for all complex $z \in \mathbb{C}$ $(Re(z) > 0)$.

For this function the reduction formula

$$\Gamma(z+1) = z\Gamma(z), \quad Re(z) > 0$$

holds. In particular, if $z = n \in \mathbb{N}_0$, then

$$\Gamma(n+1) = n!, \quad n \in \mathbb{N}_0$$

with (as usual) $0! = 1$.

Let us consider some of the starting points for a discussion of fractional calculus (see [119]). One development begins with a generalization of repeated integration. Thus if f is locally integrable on (c, ∞), then the n-fold iterated integral is given by

$$_cD_t^{-n}f(t) = \int_c^t ds_1 \int_c^{s_1} ds_2 \cdots \int_c^{s_{n-1}} f(s_n)ds_n$$

$$= \frac{1}{(n-1)!} \int_c^t (t-s)^{n-1}f(s)ds$$

for almost all t with $-\infty \le c < t < \infty$ and $n \in \mathbb{N}^+$. Writing $(n-1)! = \Gamma(n)$, an immediate generalization is the integral of f of fractional order $\alpha > 0$:

$$_cD_t^{-\alpha}f(t) = \frac{1}{\Gamma(\alpha)} \int_c^t (t-s)^{\alpha-1}f(s)ds \quad \text{(right hand)}$$

and similarly for $-\infty < t < d \le \infty$

$$_tD_d^{-\alpha}f(t) = \frac{1}{\Gamma(\alpha)} \int_t^d (s-t)^{\alpha-1}f(s)ds \quad \text{(left hand)},$$

both being defined for suitable f.

1.2.1 Definitions

Definition 1.1. *(Left and right Riemann-Liouville fractional integrals) Let $J = [a, b]$ ($-\infty < a < b < \infty$) be a finite interval of $\mathbb{R}$. The left and right Riemann-Liouville fractional integrals $_aD_t^{-\alpha}f$ and $_tD_b^{-\alpha}f$ of order $\alpha \in \mathbb{R}^+$ are defined by*

$$_aD_t^{-\alpha}f(t) = \frac{1}{\Gamma(\alpha)} \int_a^t (t-s)^{\alpha-1}f(s)ds, \quad t > a, \; \alpha > 0 \qquad (1.1)$$

and

$$_tD_b^{-\alpha}f(t) = \frac{1}{\Gamma(\alpha)} \int_t^b (s-t)^{\alpha-1}f(s)ds, \quad t < b, \; \alpha > 0, \qquad (1.2)$$

respectively, provided the right-hand sides are pointwise defined on $[a, b]$. When $\alpha = n \in \mathbb{N}^+$, definitions (1.1) and (1.2) coincide with the nth *integrals of the form*

$$_aD_t^{-n}f(t) = \frac{1}{(n-1)!}\int_a^t (t-s)^{n-1}f(s)ds$$

and

$$_tD_b^{-n}f(t) = \frac{1}{(n-1)!}\int_t^b (s-t)^{n-1}f(s)ds.$$

Definition 1.2. *(Left and right Riemann-Liouville fractional derivatives) The left and right Riemann-Liouville fractional derivatives $_aD_t^\alpha f$ and $_tD_b^\alpha f$ of order $\alpha \in \mathbb{R}^+$ are defined by*

$$_aD_t^\alpha f(t) = \frac{d^n}{dt^n}\,_aD_t^{-(n-\alpha)}f(t)$$

$$= \frac{1}{\Gamma(n-\alpha)}\frac{d^n}{dt^n}\left(\int_a^t (t-s)^{n-\alpha-1}f(s)ds\right), \quad t > a$$

and

$$_tD_b^\alpha f(t) = (-1)^n\frac{d^n}{dt^n}\,_tD_b^{-(n-\alpha)}f(t)$$

$$= \frac{1}{\Gamma(n-\alpha)}(-1)^n\frac{d^n}{dt^n}\left(\int_t^b (s-t)^{n-\alpha-1}f(s)ds\right), \quad t < b,$$

respectively, where $n = [\alpha] + 1$, $[\alpha]$ means the integral part of α. In particular, when $\alpha = n \in \mathbb{N}_0$, then

$$_aD_t^0 f(t) = \,_tD_b^0 f(t) = f(t),$$

$$_aD_t^n f(t) = f^{(n)}(t) \quad and \quad _tD_b^n f(t) = (-1)^n f^{(n)}(t),$$

where $f^{(n)}(t)$ is the usual derivative of $f(t)$ of order n. If $0 < \alpha < 1$, then

$$_aD_t^\alpha f(t) = \frac{1}{\Gamma(1-\alpha)}\frac{d}{dt}\left(\int_a^t (t-s)^{-\alpha}f(s)ds\right), \quad t > a$$

and

$$_tD_b^\alpha f(t) = -\frac{1}{\Gamma(1-\alpha)}\frac{d}{dt}\left(\int_t^b (s-t)^{-\alpha}f(s)ds\right), \quad t < b.$$

Remark 1.2. *If* $f \in C([a, b], \mathbb{R}^N)$, *it is obvious that Riemann-Liouville fractional integral of order* $\alpha > 0$ *exists on* $[a, b]$. *On the other hand, following Kilbas et al. [134], we know that Riemann-Liouville fractional derivative of order* $\alpha \in [n - 1, n)$ *exists almost everywhere on* $[a, b]$ *if* $f \in AC^n([a, b], \mathbb{R}^N)$.

The left and right Caputo fractional derivatives are defined via above Riemann-Liouville fractional derivatives.

Definition 1.3. *(Left and right Caputo fractional derivatives) The left and right Caputo fractional derivatives* ${}_a^C D_t^\alpha f(t)$ *and* ${}_t^C D_b^\alpha f(t)$ *of order* $\alpha \in \mathbb{R}^+$ *are defined by*

$$
{}_a^C D_t^\alpha f(t) = {}_a D_t^\alpha \left(f(t) - \sum_{k=0}^{n-1} \frac{f^{(k)}(a)}{k!} (t - a)^k \right)
$$

and

$$
{}_t^C D_b^\alpha f(t) = {}_t D_b^\alpha \left(f(t) - \sum_{k=0}^{n-1} \frac{f^{(k)}(b)}{k!} (b - t)^k \right),
$$

respectively, where

$$
n = [\alpha] + 1 \ \text{for} \ \alpha \notin \mathbb{N}_0; \ n = \alpha \ \text{for} \ \alpha \in \mathbb{N}_0. \tag{1.3}
$$

In particular, when $0 < \alpha < 1$, then

$$
{}_a^C D_t^\alpha f(t) = {}_a D_t^\alpha \big(f(t) - f(a) \big)
$$

and

$$
{}_t^C D_b^\alpha f(t) = {}_t D_b^\alpha \big(f(t) - f(b) \big).
$$

Riemann-Liouville fractional derivative and Caputo fractional derivative are connected with each other by the following relations.

Property 1.1.

(i) *If* $\alpha \notin \mathbb{N}_0$ *and* $f(t)$ *is a function for which Caputo fractional derivatives* ${}_a^C D_t^\alpha f(t)$ *and* ${}_t^C D_b^\alpha f(t)$ *of order* $\alpha \in \mathbb{R}^+$ *exist together with the Riemann-*

Liouville fractional derivatives $_aD_t^\alpha f(t)$ *and* $_tD_b^\alpha f(t)$, *then*

$$_a^C D_t^\alpha f(t) = {}_aD_t^\alpha f(t) - \sum_{k=0}^{n-1} \frac{f^{(k)}(a)}{\Gamma(k-\alpha+1)}(t-a)^{k-\alpha}$$

and

$$_t^C D_b^\alpha f(t) = {}_tD_b^\alpha f(t) - \sum_{k=0}^{n-1} \frac{f^{(k)}(b)}{\Gamma(k-\alpha+1)}(b-t)^{k-\alpha},$$

where $n = [\alpha] + 1$. *In particular, when* $0 < \alpha < 1$, *we have*

$$_a^C D_t^\alpha f(t) = {}_aD_t^\alpha f(t) - \frac{f(a)}{\Gamma(1-\alpha)}(t-a)^{-\alpha}$$

and

$$_t^C D_b^\alpha f(t) = {}_tD_b^\alpha f(t) - \frac{f(b)}{\Gamma(1-\alpha)}(b-t)^{-\alpha}.$$

(ii) *If* $\alpha = n \in \mathbb{N}_0$ *and the usual derivative* $f^{(n)}(t)$ *of order* n *exists, then* $_a^C D_t^n f$ *and* $_t^C D_b^n f$ *are represented by*

$$_a^C D_t^n f(t) = f^{(n)}(t) \text{ and } {}_t^C D_b^n f(t) = (-1)^{(n)} f^{(n)}(t). \tag{1.4}$$

Property 1.2. *Let* $\alpha \in \mathbb{R}^+$ *and let* n *be given by (1.3). If* $f(t) \in AC^n([a,b], \mathbb{R}^N)$, *then Caputo fractional derivatives* $_a^C D_t^\alpha f(t)$ *and* $_t^C D_b^\alpha f(t)$ *exist almost everywhere on* $[a,b]$.

(i) *If* $\alpha \notin \mathbb{N}_0$, $_a^C D_t^\alpha f(t)$ *and* $_t^C D_b^\alpha f(t)$ *are represented by*

$$_a^C D_t^\alpha f(t) = \frac{1}{\Gamma(n-\alpha)}\left(\int_a^t (t-s)^{n-\alpha-1} f^{(n)}(s)ds\right) =: {}_a^C\mathbf{D}_t^\alpha f(t)$$

and

$$_t^C D_b^\alpha f(t) = \frac{(-1)^n}{\Gamma(n-\alpha)}\left(\int_t^b (s-t)^{n-\alpha-1} f^{(n)}(s)ds\right) =: {}_t^C\mathbf{D}_b^\alpha f(t),$$

respectively, where $n = [\alpha] + 1$.
   In particular, when $0 < \alpha < 1$ *and* $f(t) \in AC([a,b], \mathbb{R}^N)$,

$$_a^C D_t^\alpha f(t) = \frac{1}{\Gamma(1-\alpha)}\left(\int_a^t (t-s)^{-\alpha} f'(s)ds\right) =: {}_a^C\mathbf{D}_t^\alpha f(t) \tag{1.5}$$

and

$$_t^C D_b^\alpha f(t) = -\frac{1}{\Gamma(1-\alpha)} \left(\int_t^b (s-t)^{-\alpha} f'(s) ds \right) =: _t^C \mathbf{D}_b^\alpha f(t). \quad (1.6)$$

(ii) *If* $\alpha = n \in \mathbb{N}_0$, *then* $_a^C D_t^\alpha f(t)$ *and* $_t^C D_b^\alpha f(t)$ *are represented by (1.4). In particular,*

$$_a^C D_t^0 f(t) = _t^C D_b^0 f(t) = f(t).$$

Remark 1.3. *If* f *is an abstract function with values in Banach space* X, *then integrals which appear in above definitions are taken in Bochner's sense.*

The fractional integrals and derivatives, defined on a finite interval $[a, b]$ of $\mathbb{R}$, are naturally extended to whole axis $\mathbb{R}$.

Definition 1.4. *(Left and right Liouville fractional integrals on the real axis) The left and right Liouville-Weyl fractional integrals* $_{-\infty} D_t^{-\alpha} f(t)$ *and* $_t D_{+\infty}^{-\alpha} f(t)$ *of order* $\alpha > 0$ *on the whole axis* $\mathbb{R}$ *are defined by*

$$_{-\infty} D_t^{-\alpha} f(t) = \frac{1}{\Gamma(\alpha)} \int_{-\infty}^t (t-s)^{\alpha-1} f(s) ds \quad (1.7)$$

and

$$_t D_{+\infty}^{-\alpha} f(t) = \frac{1}{\Gamma(\alpha)} \int_t^\infty (s-t)^{\alpha-1} f(s) ds, \quad (1.8)$$

respectively, where $t \in \mathbb{R}$ *and* $\alpha > 0$.

Definition 1.5. *(Left and right Liouville fractional derivatives on the real axis) The left and right Liouville-Weyl fractional derivatives* $_{-\infty} D_t^\alpha f(t)$ *and* $_t D_{+\infty}^\alpha f(t)$ *of order* α *on the whole axis* $\mathbb{R}$ *are defined by*

$$\begin{aligned} _{-\infty} D_t^\alpha f(t) &= \frac{d^n}{dt^n} \left(_{-\infty} D_t^{-(n-\alpha)} f(t) \right) \\ &= \frac{1}{\Gamma(n-\alpha)} \frac{d^n}{dt^n} \left(\int_{-\infty}^t (t-s)^{n-\alpha-1} f(s) ds \right) \end{aligned} \quad (1.9)$$

and

$$
\begin{aligned}
{}_t D^{\alpha}_{+\infty} f(t) &= (-1)^n \frac{d^n}{dt^n} \left({}_t D^{-(n-\alpha)}_{+\infty} f(t) \right) \\
&= \frac{(-1)^n}{\Gamma(n-\alpha)} \frac{d^n}{dt^n} \left(\int_t^{\infty} (s-t)^{n-\alpha-1} f(s) ds \right),
\end{aligned}
\tag{1.10}
$$

respectively, where $n = [\alpha] + 1$, $\alpha \geq 0$ *and* $t \in \mathbb{R}$.

In particular, when $\alpha = n \in \mathbb{N}_0$, then

$$
{}_{-\infty} D^0_t f(t) = {}_t D^0_{+\infty} f(t) = f(t),
$$

$$
{}_{-\infty} D^n_t f(t) = f^{(n)}(t) \quad \text{and} \quad {}_t D^n_{+\infty} f(t) = (-1)^n f^{(n)}(t),
$$

where $f^{(n)}(t)$ is the usual derivative of $f(t)$ of order n. If $0 < \alpha < 1$ and $t \in \mathbb{R}$, then

$$
\begin{aligned}
{}_{-\infty} D^{\alpha}_t f(t) &= \frac{1}{\Gamma(1-\alpha)} \frac{d}{dt} \left(\int_{-\infty}^t (t-s)^{-\alpha} f(s) ds \right) \\
&= \frac{\alpha}{\Gamma(1-\alpha)} \int_0^{\infty} \frac{f(t) - f(t-s)}{s^{\alpha+1}} ds
\end{aligned}
$$

and

$$
\begin{aligned}
{}_t D^{\alpha}_{+\infty} f(t) &= -\frac{1}{\Gamma(1-\alpha)} \frac{d}{dt} \left(\int_t^{\infty} (s-t)^{-\alpha} f(s) ds \right) \\
&= \frac{\alpha}{\Gamma(1-\alpha)} \int_0^{\infty} \frac{f(t) - f(t+s)}{s^{\alpha+1}} ds.
\end{aligned}
$$

Definition 1.6. *(Left and right Caputo fractional derivatives on the real axis) The left and right Caputo fractional derivatives* ${}_{-\infty}^C D^{\alpha}_t f(t)$ *and* ${}_t^C D^{\alpha}_{+\infty} f(t)$ *of order* α *(with* $\alpha > 0$ *and* $\alpha \notin \mathbb{N}$*) on the whole axis* $\mathbb{R}$ *are defined by*

$$
{}_{-\infty}^C D^{\alpha}_t f(t) = \frac{1}{\Gamma(n-\alpha)} \left(\int_{-\infty}^t (t-s)^{n-\alpha-1} f^{(n)}(s) ds \right)
\tag{1.11}
$$

and

$$
{}_t^C D^{\alpha}_{+\infty} f(t) = \frac{(-1)^n}{\Gamma(n-\alpha)} \left(\int_t^{\infty} (s-t)^{n-\alpha-1} f^{(n)}(s) ds \right),
\tag{1.12}
$$

respectively.

When $0 < \alpha < 1$, *relations (1.11) and (1.12) take the following forms:*

$$_{-\infty}^{C}\mathbf{D}_t^{\alpha} f(t) = \frac{1}{\Gamma(1-\alpha)} \left(\int_{-\infty}^{t} (t-s)^{-\alpha} f'(s) ds \right)$$

and

$$_{t}^{C}\mathbf{D}_{+\infty}^{\alpha} f(t) = -\frac{1}{\Gamma(1-\alpha)} \left(\int_{t}^{\infty} (s-t)^{-\alpha} f'(s) ds \right).$$

1.2.2 Properties

We present here some properties of the fractional integral and derivative operators that will be useful throughout this book.

Property 1.3. *If $\alpha \geq 0$ and $\beta > 0$, then*

$$_{a}D_t^{-\alpha}(t-a)^{\beta-1} = \frac{\Gamma(\beta)}{\Gamma(\beta+\alpha)}(t-a)^{\beta+\alpha-1}, \quad \alpha > 0,$$

$$_{a}D_t^{\alpha}(t-a)^{\beta-1} = \frac{\Gamma(\beta)}{\Gamma(\beta-\alpha)}(t-a)^{\beta-\alpha-1}, \quad \alpha \geq 0$$

and

$$_{t}D_b^{-\alpha}(b-t)^{\beta-1} = \frac{\Gamma(\beta)}{\Gamma(\beta+\alpha)}(b-t)^{\beta+\alpha-1}, \quad \alpha > 0,$$

$$_{t}D_b^{\alpha}(b-t)^{\beta-1} = \frac{\Gamma(\beta)}{\Gamma(\beta-\alpha)}(b-t)^{\beta-\alpha-1}, \quad \alpha \geq 0.$$

In particular, if $\beta = 1$ and $\alpha \geq 0$, then the Riemann-Liouville fractional derivatives of a constant are, in general, not equal to zero:

$$_{a}D_t^{\alpha}1 = \frac{(t-a)^{-\alpha}}{\Gamma(1-\alpha)}, \quad _{t}D_b^{\alpha}1 = \frac{(b-t)^{-\alpha}}{\Gamma(1-\alpha)}, \quad 0 < \alpha < 1.$$

On the other hand, for $j = 1, 2, ..., [\alpha] + 1$,

$$_{a}D_t^{\alpha}(t-a)^{\alpha-j} = 0, \quad _{t}D_b^{\alpha}(b-t)^{\alpha-j} = 0.$$

The semigroup properties of the fractional integration operators ${}_aD_t^{-\alpha}$ and ${}_tD_b^{-\alpha}$ are given by the following results.

Property 1.4. *If $\alpha > 0$ and $\beta > 0$, then the equations*

$$
\begin{aligned}
{}_aD_t^{-\alpha}\left({}_aD_t^{-\beta}f(t)\right) &= {}_aD_t^{-\alpha-\beta}f(t), \\
{}_tD_b^{-\alpha}\left({}_tD_b^{-\alpha}f(t)\right) &= {}_tD_b^{-\alpha-\beta}f(t)
\end{aligned}
\tag{1.13}
$$

are satisfied at almost every point $t \in [a, b]$ for $f(t) \in L^p([a, b], \mathbb{R}^N)$ $(1 \le p < \infty)$. If $\alpha + \beta > 1$, then the relations in (1.13) hold at any point of $[a, b]$.

Property 1.5.

(i) *If $\alpha > 0$ and $f(t) \in L^p([a, b], \mathbb{R}^N)$ $(1 \le p \le \infty)$, then the following equalities*

$$
{}_aD_t^{\alpha}\left({}_aD_t^{-\alpha}f(t)\right) = f(t) \quad \text{and} \quad {}_tD_b^{\alpha}\left({}_tD_b^{-\alpha}f(t)\right) = f(t)
$$

hold almost everywhere on $[a, b]$.

(ii) *If $\alpha > \beta > 0$, then, for $f(t) \in L^p([a, b], \mathbb{R}^N)$ $(1 \le p \le \infty)$, the relations*

$$
{}_aD_t^{\beta}\left({}_aD_t^{-\alpha}f(t)\right) = {}_aD_t^{-\alpha+\beta}f(t)
$$

and

$$
{}_tD_b^{\beta}\left({}_tD_b^{-\alpha}f(t)\right) = {}_tD_b^{-\alpha+\beta}f(t)
$$

hold almost everywhere on $[a, b]$.

In particular, when $\beta = k \in \mathbb{N}^+$ and $\alpha > k$, then

$$
{}_aD_t^{k}\left({}_aD_t^{-\alpha}f(t)\right) = {}_aD_t^{-\alpha+k}f(t)
$$

and

$$
{}_tD_b^{k}\left({}_tD_b^{-\alpha}f(t)\right) = (-1)^k {}_tD_b^{-\alpha+k}f(t).
$$

To present the next property, we use the spaces of functions $_aD_t^{-\alpha}(L^p)$ and $_tD_b^{-\alpha}(L^p)$ defined for $\alpha > 0$ and $1 \leq p \leq \infty$ by

$$_aD_t^{-\alpha}(L^p) = \{f : f = {_aD_t^{-\alpha}}\varphi, \; \varphi \in L^p([a,b], \mathbb{R}^N)\}$$

and

$$_tD_b^{-\alpha}(L^p) = \{f : f = {_tD_b^{-\alpha}}\phi, \; \phi \in L^p([a,b], \mathbb{R}^N)\},$$

respectively. The composition of the fractional integration operator $_aD_t^{-\alpha}$ with the fractional differentiation operator $_aD_t^{\alpha}$ is given by the following result.

Property 1.6. *Set $\alpha > 0$, $n = [\alpha] + 1$. Let $f_{n-\alpha}(x) = {_aD_t^{-(n-\alpha)}}f(t)$ be the fractional integral (1.1) of order $n - \alpha$.*

(i) *If $1 \leq p \leq \infty$ and $f(t) \in {_aD_t^{-\alpha}}(L^p)$, then*

$$_aD_t^{-\alpha}\left(_aD_t^{\alpha}f(t)\right) = f(t).$$

(ii) *If $f(t) \in L^1([a,b], \mathbb{R}^N)$ and $f_{n-\alpha}(t) \in AC^n([a,b], \mathbb{R}^N)$, then the equality*

$$_aD_t^{-\alpha}\left(_aD_t^{\alpha}f(t)\right) = f(t) - \sum_{j=1}^{n} \frac{f_{n-\alpha}^{(n-j)}(a)}{\Gamma(\alpha - j + 1)}(t - a)^{\alpha - j}$$

holds almost everywhere on $[a,b]$.

Property 1.7. *Let $\alpha > 0$ and $n = [\alpha] + 1$. Also let $g_{n-\alpha}(t) = {_tD_b^{\alpha-n}}g(t)$ be the fractional integral (1.2) of order $n - \alpha$.*

(i) *If $1 \leq p \leq \infty$ and $g(t) \in {_tD_b^{-\alpha}}(L^p)$, then*

$$_tD_b^{-\alpha}\left(_tD_b^{\alpha}g(t)\right) = g(t).$$

(ii) *If $g(t) \in L^1(a,b)$ and $g_{n-\alpha}(t) \in AC^n([a,b], \mathbb{R}^N)$, then the equality*

$$_tD_b^{-\alpha}\left(_tD_b^{\alpha}g(t)\right) = g(t) - \sum_{j=1}^{n} \frac{(-1)^{n-j}g_{n-\alpha}^{(n-j)}(a)}{\Gamma(\alpha - j + 1)}(b - t)^{\alpha - j}$$

holds almost everywhere on $[a,b]$.

In particular, if $0 < \alpha < 1$, then

$$_tD_b^{-\alpha}\left(_tD_b^\alpha g(t)\right) = g(t) - \frac{g_{1-\alpha}(a)}{\Gamma(\alpha)}(b-t)^{\alpha-1},$$

where $g_{1-\alpha}(t) = {_tD_b^{\alpha-1}}g(t)$ while for $\alpha = n \in \mathbb{N}^+$, the following equality holds:

$$_tD_b^{-n}\left(_tD_b^n g(t)\right) = g(t) - \sum_{k=0}^{n-1}\frac{(-1)^k g^{(k)}(a)}{k!}(b-t)^k.$$

Property 1.8. *Let* $\alpha > 0$ *and let* $y(t) \in L^\infty([a,b], \mathbb{R}^N)$ *or* $y(t) \in C([a,b], \mathbb{R}^N)$. *Then*

$$_a^C D_t^\alpha\left(_aD_t^{-\alpha}y(t)\right) = y(t) \text{ and } _t^C D_b^\alpha\left(_tD_b^{-\alpha}y(t)\right) = y(t).$$

Property 1.9. *Let* $\alpha > 0$ *and let* n *be given by (1.3). If* $y(t) \in AC^n([a,b], \mathbb{R}^N)$ *or* $y(t) \in C^n([a,b], \mathbb{R}^N)$, *then*

$$_aD_t^{-\alpha}\left(_a^C D_t^\alpha y(t)\right) = y(t) - \sum_{k=0}^{n-1}\frac{y^{(k)}(a)}{k!}(t-a)^k$$

and

$$_tD_b^{-\alpha}\left(_t^C D_b^\alpha y(t)\right) = y(t) - \sum_{k=0}^{n-1}\frac{(-1)^k y^{(k)}(b)}{k!}(b-t)^k.$$

In particular, if $0 < \alpha \le 1$ and $y(t) \in AC([a,b], \mathbb{R}^N)$ or $y(t) \in C([a,b], \mathbb{R}^N)$, then

$$_aD_t^{-\alpha}\left(_a^C D_t^\alpha y(t)\right) = y(t) - y(a) \text{ and } _tD_b^{-\alpha}\left(_t^C D_b^\alpha y(t)\right) = y(t) - y(b).$$

Lemma 1.3. *[222]*

(i) *Let* $\xi, \eta \in \mathbb{R}$ *such that* $\eta > -1$. *If* $t > 0$, *then*

$$_0D_t^{-\xi}\frac{t^\eta}{\Gamma(\eta+1)} = \begin{cases} \dfrac{t^{\xi+\eta}}{\Gamma(\xi+\eta+1)}, & \text{if } \xi+\eta \ne -n, \\ 0, & \text{if } \xi+\eta = -n, \end{cases} \quad (n \in \mathbb{N}^+).$$

(ii) *Let $\xi > 0$ and $\varphi \in L((0, a), X)$. Define*

$$G_\xi(t) = {}_0D_t^{-\xi}\varphi, \quad \text{for } t \in (0, a),$$

then

$${}_0D_t^{-\eta}G_\xi(t) = {}_0D_t^{-(\xi+\eta)}\varphi(t), \quad \eta > 0, \quad \text{almost all } t \in [0, a].$$

At the end of this subsection, we present some properties of two special functions.

Definition 1.7. *[119, 202] The generalized Mittag-Leffler special function $E_{\alpha,\beta}$ is defined by*

$$E_{\alpha,\beta}(z) = \sum_{k=0}^{\infty} \frac{z^k}{\Gamma(\alpha k + \beta)} = \frac{1}{2\pi i}\int_\Upsilon \frac{\lambda^{\alpha-\beta}e^\lambda}{\lambda^\alpha - z}d\lambda, \quad \alpha, \beta > 0, z \in \mathbb{C},$$

where Υ is a contour which starts and ends as $-\infty$ and encircles the disc $|\lambda| \leq |z|^{1/\alpha}$ counterclockwise.

If $0 < \alpha < 1$, $\beta > 0$, then the asymptotic expansion of $E_{\alpha,\beta}$ as $z \to \infty$ is given by

$$E_{\alpha,\beta}(z) = \begin{cases} \dfrac{1}{\alpha}z^{(1-\beta)/\alpha}\exp(z^{1/\alpha}) + \varepsilon_{\alpha,\beta}(z), & |\arg z| \leq \dfrac{1}{2}\alpha\pi, \\[2ex] \varepsilon_{\alpha,\beta}(z), & |\arg(-z)| < \left(1 - \dfrac{1}{2}\alpha\right)\pi, \end{cases}$$

where

$$\varepsilon_{\alpha,\beta}(z) = -\sum_{n=1}^{N-1}\frac{z^{-n}}{\Gamma(\beta - \alpha n)} + O(|z|^{-N}), \quad \text{as } z \to \infty.$$

For short, set

$$E_\alpha(z) = E_{\alpha,1}(z), \quad e_\alpha(z) = E_{\alpha,\alpha}(z).$$

Then Mittag-Leffler have the following properties.

Property 1.10. *For $\alpha \in (0, 1)$ and $t \in \mathbb{R}$,*

(i) $E_\alpha(t), e_\alpha(t) > 0$;

(ii) $(E_\alpha(t))' = \frac{1}{\alpha}e_\alpha(t)$;

(iii) $\lim\limits_{t \to -\infty} E_\alpha(t) = \lim\limits_{t \to -\infty} e_\alpha(t) = 0$;

(iv) $^C_0D^\alpha_t E(\omega t^\alpha) = \omega E(\omega t^\alpha)$, $_0D^{\alpha-1}_t(t^{\alpha-1}e_\alpha(\omega t^\alpha)) = E_\alpha(\omega t^\alpha)$, $\omega \in \mathbb{C}$.

Definition 1.8. *[175] The Wright function Ψ_α is defined by*

$$\Psi_\alpha(\theta) = \sum_{n=0}^{\infty} \frac{(-\theta)^n}{n!\Gamma(-\alpha n + 1 - \alpha)}$$

$$= \frac{1}{\pi}\sum_{n=1}^{\infty} \frac{(-\theta)^n}{(n-1)!}\Gamma(n\alpha)\sin(n\pi\alpha), \quad \theta \in \mathbb{C}$$

with $0 < \alpha < 1$.

Remark 1.4. *If $\theta \in \mathbb{R}^+$, then*

$$\Psi_\alpha(\theta) = \frac{1}{\pi\alpha}\sum_{n=1}^{\infty}(-\theta)^{n-1}\frac{\Gamma(1+\alpha n)}{n!}\sin(n\pi\alpha), \quad \alpha \in (0,1).$$

Property 1.11. *[175]*

(i) $\Psi_\alpha(t) \geq 0$, $\quad t \in (0, \infty)$;

(ii) $\displaystyle\int_0^\infty \frac{\alpha}{t^{\alpha+1}}\Psi_\alpha(t^{-\alpha})e^{-\lambda t}dt = e^{-\lambda^\alpha}$, $\quad Re(\lambda) \geq 0$;

(iii) $\displaystyle\int_0^\infty \Psi_\alpha(t)t^r dt = \frac{\Gamma(1+r)}{\Gamma(1+\alpha r)}$, $\quad r \in (-1, \infty)$;

(iv) $\displaystyle\int_0^\infty \Psi_\alpha(t)e^{-zt}dt = E_\alpha(-z)$, $\quad z \in \mathbb{C}$;

(v) $\displaystyle\int_0^\infty \alpha t\Psi_\alpha(t)e^{-zt}dt = e_\alpha(-z)$, $\quad z \in \mathbb{C}$.

1.3 SEMIGROUPS AND ALMOST SECTORIAL OPERATORS

1.3.1 C_0-semigroup

Let us recall the definitions and properties of operator semigroups, for details see Banasiak et al. [34] and Pazy [199].

Let X be a Banach space and $\mathscr{L}(X)$ be the Banach space of linear bounded operators.

Definition 1.9. *A one parameter family* $\{T(t)\}_{t\geq 0} \subset \mathscr{L}(X)$ *is a semigroup of bounded linear operators on X if*

(i) $T(t)T(s) = T(t + s)$, *for* $t, s \geq 0$;
(ii) $T(0) = I$; *here, I denotes the identity operator in X.*

Definition 1.10. *A semigroup of bounded linear operators* $\{T(t)\}_{t\geq 0}$ *is uniformly continuous if*

$$\lim_{t\to 0+} \|T(t) - I\| = 0.$$

From the definition it is clear that if $\{T(t)\}_{t\geq 0}$ is a uniformly continuous semigroup of bounded linear operators, then

$$\lim_{s\to t} \|T(s) - T(t)\| = 0.$$

Definition 1.11. *We say that the semigroup* $\{T(t)\}_{t\geq 0}$ *is strongly continuous (or a C_0-semigroup) if the map $t \to T(t)x$ is strongly continuous, for each $x \in X$, i.e.,*

$$\lim_{t\to 0+} T(t)x = x, \ \forall \, x \in X.$$

Definition 1.12. *Let* $\{T(t)\}_{t\geq 0}$ *be a C_0-semigroup defined on X. The linear operator A is the infinitesimal generator of $\{T(t)\}_{t\geq 0}$ defined by*

$$Ax = \lim_{t\to 0+} \frac{T(t)x - x}{t}, \ \text{for } x \in D(A),$$

where $D(A) = \left\{ x \in X : \lim_{t\to 0+} \frac{T(t)x-x}{t} \ \text{exists in } X \right\}$.

If there are $M \geq 0$ and $\nu \in \mathbb{R}$ such that $\|T(t)\| \leq Me^{\nu t}$, then

$$(\lambda I - A)^{-1}x = \int_0^\infty e^{-\lambda t}T(t)x\,dt, \quad Re(\lambda) > \nu, \ x \in X. \tag{1.14}$$

A C_0-semigroup $\{T(t)\}_{t\geq 0}$ is called exponentially stable if there exist constants $M > 0$ and $\delta > 0$ such that

$$\|T(t)\| \leq Me^{-\delta t}, \quad t \geq 0. \tag{1.15}$$

The growth bound ν_0 of $\{T(t)\}_{t\geq 0}$ is defined by

$$\nu_0 = \inf\{\delta \in \mathbb{R} : \text{ there exists } M_\delta > 0 \\ \text{ such that } \|T(t)\| \leq M_\delta e^{\delta t}, \forall\, t \geq 0\}. \tag{1.16}$$

Furthermore, ν_0 can also be obtained by the following formula:

$$\nu_0 = \limsup_{t \to +\infty} \frac{\ln \|T(t)\|}{t}. \tag{1.17}$$

Definition 1.13. *A C_0-semigroup $\{T(t)\}_{t\geq 0}$ is called uniformly bounded if there exists a constant $M > 0$ such that*

$$\|T(t)\| \leq M, \quad t \geq 0. \tag{1.18}$$

Definition 1.14. *A C_0-semigroup $\{T(t)\}_{t\geq 0}$ is called compact if $T(t)$ is compact for $t > 0$.*

Property 1.12. *If $\{T(t)\}_{t\geq 0}$ is compact, then $\{T(t)\}$ is equicontinuous for $t > 0$.*

Definition 1.15. *A C_0-semigroup $\{T(t)\}_{t\geq 0}$ is called positive if $T(t)x \geq \theta$ for all $x \geq \theta$ and $t \geq 0$.*

1.3.2 Almost Sectorial Operators

We first introduce some special functions and classes of functions which will be used in the following, for more details, we refer to Markus [177] and Periago and Straub [200].

Let $-1 < p < 0$ and S_μ^0 with $0 < \mu < \pi$ be the open sector

$$\{z \in \mathbb{C}\backslash\{0\} : |\arg z| < \mu\}$$

and S_μ be its closure, that is,

$$S_\mu = \{z \in \mathbb{C}\backslash\{0\} : |\arg z| \leq \mu\} \cup \{0\}.$$

Set

$$\mathcal{F}_0^\gamma(S_\mu^0) = \bigcup_{s<0} \mathcal{G}_s^\gamma(S_\mu^0) \bigcup \mathcal{G}_0(S_\mu^0),$$
$$\mathcal{F}(S_\mu^0) = \{f \in \mathcal{H}(S_\mu^0) : \text{there exist } k, n \in \mathbb{N} \text{ such that } f\psi_n^k \in \mathcal{F}_0(S_\mu^0)\},$$

where

$$\mathcal{H}(S_\mu^0) = \{f : S_\mu^0 \mapsto \mathbb{C} : f \text{ is holomorphic}\},$$
$$\mathcal{H}^\infty(S_\mu^0) = \{f \in \mathcal{H}(S_\mu^0) : f \text{ is bounded}\},$$
$$\varphi_0(z) = \frac{1}{1+z}, \quad \psi_n(z) = \frac{z}{(1+z)^n}, \quad z \in \mathbb{C}\backslash\{-1\}, n \in \mathbb{N} \cup \{0\},$$
$$\mathcal{G}_0(S_\mu^0) = \left\{f \in \mathcal{H}(S_\mu^0) : \sup_{z \in S_\mu^0}\left|\frac{f(z)}{\varphi_0(z)}\right| < \infty\right\},$$

and for each $s < 0$,

$$\mathcal{G}_s^\gamma(S_\mu^0) := \left\{f \in \mathcal{H}(S_\mu^0) : \sup_{z \in S_\mu^0} |\psi_n^s(z)f(z)| < \infty\right\},$$

where n is the smallest integer such that $n \geq 2$ and $\gamma + 1 < -(n-1)s$.

Since for each $\beta \in \mathbb{C}$, $z^\beta \in \mathcal{F}(S_\mu^0)$ ($z \in \mathbb{C} \backslash (-\infty, 0]$, $0 < \mu < \pi$), one can define, via the triple $(\mathcal{F}_0^\gamma(S_\mu^0), \mathcal{F}(S_\mu^0), f(A))$, the complex powers of A which are closed by

$$A^\beta = z^\beta(A), \quad \beta \in \mathbb{C}.$$

However, in difference to the case of sectorial operators, having $0 \in \rho(A)$ does not imply that the complex powers $A^{-\beta}$ with $Re(\beta) > 0$ are bounded. The operator $A^{-\beta}$ belongs to $\mathscr{L}(X)$ whenever $Re(\beta) > 1 + \gamma$. So, in this situation, the linear space $X^\beta = D(A^\beta)$, $\beta > 1 + \gamma$, endowed with the graph norm $|x|_\beta = |A^\beta x|$, $x \in X^\beta$, is a Banach space.

We state the concept of almost sectorial operators as follows.

Definition 1.16. *[55] Let* $-1 < p < 0$ *and* $0 < \omega < \frac{\pi}{2}$*. By* $\Theta_\omega^p(X)$ *we denote the family of all linear closed operators* $A : D(A) \subset X \to X$ *which satisfy*

(i) $\sigma(A) \subset S_\omega = \{z \in \mathbb{C} \setminus \{0\} : |\arg z| \leq \omega\} \cup \{0\}$;
(ii) *for every* $\omega < \mu < \pi$*, there exists a constant* C_μ *such that*

$$\|R(z; A)\| \leq C_\mu |z|^p, \ \text{for all } z \in \mathbb{C} \setminus S_\mu,$$

where the family $R(z; A) = (zI - A)^{-1}$, $z \in \rho(A)$*, of bounded linear operators is the resolvent of* A*.*

A linear operator A *will be called an almost sectorial operator on* X *if* $A \in \Theta_\omega^p(X)$*.*

Remark 1.5. *Let* $A \in \Theta_\omega^p(X)$*. Then the definition implies that* $0 \in \rho(A)$*.*

We denote the semigroup associated with A by $\{Q(t)\}_{t \geq 0}$. For $t \in S_{\frac{\pi}{2} - \omega}^0$,

$$Q(t) = e^{-tz}(A) = \frac{1}{2\pi i} \int_{\Gamma_\theta} e^{-tz} R(z; A) dz,$$

where the integral contour $\Gamma_\theta = \{\mathbb{R}^+ e^{i\theta}\} \cup \{\mathbb{R}^+ e^{-i\theta}\}$ is oriented counterclockwise and $\omega < \theta < \mu < \frac{\pi}{2} - |\arg t|$ forms an analytic semigroup of growth order $1 + p$.

Remark 1.6. *From Periago and Straub [200], note that if* $A \in \Theta_\omega^p(X)$*, then* A *generates a semigroup* $Q(t)$ *with a singular behavior at* $t = 0$ *in a sense, called semigroup of growth* $1 + p$*. Moreover, the semigroup* $Q(t)$ *is analytic in an open sector of the complex plane* $\mathbb{C}$*, but the strong continuity fails at* $t = 0$ *for data which are not sufficiently smooth.*

Property 1.13. *[200] Let* $A \in \Theta_\omega^p(X)$ *with* $-1 < p < 0$ *and* $0 < \omega < \frac{\pi}{2}$*. Then the following properties remain true:*

(i) $Q(t)$ *is analytic in* $S_{\frac{\pi}{2} - \omega}^0$ *and* $\frac{d^n}{dt^n} Q(t) = (-A)^n Q(t) \ (t \in S_{\frac{\pi}{2} - \omega}^0)$;
(ii) *the functional equation* $Q(s + t) = Q(s)Q(t)$ *for all* $s, t \in S_{\frac{\pi}{2} - \omega}^0$ *holds*;
(iii) *there is a constant* $c_0 = c_0(p) > 0$ *such that* $\|Q(t)\| \leq c_0 t^{-p-1} (t > 0)$;
(iv) *if* $\beta > 1 + p$*, then* $D(A^\beta) \subset \Sigma_Q = \{x \in X : \lim_{t \to 0+} Q(t)x = x\}$;

(v) $R(\lambda, A) = \int_0^\infty e^{-\lambda t} Q(t) dt$ for every $\lambda \in \mathbb{C}$ with $Re(\lambda) > 0$.

1.4 SPACES OF ASYMPTOTICALLY PERIODIC FUNCTIONS

This subsection is devoted to some preliminary facts needed in the sequel. Let X be an ordered Banach space with norm $|\cdot|$ and partial order $\leq$, whose positive cone $P = \{y \in X : y \geq \theta\}$ (θ is the zero element of X) is normal with normal constant N. The notation $C_b(X)$ stands for the Banach space of all bounded continuous functions from $\mathbb{R}$ into X equipped with the sup-norm $\|\cdot\|_\infty$, that is,

$$C_b(X) = \{x : \mathbb{R} \to X : x \text{ is continuous, and } \|x\|_\infty = \sup_{t \in \mathbb{R}} |x(t)|\}.$$

For $x, y \in C_b(X)$, $x \leq y$ if $x(t) \leq y(t)$ for all $t \in \mathbb{R}$. $P_\omega(X)$ stands for the subspace of $C_b(X)$ consisting of all X-valued continuous ω-periodic functions. Set

$$SAP_\omega(X) = \{f \in C_b(X) : \exists\, \omega > 0, |f(t + \omega) - f(t)| \to 0 \text{ as } t \to \infty\}.$$

The class of functions in $SAP_\omega(X)$ is called S-asymptotically ω-periodic (see [116] for a discussion of qualitative properties of this class of functions). We note that $P_\omega(X)$ and $SAP_\omega(X)$ are Banach spaces (see [116]), and $P_\omega(X) \subset SAP_\omega(X)$.

We recall that a function $f \in C_b(\mathbb{X})$ is said to be almost periodic (in the sense of Bohr) if for any $\epsilon > 0$, it is possible to find a real number $\omega = \omega(\epsilon) > 0$ for any interval of length $\omega(\varepsilon)$, there exists a number $\tau = \tau(\varepsilon)$ in this interval such that $\|f(t + \tau) - f(t)\|_\infty < \epsilon$ for all $t \in \mathbb{R}$. We denote by $AP(X)$ the set of all these functions. The space of almost automorphic functions (resp., compact almost automorphic functions) will be denoted by $AA(X)$ (resp., $AA_c(X)$). Recall that function $f \in C_b(X)$ belongs to $AA(X)$ (resp., $AA_c(X)$) if and only if for all sequence $\{s'_n\}_{n \in \mathbb{N}}$ of real numbers there exists a subsequence $\{s_n\}_{n \in \mathbb{N}} \subset \{s'_n\}_{n \in \mathbb{N}}$ such that $g(t) := \lim_{n \to \infty} f(t + s_n)$ and $f(t) = \lim_{n \to \infty} g(t - s_n)$ for each $t \in \mathbb{R}$ (resp., uniformly on compact subsets of $\mathbb{R}$). Clearly the function g above is continuous on $\mathbb{R}$. Now we consider the set $C_0(X) = \{f \in C_b(X) : \lim_{|t| \to \infty} |f(t)| = 0\}$ and define the space of asymptotically periodic functions as $AP_\omega(X) = P_\omega(X) \oplus C_0(X)$. Analogously, we define the space of asymptotically almost periodic functions,

$$AAP(X) = AP(X) \oplus C_0(X),$$

the space of asymptotically compact almost automorphic functions,

$$AAA_c(X) = AA_c(X) \oplus C_0(X),$$

and the space of asymptotically almost automorphic functions,

$$AAA(X) = AA(X) \oplus C_0(X).$$

Next, we consider the set

$$P_0(X) = \left\{ f \in C_b(X) : \lim_{T \to \infty} \frac{1}{2T} \int_{-T}^{T} |f(s)| ds = 0 \right\},$$

and define the following classes of spaces: the space of pseudo-periodic functions,

$$PP_\omega(X) = P_\omega(X) \oplus P_0(X),$$

the space of pseudo-almost periodic functions,

$$PAP(X) = AP(X) \oplus P_0(X),$$

the space of pseudo-compact almost automorphic functions,

$$PAA_c(X) = AA_c(X) \oplus P_0(X),$$

and the space of pseudo-almost automorphic functions,

$$PAA(X) = AA(X) \oplus P_0(X).$$

We have the following diagram that summarizes the relation of the different classes of subspaces defined previously (see [163]):

$$
\begin{array}{ccccc}
AA(X) & \subset & AAA(X) & \subset & PAA(X) \\
\cup & & \cup & & \cup \\
AA_c(X) & \subset & AAA_c(X) & \subset & PAA_c(X) \\
\cup & & \cup & & \cup \\
AP(X) & \subset & AAP(X) & \subset & PAP(X) \\
\cup & & \cup & & \cup \\
P_\omega(X) & \subset & AP_\omega(X) & \subset & PP_\omega(X) \\
& & \cap & & \\
& & SAP_\omega(X) & &
\end{array}
$$

Denote by $\mathcal{M}(\mathbb{R}, X)$ or simply $\mathcal{M}(X)$ the following function spaces

$$
\begin{aligned}
\mathcal{M}(X) = \{ & P_\omega(X), AP(X), AA_c(X), AA(X), AP_\omega(X), \\
& AAP(X), AAA_c(X), AAA(X), PP_\omega(X), \\
& PAP(X), PAA_c(X), PAA(X), SAP_\omega(X) \}.
\end{aligned}
$$

We define the set $\mathcal{M}(\mathbb{R} \times X, X)$ which consists of all functions $f : \mathbb{R} \times X \to X$ such that $f(\cdot, x) \in \mathcal{M}(X)$ uniformly for each $x \in K$, where K is any bounded subset of X. If $\Omega(X) \in \mathcal{M}(X) \cup \{C_0(X), P_0(X)\}$ is fixed, then, given $x \in \Omega(X)$ and $f \in C_b(\mathbb{R} \times X, X)$, sufficient conditions to ensure that $f(\cdot, x(\cdot))$ belongs to $\Omega(X)$ should be chosen among the following:

(A_1) $f(t, \cdot)$ is uniformly continuous with respect to t on $\mathbb{R}$ for each bounded subset of X. More precisely, given $\varepsilon > 0$ and $K \subset X$, there exists $\delta > 0$ such that $x, y \in K$ and $|x - y| < \delta$ imply that $|f(t, x) - f(t, y)| < \varepsilon$;

(A_2) $\{f(t, x) : t \in \mathbb{R}, x \in K\}$ is bounded for all bounded subset and $K \subset X$;

(A_3) if $f = f_1 + f_2$, where $f_1 \in \{C_0(X), P_0\} \setminus \{0\}$, $f_2 \in \Phi(X) = \{P_\omega(X), AP(X), AA_c(X), AA(X)\}$, and $f_2(t, \cdot)$ is uniformly continuous with respect to t on $\mathbb{R}$ for each bounded subset of X.

Then the following results hold (see [163]):

$\Omega(X)$	(A_1)	(A_2)	(A_3)
P_w			
AP			
AA_c	•		
AA	•		
AP_ω	•		
AAP	•		
AAA_c	•		•
AAA	•		•
PP_ω		•	•
PAP		•	•
PAA_c	•	•	•
PAA	•	•	•
SAP_ω	•		
C_0	•		
P_0	•		

$$(1.19)$$

Corollary 1.1. *[163] Let $\Omega(X) \in \mathcal{M}(X)$ and $f \in \Omega(\mathbb{R} \times X, X) \in \mathcal{M}(\mathbb{R} \times X, X)$ be given and fixed. Assume that there exists a constant $L_f > 0$ such that*

$$|f(t,x) - f(t,y)| \leq L_f|x - y|$$

for all $t \in \mathbb{R}$ and $x, y \in X$. Let $x \in \Omega(X)$, then $f(\cdot, x(\cdot)) \in \Omega(X)$.

1.5 WEAK COMPACTNESS OF SETS AND OPERATORS

Let X be a real Banach space with norm $|\cdot|$ and X^* be its topological dual, i.e., the vector space of all linear continuous functionals from X to $\mathbb{R}$, which, endowed with the dual norm $\|f\| = \sup_{|x| \leq 1} |\langle f, x \rangle|$, for $f \in X^*$, is, in its turn, a real Banach space too, here $\langle \cdot, \cdot \rangle$ denotes duality product. Thereafter, if $x \in X$ and $f \in X^*$, $\langle f, x \rangle$ denotes $f(x)$. We denote by $\mathrm{Fin}(X^*)$ the class of all finite subsets in X^*. Let $F \in \mathrm{Fin}(X^*)$. Then, the function $|\cdot|_F : X \to \mathbb{R}$, defined by

$$|x|_F = \max\{|\langle f, x \rangle| : f \in F\}$$

for each $x \in X$, is a seminorm on X.

The family of seminorms $\{|\cdot|_F : F \in \text{Fin}(X^*)\}$ denotes the so-called weak topology and X, endowed with this topology, denoted by X_w, is a locally convex topological vector space (see [54]).

A subset A of a Banach space X is called weakly closed if it is closed in the weak topology. The symbol $\overline{D}^w$ denotes the weak closure of D.

We will say that $\{x_n\} \subset X$ converges weakly to $x_0 \in X$, and we write $x_n \rightharpoonup x_0$, if for each $f \in X^*$, $f(x) \to f(x_0)$. We recall (see [43]) that a sequence $\{x_n\} \subset C([0, b], X)$ weakly converges to an element $x \in C([0, b], X)$ if and only if

(i) there exists $N > 0$ such that, for every $n \in \mathbb{N}^+$ and $t \in [0, b]$, $|x_n(t)| \leq N$;

(ii) for every $t \in [0, b]$, $x_n(t) \rightharpoonup x(t)$.

Definition 1.17.

(i) *A subset A of a normed space X is said to be (relatively) weakly compact if (the weak closure of) A is compact in the weak topology of X.*

(ii) *A subset A of a Banach space X is weakly sequentially compact if any sequence in A has a subsequence which converges weakly to an element of X.*

Definition 1.18. *Suppose that X and Y are Banach spaces. A linear operator T from X into Y is weakly compact if $T(B)$ is a relatively weakly compact subset of Y whenever B is a bounded subset of X.*

We mention also two results that are contained in the so-called Eberlein-Smulian theory.

Theorem 1.3. *[132] Let Ω be a subset of a Banach space X. The following statements are equivalent:*

(i) *Ω is relatively weakly compact;*
(ii) *Ω is relatively weakly sequentially compact.*

Corollary 1.2. *[132] Let Ω be a subset of a Banach space X. The following statements are equivalent:*

(i) *Ω is weakly compact;*

(ii) Ω *is weakly sequentially compact.*

We recall Krein-Smulian theorem and Pettis measurability theorem.

Theorem 1.4. *[84] The convex hull of a weakly compact set in a Banach space X is weakly compact.*

Theorem 1.5. *[201] Let (E, Σ) be a measure space, X be a separable Banach space. Then a function $f : E \to X$ is measurable if and only if for every $x^* \in X^*$ the function $x^* \circ f : E \to \mathbb{R}$ is measurable with respect to Σ and the Borel σ-algebra in $\mathbb{R}$.*

We recall that a bounded subset in a reflexive Banach space is relatively weakly compact.

Lemma 1.4. *[280] Let X be reflexive and $1 < p < \infty$. A subset $K \subset L^p([0, b], X)$ is relatively weakly sequentially compact in $L^p([0, b], X)$ if and only if K is bounded in $L^p([0, b], X)$.*

1.6 MULTIVALUED ANALYSIS

1.6.1 Multivalued Maps

Multivalued maps play a significant role in the description of processes in control theory since the presence of controls provides an intrinsic multivalence in the evolution of the system. In this subsection, we introduce some general properties on multivalued maps. The material in this subsection is taken from Kamenskii et al. [131].

Let Y and Z be metric spaces. $P(Y)$ stands for the collection of all nonempty subsets of Y. As usual, we denote

$P_{cl}(Y) = \{D \in P(Y), \text{ closed }\}$;
$P_{cp}(Y) = \{D \in P(Y), \text{ compact }\}$;
$P_{cl,cv}(Y) = \{D \in P(Y), \text{ closed and convex }\}$;
$P_{cp,cv}(Y) = \{D \in P(Y), \text{ compact and convex }\}$;
co(D) (resp., $\overline{\text{co}}$(D)) be the convex hull (resp., convex closed hull in D) of a subset D.

A multivalued map φ of Y into Z is a correspondence which associates to every $y \in Y$ a nonempty subset $\varphi(y) \subseteq Z$, called the value of y. We write this correspondence as $\varphi : Y \to P(Z)$. If $D \subseteq Y$, then the set $\varphi(D) = \bigcup_{y \in D} \varphi(y)$ is called the image of D under φ. The set $\mathrm{Gra}(\varphi) \subseteq Y \times Z$, defined by $\mathrm{Gra}(\varphi) = \{(y, z) : y \in Y, z \in \varphi(y)\}$, is the graph of φ.

Definition 1.19. *Let* $\varphi : Y \to P(Z)$ *be a multivalued map and* D *be a subset of* Z. *The complete preimage* $\varphi^{-1}(D)$ *of a set* D *is the set*

$$\varphi^{-1}(D) = \{y \in Y : \varphi(y) \cap D \neq \emptyset\}.$$

Definition 1.20. *A multivalued map* $\varphi : Y \to P(Z)$ *is said to be*

(i) *closed if its graph* $\mathrm{Gra}(\varphi)$ *is closed subset of the space* $Y \times Z$;
(ii) *upper semicontinuous (shortly, u.s.c.) if the set* $\varphi^{-1}(D)$ *is closed for every closed set* $D \subset Z$.

Definition 1.21. *Let* D *be nonempty subset of a Banach space* Y *and* $\varphi : D \to P(Y)$ *be a multivalued map:*

(i) φ *is said to have weakly sequentially closed graph if for every sequence* $\{x_n\} \subset D$ *with* $x_n \rightharpoonup x$ *in* D *and for every sequence* $\{x_n\}$ *with* $y_n \in \varphi(x_n)$, $\forall n \in \mathbb{N}$, $y_n \rightharpoonup y$ *in* Y *implies* $y \in \varphi(x)$;
(ii) φ *is called weakly upper semicontinuous (shortly, weakly u.s.c.) if* $\varphi^{-1}(A)$ *is closed for all weakly closed* $A \subset Y$;
(iii) β *is* ϵ-δ *u.s.c. if for every* $w_0 \in Y$ *and* $\epsilon > 0$ *there exists* $\delta > 0$ *such that* $\beta(y) \subset \beta(w_0) + B_\epsilon(0)$ *for all* $y \in B_\delta(w_0) \cap D$.

Lemma 1.5. *[44] Let* $\varphi : D \subset Y \to P(Z)$ *be a multivalued map with weakly compact values. Then*

(i) φ *is weakly u.s.c. if* φ *is* ϵ-δ *u.s.c., and*
(ii) *suppose further that* φ *has convex values and* Z *is reflexive. Then* φ *is weakly u.s.c. if and only if* $\{x_n\} \subset D$ *with* $x_n \to x_0 \in D$ *and* $y_n \in \varphi(x_n)$ *implies* $y_n \rightharpoonup y_0 \in \varphi(x_0)$, *up to a subsequence.*

Definition 1.22. *A multivalued map* $\varphi : Y \to P(Z)$ *is*

(i) *compact if its range* $\varphi(Y)$ *is relatively compact in* Z, *i.e.,* $\overline{\varphi(Y)}$ *is compact in* Z;

(ii) *locally compact if every point* $y \in Y$ *has a neighborhood* $V(y)$ *such that the restriction of* φ *to* $V(y)$ *is compact;*

(iii) *quasicompact if* $\varphi(D)$ *is relatively compact for each compact set* $D \subset Y.$

It is clear that (i) $\Longrightarrow$ (ii) $\Longrightarrow$ (iii). The following facts will be used.

Lemma 1.6. *Let* Y *be a topological spaces,* Z *a regular topological space and* $\varphi : Y \to P_{cl}(Z)$ *an u.s.c. multivalued map. Then* φ *is closed.*

The inverse relation between u.s.c. maps and closed ones is expressed in the following lemma.

Lemma 1.7. *Let* Y *and* Z *be metric spaces and* $\varphi : Y \to P_{cp}(Z)$ *a closed quasicompact multivalued map. Then* φ *is u.s.c.*

Let us consider some properties of closed and u.s.c. multivalued map.

Lemma 1.8. *Let* $\varphi : Y \to P_{cl}(Z)$ *be a closed multivalued map. If* $D \subset Y$ *is a compact set then its image* $\varphi(D)$ *is a closed subset of* $Z.$

Lemma 1.9. *Let* $\varphi : Y \to P_{cp}(Z)$ *be an u.s.c. multivalued map. If* $D \subset Y$ *is a compact set then its image* $\varphi(D)$ *is a compact subset of* $Z.$

Lemma 1.10. *Let* Y *and* Z *be Banach space, and let the multivalued map* $\varphi : [0, b] \times Y \to P_{cp}(Z)$ *be such that*

(i) *for every* $x \in Y$ *the multifunction* $\varphi(\cdot, x) : [0, b] \to P_{cp}(Z)$ *has a strongly measurable selection;*

(ii) *for a.e.* $t \in [0, b]$ *the multivalued map* $\varphi(t, \cdot) : Y \to P_{cp}(Z)$ *is u.s.c.*

Then for every strongly measurable function $q : [0, b] \to Y$ *there exists a strongly measurable selection* $g : [0, b] \to Z$ *of the multifunction* $G : [0, b] \to P_{cp}(Z),$ $G(t) = \varphi(t, q(t)).$

Theorem 1.6. *[196] Let* X *be a metrizable locally convex linear topological space and let* D *be a weakly compact, convex subset of* $X.$ *Suppose*

$\varphi : D \rightarrow P_{cl,cv}(D)$ *has weakly sequentially closed graph. Then* φ *has a fixed point.*

1.6.2 Measure of Noncompactness

We recall here some definitions and properties of measure of noncompactness and condensing maps. For more details, we refer the reader to Akhmerov et al. [9], Banaš and Goebel [33], Deimling [77], Heinz [112], Kamenskii et al. [131], and Lakshmikantham and Leela [143].

Definition 1.23. *Let* Y^+ *be the positive cone of an order Banach space* $(Y, \leq)$. *A function* α *defined on the set of all bounded subsets of the Banach space* X *with values in* Y^+ *is called a measure of noncompactness (MNC) on* X *if* $\alpha(\overline{co}\Omega) = \alpha(\Omega)$ *for all bounded subsets* $\Omega \subset X$.

The MNC α is said to be:

(i) *Monotone* if for all bounded subsets B_1, B_2 of X, $B_1 \subseteq B_2$ implies $\alpha(B_1) \leq \alpha(B_2)$;
(ii) *Nonsingular* if $\alpha(\{x\} \cup B) = \alpha(B)$ for every $x \in X$ and every nonempty subset $B \subseteq X$;
(iii) *Regular* $\alpha(B) = 0$ if and only if B is relatively compact in X.

One of the most important examples of MNC is Hausdorff MNC α defined on each bounded subset B of X by

$$\alpha(B) = \inf \left\{ \varepsilon > 0 : B \subset \bigcup_{j=1}^{m} B_\varepsilon(x_j) \text{ where } x_j \in X \right\},$$

where $B_\varepsilon(x_j)$ is a ball of radius $\leq \varepsilon$ centered at x_j, $j = 1, 2, ..., m$. Without confusion, Kuratowski MNC α_1 is defined on each bounded subset B of X by

$$\alpha_1(B) = \inf \left\{ \varepsilon > 0 : B \subset \bigcup_{j=1}^{m} M_j \text{ and } \text{diam}(M_j) \leq \varepsilon \right\},$$

where the diameter of M_j is defined by $\text{diam}(M_j) = \sup\{|x - y| : x, y \in M_j\}$, $j = 1, 2, ..., m$.

It is well known that Hausdorff MNC α and Kuratowski MNC α_1 enjoy the above properties (i)-(iii) and other properties.

(iv) $\alpha(B_1 + B_2) \leq \alpha(B_1) + \alpha(B_2)$, where $B_1 + B_2 = \{x + y : x \in B_1,\, y \in B_2\}$;

(v) $\alpha(B_1 \cup B_2) \leq \max\{\alpha(B_1), \alpha(B_2)\}$;

(vi) $\alpha(\lambda B) \leq |\lambda| \alpha(B)$ for any $\lambda \in \mathbb{R}$.

In particular, the relationship between Hausdorff MNC α and Kuratowski MNC α_1 is given by

(vii) $\alpha(B) \leq \alpha_1(B) \leq 2\alpha(B)$.

In the following, several examples of useful measures of noncompactness in spaces of continuous functions are presented.

Example 1.1. We consider general example of MNC in the space of continuous functions $C([a, b], X)$. For $\Omega \subset C([a, b], X)$ define

$$\phi(\Omega) = \sup_{t \in [a,b]} \alpha(\Omega(t)),$$

where α is Hausdorff MNC in X and $\Omega(t) = \{y(t) : y \in \Omega\}$.

Example 1.2. Consider another useful MNC in the space $C([a, b], X)$. For a bounded $\Omega \subset C([a, b], X)$, set

$$\nu(\Omega) = \left(\sup_{t \in [a,b]} \alpha(\Omega(t)), \mathrm{mod}_C(\Omega) \right);$$

here, the modulus of equicontinuity of the set of functions $\Omega \subset C([a, b], X)$ has the following form:

$$\mathrm{mod}_C(\Omega) = \lim_{\delta \to 0} \sup_{x \in \Omega} \max_{|t_1 - t_2| \leq \delta} |x(t_1) - x(t_2)|. \tag{1.20}$$

Example 1.3. We consider one more MNC in the space $C([a, b], X)$. For a bounded $\Omega \subset C([a, b], X)$, set

$$\nu(\Omega) = \max_{D \in \Delta(\Omega)} \left(\sup_{t \in [a,b]} \exp^{-Lt} \alpha(D(t)), \mathrm{mod}_C(D) \right),$$

where $\Delta(\Omega)$ is the collection of all denumerable subsets of Ω, L is a constant, and $\mathrm{mod}_C(D)$ is given in formula (1.20).

Let $J = [0, b], b \in \mathbb{R}^+$. For any $W \subset C(J, X)$, we define

$$\int_0^t W(s)ds = \left\{ \int_0^t u(s)ds : u \in W \right\}, \text{ for } t \in [0, b],$$

where $W(s) = \{u(s) \in X : u \in W\}$.

We present here some useful properties.

Property 1.14. *If $W \subset C(J, X)$ is bounded and equicontinuous, then $\overline{co}W \subset C(J, X)$ is also bounded and equicontinuous.*

Property 1.15. *If $W \subset C(J, X)$ is bounded and equicontinuous, then $t \to \alpha(W(t))$ is continuous on J, and*

$$\alpha(W) = \max_{t \in J} \alpha(W(t)), \quad \alpha\left(\int_0^t W(s)ds\right) \le \int_0^t \alpha(W(s))ds,$$

for $t \in [0, b]$.

Property 1.16. *Let $\{u_n\}_{n=1}^\infty$ be a sequence of Bochner integrable functions from J into X with $|u_n(t)| \le \tilde{m}(t)$ for almost all $t \in J$ and every $n \ge 1$, where $\tilde{m} \in L(J, \mathbb{R}^+)$, then the function $\psi(t) = \alpha(\{u_n(t)\}_{n=1}^\infty)$ belongs to $L(J, \mathbb{R}^+)$ and satisfies*

$$\alpha\left(\left\{\int_0^t u_n(s)ds : n \ge 1\right\}\right) \le 2\int_0^t \psi(s)ds.$$

Property 1.17. *If W is bounded, then for each $\varepsilon > 0$, there is a sequence $\{u_n\}_{n=1}^\infty \subset W$, such that*

$$\alpha(W) \le 2\alpha(\{u_n\}_{n=1}^\infty) + \varepsilon.$$

Consider an abstract operator $\mathcal{L} : L^1([0, b], X) \to C([0, b], X)$ satisfying the following conditions:

(L_1) there exists a constant $C > 0$ such that

$$|(\mathcal{L}g_1)(t) - (\mathcal{L}g_2)(t)| \le C \int_0^t |g_1(s) - g_2(s)|ds$$

for all $g_1, g_2 \in L^1([0, b], X)$, $t \in [0, b]$;

(L_2) for each compact set $K \subset X$ and sequence $\{g_n\} \subset L^1([0,b], X)$ such that $\{g_n(t)\} \subset K$ for a.e. $t \in [0,b]$, the weak convergence $g_n \rightharpoonup g_0$ implies $\mathcal{L}(g_n) \to \mathcal{L}(g_0)$ strongly in $C([0,b], X)$.

Remark 1.7. *A typical example for $\mathcal{L}$ is the Cauchy operator*

$$(Gg)(t) = \int_0^t g(s)ds, \quad g \in L^1([0,b], X),$$

which satisfies conditions (L_1)-(L_2) with $C = 1$ (see [131]).

Property 1.18. *Let $\mathcal{L}$ satisfy (L_1)-(L_2) and a sequence $\{g_n\} \subset L^1([0,b], X)$ be integrably bounded, i.e.,*

$$|g_n(t)| \le \varrho(t) \ for \ a.e. \ t \in [0,b],$$

where $\varrho \in L^1([0,b], \mathbb{R}^+)$. Assume that there exists $\varpi \in L^1([0,b], \mathbb{R}^+)$ such that

$$\alpha(\{g_n(t)\}) \le \varpi(t) \ for \ a.e. \ t \in [0,b].$$

Then

$$\alpha(\{(\mathcal{L}g_n)(t)\}) \le 2C \int_0^t \varpi(s)ds$$

for each $t \in [0,b]$.

The following α-estimate, which is similar to Theorem 4.2.3 in [131], will be used in the sequel.

Property 1.19. *Assume that X is a separable Banach space. Let $F : [0,b] \to P(X)$ be $L^p(p \ge 1)$-integrable bounded multifunction such that*

$$\alpha(F(t)) \le q(t),$$

for a.e. $t \in [0,b]$; here, $q(t) \in L^p([0,b], \mathbb{R}^+)$. Then

$$\alpha\left(\int_0^t F(\tau)d\tau\right) \le \int_0^t q(\tau)d\tau,$$

for a.e. $t \in [0,b]$. In particular, if the multifunction $F : [0,b] \to P_{cp}(X)$ is measurable and L^p-integrably bounded, then the function $\alpha(F(\cdot))$ is *integrable and, moreover,*

$$\alpha\left(\int_0^t F(\tau)d\tau\right) \le \int_0^t \alpha(F(\tau))d\tau,$$

for a.e. $t \in [0, b]$.

We also recall definition of condensing maps and fixed point theorems via condensing maps, see, e.g., Akhmerov et al. [9] and Kamenskii et al. [131].

Definition 1.24. *A multivalued map $\varphi : X \to P_{cp}(X)$ is said to be condensing with respect to an MNC β (β-condensing) if for every bounded set $D \subset X$ that is not relatively compact, we have $\beta(\varphi(D)) \not\ge \beta(D)$.*

Theorem 1.7. *Let Ω be a bounded convex closed subset of X and $\mathscr{H} : \Omega \to \Omega$ a β-condensing map. Then $\mathrm{Fix}\mathscr{H} = \{x : x = \mathscr{H}(x)\}$ is nonempty.*

Theorem 1.8. *Let $\Omega \subset X$ be a bounded open neighborhood of zero and $\mathscr{H} : \overline{\Omega} \to X$ a β-condensing map satisfying the boundary condition $x \ne \hat{\lambda}\mathscr{H}(x)$ for all $x \in \partial\Omega$ and $\hat{\lambda} \in (0, 1]$. Then $\mathrm{Fix}\mathscr{H}$ is a nonempty compact set.*

Theorem 1.9. *Let Ω be a bounded convex closed subset of a Banach space X, and $\varphi : \Omega \to P_{cp,cv}(\Omega)$ an u.s.c. β-condensing multivalued map. Then the fixed point set $\mathrm{Fix}\varphi = \{x : x \in \varphi(x)\}$ is a nonempty compact set.*

1.6.3 R_δ-set

In the study of the topological structure of the solution sets for differential equations and inclusions, an important aspect is the R_δ-property. Recall that a subset D of a metric space is an R_δ-set if there exists a decreasing sequence $\{D_n\}_{n=1}^\infty$ of compact and contractible sets such that $D = \bigcap_{n=1}^\infty D_n$ (see Definition 1.27 below). This means that an R_δ-set is acyclic (in particular, nonempty, compact, and connected) and may not be a singleton but, from the point of view of algebraic topology, it is equivalent to a point, in the sense that it has the same homology groups as one point space.

Definition 1.25. *[121]*

(i) X *is called an absolute retract (AR space) if for any metric space Y and any closed subset $D \subset Y$, there exists a continuous function $h : D \to X$ which can be extended to a continuous function $\tilde{h} : Y \to X$.*

(ii) X *is called an absolute neighborhood retract (ANR space) if for any metric space Y, closed subset $D \subset Y$ and continuous function $h : D \to X$, there exists a neighborhood $U \supset D$ and a continuous extension $\tilde{h} : U \to X$ of h.*

Obviously, if X is an AR space then it is an ANR space. Furthermore, as in [83], if D is a convex set in a locally convex linear space then it is an AR space. This yields that each convex set of a Fréchet space is an AR space, since every Fréchet space is locally convex. In particular, every Banach space is an AR pace.

Definition 1.26. *A nonempty subset D of a metric space Y is said to be contractible if there exists a point $y_0 \in D$ and a continuous function $h : D \times [0,1] \to D$ such that $h(y,1) = y_0$ and $h(y,0) = y$ for every $y \in D$.*

Definition 1.27. *A subset D of a metric space is called an R_δ-set if there exists a decreasing sequence $\{D_n\}$ of compact and contractible sets such that*

$$D = \bigcap_{n=1}^{\infty} D_n.$$

Note that any R_δ-set is nonempty, compact, and connected. What followed is the hierarchy for nonempty subsets of a metric space:

$$\text{compact+convex} \subset \text{compact } AR \text{ space}$$
$$\subset \text{compact+ contractible} \subset R_\delta\text{-set,} \tag{1.21}$$

and all the above inclusions are proper.

Theorem 1.10. *[44] Let X be a complete metric space, α denote Hausdorff MNC in X, and let $\emptyset \neq D \subset X$. Then the following statements are equivalent:*

(i) D *is an R_δ-set;*

(ii) D *is an intersection of a decreasing sequence* $\{D_n\}$ *of closed contractible spaces with* $\alpha(D_n) \to 0$;

(iii) D *is compact and absolutely neighborhood contractible, i.e.,* D *is contractible in each neighborhood in* $Y \in ANR$.

Definition 1.28. *A multivalued map* $\varphi : Y \to P(Z)$ *is an* R_δ-*map if* φ *is u.s.c. and* $\varphi(y)$ *is an* R_δ-*set for each* $y \in Y$.

It is clear that every u.s.c. multivalued map with contractible values can be seen as an R_δ-map. In particular, every single-valued continuous map is an R_δ-map.

The following theorem presents a sufficient condition for a set of being R_δ.

Theorem 1.11. *[48] Let* Y *be a metric space and* E *a Banach space. Suppose that* $\mathcal{F} : Y \to E$ *is a proper map, i.e.,* $\mathcal{F}$ *is continuous and* $\mathcal{F}^{-1}(K)$ *is compact for each compact set* $K \subset E$. *In addition, if there exists a sequence* $\{\mathcal{F}_n\}$ *of mappings from* Y *into* E *such that*

(i) $\mathcal{F}_n$ *is proper and* $\{\mathcal{F}_n\}$ *converges to* $\mathcal{F}$ *uniformly on* Y, *and*

(ii) *for a given point* $y_0 \in E$ *and for all* y *in a neighborhood* $U(y_0)$ *of* y_0 *in* E, *there exists exactly one solution* x_n *of the equation* $\mathcal{F}_n(x) = y$,

then $\mathcal{F}^{-1}(y_0)$ *is an* R_δ-*set.*

We need the following fixed point theorem which is due to [106].

Theorem 1.12. *Let* Y *be an* ANR *space. Assume that* $\varphi : Y \to P(Y)$ *can be factorized as*

$$\varphi = \varphi_n \circ \varphi_{n-1} \circ \cdots \circ \varphi_1,$$

where $\varphi_i : Y_{i-1} \to P(Y_i)$, $i = 1, ..., n$, *are* R_δ-*maps and* Y_i, $i = 1, ..., n-1$, *are* ANR *spaces and* $Y_0 = Y_n = Y$ *are* AR *spaces. If there exists a compact subset* $K \subset Y$ *satisfying* $\varphi(Y) \subset K$, *then* φ *has a fixed point in* Y.

We also need the following result, which can be seen from the inclusion relation (1.21) and Theorem 1.12.

Theorem 1.13. *Let X be a Banach space and $D \subset X$ be a nonempty compact convex subset. If the multivalued map $\varphi : D \to P(D)$ is u.s.c. with contractible values, then φ has a fixed point.*

1.7 STOCHASTIC PROCESS

We present some important concepts and results of stochastic process in this section. The material is taken from Arnold [15], Gawarecki et al. [102], and Prato et al. [205].

1.7.1 Random Variables

Let Ω be a sample space and $\mathscr{F}$ a σ-algebra of the subset of Ω. A function $\mathbb{P}(\cdot)$ defined on $\mathscr{F}$ and taking values in the unit interval $[0, 1]$ is called a probability measure, if

(i) $\mathbb{P}(\Omega) = 1$;

(ii) $\mathbb{P}(A) \geq 0$ for all $A \in \mathscr{F}$;

(iii) for an at most countable family $\{A_n, n \geq 1\}$ of mutually disjoint event, we have

$$\mathbb{P}\left\{ \bigcup_{n \geq 1} A_n \right\} = \sum_{n \geq 1} \mathbb{P}(A_n).$$

The triple $(\Omega, \mathscr{F}, \mathbb{P})$ is a probability space.

$\mathbb{F} = (\mathscr{F}_t)_{t \geq 0}$ is a family of sub-σ-algebras $\mathscr{F}_t$ of σ-algebra $\mathscr{F}$ such that $\mathscr{F}_s \subset \mathscr{F}_t$ for $0 \leq s < t < \infty$. $\mathbb{P}_{\mathbb{F}} = (\Omega, \mathscr{F}, \mathbb{F}, \mathbb{P})$ is said to be a filtered probability space. We say that a filtration $\mathscr{F}_t$ satisfies the usual conditions if $\mathscr{F}_0$ contains all $\mathbb{P}$-null sets of $\mathscr{F}$ and $\mathscr{F}_t = \bigcap_{\varepsilon > 0} \mathscr{F}_{t+\varepsilon}$ for every $t \geq 0$. If the last condition is satisfied, we say that a filtration $\mathscr{F}$ is right continuous.

Let $(X, \mathscr{B}_X)$ be measurable space, we mean an $(\mathscr{F}, \mathscr{B}_X)$-measurable mapping $x : \Omega \to X$, i.e., such that $x^{-1}(A) \in \mathscr{F}$ for every $A \in \mathscr{B}_X$, where as usual, $\mathscr{B}_X$ denotes the Borel σ-algebra on X and $x^{-1}(A) = \{\omega \in \Omega : x(\omega) \in A\}$. We shall also say that x is a random variable on Ω with values at X.

The integral of an integrable random variable x is called its mean value or expectation and is denoted by

$$\mathbb{E}(x) = \int x(w)d\mathbb{P}.$$

Let $\mathcal{K}$ and $\mathcal{H}$ be separable Hilbert spaces, and Q be either a symmetric nonnegative definite trace-class operator on $\mathcal{K}$ or $Q = I_{\mathcal{K}}$, the identity operator on $\mathcal{K}$. In case Q is trace-class, we will always assume that its all eigenvalues $\lambda_j > 0, ...$; otherwise we can start with the Hilbert space $\ker(Q)^\perp$ instead of $\mathcal{K}$. The associated eigenvectors forming an orthonormal basis (ONB) in $\mathcal{K}$ will be denoted by e_j.

Denote $\mathscr{L}(\mathcal{K}, \mathcal{H})$ by all bounded linear operators from $\mathcal{K}$ to $\mathcal{H}$. Then the space of Hilbert-Schmidt operators from $\mathcal{K}$ to $\mathcal{H}$ is defined as

$$\mathscr{L}_2(\mathcal{K}, \mathcal{H}) = \left\{ \Phi \in \mathscr{L}(\mathcal{K}, \mathcal{H}) : \sum_{i=1}^{\infty} |\Phi e_i|_{\mathcal{H}}^2 < \infty \right\}.$$

It is well known (see [214]) that $\mathscr{L}_2(\mathcal{K}, \mathcal{H})$ equipped with the norm

$$\|\Phi\|_{\mathscr{L}_2(\mathcal{K}, \mathcal{H})} = \sum_{i=1}^{\infty} |\Phi e_i|_{\mathcal{H}}^2$$

is a Hilbert space.

On the other hand, the space $Q^{\frac{1}{2}}\mathcal{K}$ equipped with the scalar product

$$\langle u, v \rangle_{Q^{\frac{1}{2}}\mathcal{K}} = \sum_{j=1}^{\infty} \frac{1}{\lambda_j} (u, e_j)_{\mathcal{K}} (v, e_j)_{\mathcal{K}}$$

is a separable Hilbert space with an ONB $\{\lambda_j^{\frac{1}{2}} e_j\}_{j=1}^{\infty}$.

Consider $\mathscr{L}_2^0 = \mathscr{L}_2(Q^{\frac{1}{2}}\mathcal{K}, \mathcal{H})$, the space of Hilbert-Schmidt operators from $Q^{\frac{1}{2}}\mathcal{K}$ to $\mathcal{H}$. If $\{\tilde{e}_j\}_{j=1}^{\infty}$ is an ONB in $\mathcal{H}$, then the Hilbert-Schmidt norm of an operator $\Phi \in \mathscr{L}_2^0$ is given by

$$\|\Phi\|_{\mathscr{L}_2^0} = \sum_{i,j=1}^{\infty} (\Phi(\lambda_j^{\frac{1}{2}} e_j), \tilde{e}_i)_{\mathcal{H}}^2 = \sum_{i,j=1}^{\infty} (\Phi(Q^{\frac{1}{2}} e_j), \tilde{e}_i)_{\mathcal{H}}^2$$
$$= \left\| \Phi Q^{\frac{1}{2}} \right\|_{\mathscr{L}_2(\mathcal{K}, \mathcal{H})}^2 = \mathrm{tr}(\Phi Q^{\frac{1}{2}})(\Phi Q^{\frac{1}{2}})^*.$$

1.7.2 Stochastic Calculus

An X-valued stochastic process (briefly, an X-valued process) indexed by a set I is a family of X-valued random variables $\{X(i), i \in I\}$ defined on some underlying probability space $(\Omega, \mathscr{F}, \mathbb{P})$.

Definition 1.29. *An X-valued process $\{X(i), i \in I\}$ is called Gaussian, if for all $N > 1$ and $i_1, ..., i_N \in I$ the X^N-valued random variable $(X(i_1), ..., X(i_N))$ is Gaussian.*

Definition 1.30. *A real-valued process $\{W(t), t \in [0, T]\}$ is called a Brownian motion, if it enjoys the following properties:*

(i) $W(0) = 0$;
(ii) $W(t) - W(s)$ *is independent of* $\{W(r), r \in [0, s]\}$ *for* $0 \leq s \leq t \leq T$;
(iii) $W(t) - W(s)$ *is Gaussian with variance* $(t - s)$.

Definition 1.31.

(a) *For an $\mathscr{L}(\mathcal{H}, X)$-valued step function of the form $\Phi(t, \omega) = \phi_1(\omega) I_{[t_0, t_1]}(t) + \sum_{i=2}^{n} \phi_i(\omega) I_{(t_{i-1}, t_i]}(t)$, where $0 = t_0 < t_1 < \cdots < t_n = T$ and ϕ_i, $i = 1, ..., n$, are, respectively, $\mathscr{F}_0$-measurable and $\mathscr{F}_{t_i}$-measurable $\mathscr{L}_2(\mathcal{K}, \mathcal{H})$-valued random variables such that $\phi_i(\omega) \in \mathscr{L}(\mathcal{K}, \mathcal{H})$, $i = 1, ..., n$. We define the stochastic integral process $\int_0^t \Phi(s) dW(s), 0 \leq t \leq T$, by*

$$\int_0^t \Phi(s) dW(s) = \sum_{i=1}^{n} \phi_i (W(t_i) - W(t_{i-1})).$$

(b) *A function $\Phi : [0, T] \to \mathscr{L}(\mathcal{H}, X)$ is said to be stochastically integrable with respect to the $\mathcal{H}$-cylindrical Brownian motion W if there exists a sequence of finite rank step functions $\Phi_n : [0, T] \to \mathscr{L}(\mathcal{H}, X)$ such that:*
(i) *for all $h \in \mathcal{H}$, we have $\lim_{n \to \infty} \Phi_n h = \Phi h$ in measure;*
(ii) *there exists an X-valued random variable x such that*

$$\lim_{n \to \infty} \int_0^t \Phi_n(s) dW(s) = x$$

in probability.

The stochastic integral of a stochastically integrable function $x : [0, T] \to$
$\mathscr{L}(\mathcal{H}, X)$ is then defined as the limit in probability

$$\int_0^t \Phi(s)dW(s) = \lim_{n\to\infty} \int_0^t \Phi_n(s)dW(s).$$

The relationship

$$x(t, \omega) = \int_0^t \Phi(s, \omega)dW(s, \omega)$$

can also be written as

$$dx(t) = \Phi(t)dW(t).$$

This is a special so-called stochastic differential. Let us look at a somewhat
more general stochastic process of the form

$$x(t, \omega) = x(0, \omega) + \int_0^t f(s, \omega)ds + \int_0^t \Phi(s, \omega)dW(s, \omega); \qquad (1.22)$$

here, $\int_0^t f(s, \omega)ds$ is the usual Lebesgue or possibly Riemann integral.

Definition 1.32. *We shall say that a stochastic process $x(t)$ defined by*
equation (1.22) possesses the stochastic differential $f(t)dt + \Phi(t)dW(t)$
and we shall write

$$\begin{aligned} dx(t) &= f(t)dt + \Phi(t)dW(t) \\ &= fdt + \Phi dW. \end{aligned}$$

CHAPTER 2

Fractional Evolution Equations

Abstract   In this chapter, we first study the existence of Cauchy problems for fractional evolution equations. The suitable mild solutions of fractional Cauchy problems with Riemann-Liouville derivative and Caputo derivative are introduced, respectively. By using fixed point theorems and Hausdorff measure of noncompactness, we give existence results of mild solutions in the cases that the almost sectorial operator is compact and noncompact, respectively. In Section 2.2, we discuss the existence and uniqueness of the bounded solutions on real axis for fractional evolution equations with Liouville fractional derivative of order $q \in (0, 1)$ with the lower limit $-\infty$. Some sufficient conditions are established for the existence and uniqueness of periodic solutions, S-asymptotically periodic solutions, and other types of bounded solutions.

Keywords   Cauchy problems, Mild solutions, Bounded solutions, Periodic solutions, Existence, Uniqueness, Laplace transform, Fourier transforms, Fixed point theorem, Hausdorff measure of noncompactness.

2.1 CAUCHY PROBLEMS

2.1.1 Introduction

In this section, we assume that X is a Banach space with the norm $|\cdot|$. Let $a \in \mathbb{R}^+$, $J = [0, a]$, and $J' = (0, a]$. Let $C(J, X)$ be the Banach space of continuous functions from J into X with the norm $\|x\| = \sup_{t \in [0,a]} |x(t)|$, where $x \in C(J, X)$, and $\mathscr{L}(X)$ be the space of all bounded linear operators from X to X with the norm $\|T\|_{\mathscr{L}(X)} = \sup\{|T(x)| : |x| = 1\}$, where $T \in \mathscr{L}(X)$ and $x \in X$.

Consider the following Cauchy problems of fractional evolution equation with Riemann-Liouville derivative

$$\begin{cases} {}_0D_t^q x(t) = Ax(t) + f(t, x(t)), & \text{a.e. } t \in [0, a], \\ {}_0D_t^{q-1} x(0) = x_0, \end{cases} \tag{2.1}$$

Fractional Evolution Equations and Inclusions: Analysis and Control. http://dx.doi.org/10.1016/B978-0-12-804277-9.50002-X

and fractional evolution equation with Caputo derivative

$$\begin{cases} {}_0^C D_t^q x(t) = Ax(t) + f(t, x(t)), & \text{a.e. } t \in [0, a], \\ x(0) = x_0, \end{cases} \tag{2.2}$$

where $0 < q < 1$, $f : J \times X \to X$ is a given function, A is an almost sectorial operator on a complex Banach space, that is, $A \in \Theta_\omega^p(X)$ ($-1 < p < 0, 0 < \omega < \frac{\pi}{2}$). We denote the semigroup associated with A by $\{Q(t)\}_{t \geq 0}$.

Subsection 2.1.2 is devoted to obtaining the appropriate definition on the mild solutions of problems (2.1) and (2.2) by considering an integral equation which is given in terms of Wright function. In Subsection 2.1.3, we introduce various existence and uniqueness results of mild solutions for Cauchy problem with Riemann-Liouville derivative. In Subsection 2.1.4, we establish various existence and uniqueness results of mild solutions for Cauchy problem with Caputo derivative.

2.1.2 Definition of Mild Solutions

Define operator families $\{S_q(t)\}|_{t \in S_{\frac{\pi}{2}-\omega}^0}$, $\{P_q(t)\}|_{t \in S_{\frac{\pi}{2}-\omega}^0}$ by

$$S_q(t) = \int_0^\infty \Psi_q(\theta) Q(t^q \theta) d\theta, \quad \text{for } t \in S_{\frac{\pi}{2}-\omega}^0,$$

$$P_q(t) = \int_0^\infty q\theta \Psi_q(\theta) Q(t^q \theta) d\theta, \quad \text{for } t \in S_{\frac{\pi}{2}-\omega}^0,$$

where $\Psi_q(\theta)$ is the Wright function (see Definition 1.8).

Proposition 2.1. *For each fixed* $t \in S_{\frac{\pi}{2}-\omega}^0$, $S_q(t)$ *and* $P_q(t)$ *are linear and bounded operators on X. Moreover, for all* $t > 0$

$$\|S_q(t)\|_{\mathscr{L}(X)} \leq C_1 \, t^{-q(1+p)}, \quad \|P_q(t)\|_{\mathscr{L}(X)} \leq C_2 \, t^{-q(1+p)}, \tag{2.3}$$

where $-1 < p < 0$, *and*

$$C_1 = c_0 \frac{\Gamma(-p)}{\Gamma(1 - q(1+p))}, \quad C_2 = qc_0 \frac{\Gamma(1-p)}{\Gamma(1-qp)},$$

c_0 *is definited as in Property 1.13.*

Proof. By Property 1.13(iii) and Property 1.11(iii), we have

$$|S_q(t)x| \leq c_0 \int_0^\infty \Psi_q(\theta)\theta^{-(1+p)}t^{-q(1+p)}|x|d\theta$$

$$\leq c_0 \frac{\Gamma(-p)}{\Gamma(1-q(1+p))}t^{-q(1+p)}|x|, \quad t > 0, \ x \in X,$$

$$|P_q(t)x| \leq qc_0 \int_0^\infty \Psi_q(\theta)\theta^{-p}t^{-q(1+p)}|x|d\theta$$

$$\leq qc_0 \frac{\Gamma(1-p)}{\Gamma(1-qp)}t^{-q(1+p)}|x|, \quad t > 0, \ x \in X.$$

Therefore, the estimates in (2.3) hold. The proof is completed. $\qquad\square$

Proposition 2.2. *For* $t > 0$, $S_q(t)$ *and* $P_q(t)$ *are strongly continuous, which means that, for any* $x \in X$ *and* $0 < t' < t'' \leq a$, *we have*

$$|S_q(t'')x - S_q(t')x| \to 0, \quad |P_q(t'')x - P_q(t')x| \to 0, \quad as \ t'' \to t'.$$

Proposition 2.3. *Let* $\beta > 1 + p$. *For all* $x \in D(A^\beta)$, *we have*

$$\lim_{t\to 0+} S_q(t)x = x \ and \ \lim_{t\to 0+} P_q(t)x = \frac{x}{\Gamma(q)}.$$

Proof. For any $x \in X$, we have

$$S_q(t)x - x = \int_0^\infty \Psi_q(\theta)(Q(t^q\theta)x - x)d\theta$$

and

$$P_q(t)x - \frac{x}{\Gamma(q)} = \int_0^\infty q\theta\Psi_q(\theta)(Q(t^q\theta)x - x)d\theta.$$

On the other hand, from (iv) of Property 1.13, it follows that $D(A^\beta) \subset \Sigma_Q$ in view of $\beta > 1 + p$. Therefore, we deduce, using (iii) of Property 1.13, that for any $x \in D(A^\beta)$, there exists a function $\eta(\theta) \in L((0,\infty),\mathbb{R}^+)$ depending on $\Psi_q(\theta)$ such that

$$|\Psi_q(\theta)(Q(t^q\theta)x - x)| \leq \eta(\theta).$$

Hence, by means of Theorem 1.1 we obtain

$$S_q(t)x - x \to 0 \ and \ P_q(t)x - \frac{x}{\Gamma(q)} \to 0, \quad as \ t \to 0+.$$

The proof is completed.   □

Before presenting the definition of mild solution of problem (2.1), we first give the following lemmas.

Lemma 2.1. *Cauchy problem (2.1) is equivalent to the integral equation*

$$x(t) = \frac{x_0}{\Gamma(q)} t^{q-1} + \frac{1}{\Gamma(q)} \int_0^t (t-s)^{q-1} \big(Ax(s) + f(s, x(s))\big) ds, \quad (2.4)$$

for $t \in (0, a]$, provided that the integral in (2.4) exists.

Proof. Suppose (2.4) is true. By Lemma 1.3, we obtain

$$_0D_t^{q-1}x(t) = x_0 + \int_0^t \big(Ax(s) + f(s, x(s))\big) ds, \quad \text{a.e. } t \in [0, a],$$

and this proves that $_0D_t^{q-1}x(t)$ is absolutely continuous on $[0, a]$. Then we have

$$_0D_t^q x(t) = \frac{d}{dt}\, _0D_t^{q-1}x(t) = Ax(t) + f(t, x(t)), \quad \text{a.e. } t \in [0, a]$$

and

$$_0D_t^{q-1}x(0) = x_0.$$

The proof of the converse is given as follows. Suppose (2.1) is true, then

$$_0D_t^{-q}\big(_0D_t^q x(t)\big) = {_0D_t^{-q}}\big(Ax(t) + f(t, x(t))\big).$$

Since

$$_0D_t^{-q}\big(_0D_t^q x(t)\big) = x(t) - \frac{t^{q-1}}{\Gamma(q)}\, _0D_t^{q-1}x(0)$$

$$= x(t) - \frac{t^{q-1}}{\Gamma(q)} x_0, \quad \text{for } t \in (0, a],$$

then we have

$$x(t) = \frac{t^{q-1}}{\Gamma(q)} x_0 + {_0D_t^{-q}}\big(Ax(t) + f(t, x(t))\big)$$

$$= \frac{t^{q-1}}{\Gamma(q)} x_0 + \frac{1}{\Gamma(q)} \int_0^t (t-s)^{q-1} \big(Ax(s) + f(s, x(s)) \big) ds,$$

for $t \in (0, a]$. The proof is completed.   □

Lemma 2.2. *If*

$$x(t) = \frac{x_0}{\Gamma(q)} t^{q-1} + \frac{1}{\Gamma(q)} \int_0^t (t-s)^{q-1} \big(Ax(s) + f(s, x(s)) \big) ds, \quad for \ t > 0$$

(2.5)

holds, then we have

$$x(t) = t^{q-1} P_q(t) x_0 + \int_0^t (t-s)^{q-1} P_q(t-s) f(s, x(s)) ds, \quad for \ t > 0.$$

Proof. Let $\lambda > 0$. Applying the Laplace transform

$$\nu(\lambda) = \int_0^\infty e^{-\lambda s} x(s) ds \ \text{ and } \ \omega(\lambda) = \int_0^\infty e^{-\lambda s} f(s, x(s)) ds, \ \text{ for } \lambda > 0,$$

to (2.5), we have

$$\begin{aligned}
\nu(\lambda) &= \frac{1}{\lambda^q} x_0 + \frac{1}{\lambda^q} A\nu(\lambda) + \frac{1}{\lambda^q} \omega(\lambda) \\
&= (\lambda^q I - A)^{-1} x_0 + (\lambda^q I - A)^{-1} \omega(\lambda) \\
&= \int_0^\infty e^{-\lambda^q s} Q(s) x_0 ds + \int_0^\infty e^{-\lambda^q s} Q(s) \omega(\lambda) ds,
\end{aligned}$$

(2.6)

provided that the integrals in (2.6) exist, where I is the identity operator defined on X.

Set

$$\psi_q(\theta) = \frac{q}{\theta^{q+1}} \Psi_q(\theta^{-q}),$$

whose Laplace transform is given by Property 1.11:

$$\int_0^\infty e^{-\lambda \theta} \psi_q(\theta) d\theta = e^{-\lambda^q},$$

(2.7)

where $q \in (0, 1)$. Using (2.7), we get

$$
\int_0^\infty e^{-\lambda^q s} Q(s) x_0 ds
$$
$$
= \int_0^\infty q t^{q-1} e^{-(\lambda t)^q} Q(t^q) x_0 dt
$$
$$
= \int_0^\infty \int_0^\infty q \psi_q(\theta) e^{-(\lambda t \theta)} Q(t^q) t^{q-1} x_0 d\theta dt \tag{2.8}
$$
$$
= \int_0^\infty \int_0^\infty q \psi_q(\theta) e^{-\lambda t} Q\left(\frac{t^q}{\theta^q}\right) \frac{t^{q-1}}{\theta^q} x_0 d\theta dt
$$
$$
= \int_0^\infty e^{-\lambda t} \left[q \int_0^\infty \psi_q(\theta) Q\left(\frac{t^q}{\theta^q}\right) \frac{t^{q-1}}{\theta^q} x_0 d\theta \right] dt
$$

and

$$
\int_0^\infty e^{-\lambda^q s} Q(s) \omega(\lambda) ds
$$
$$
= \int_0^\infty \int_0^\infty q t^{q-1} e^{-(\lambda t)^q} Q(t^q) e^{-\lambda s} f(s, x(s)) ds dt
$$
$$
= \int_0^\infty \int_0^\infty \int_0^\infty q \psi_q(\theta) e^{-(\lambda t \theta)} Q(t^q) e^{-\lambda s} t^{q-1} f(s, x(s)) d\theta ds dt
$$
$$
= \int_0^\infty \int_0^\infty \int_0^\infty q \psi_q(\theta) e^{-\lambda(t+s)} Q\left(\frac{t^q}{\theta^q}\right) \frac{t^{q-1}}{\theta^q} f(s, x(s)) d\theta ds dt \tag{2.9}
$$
$$
= \int_0^\infty e^{-\lambda t} \left[q \int_0^t \int_0^\infty \psi_q(\theta) \right.
$$
$$
\left. \times Q\left(\frac{(t-s)^q}{\theta^q}\right) \frac{(t-s)^{q-1}}{\theta^q} f(s, x(s)) d\theta ds \right] dt.
$$

According to (2.8) and (2.9), we have

$$
\nu(\lambda) = \int_0^\infty e^{-\lambda t} \left[q \int_0^\infty \psi_q(\theta) Q\left(\frac{t^q}{\theta^q}\right) \frac{t^{q-1}}{\theta^q} x_0 d\theta, \right.
$$
$$
\left. + q \int_0^t \int_0^\infty \psi_q(\theta) Q\left(\frac{(t-s)^q}{\theta^q}\right) \frac{(t-s)^{q-1}}{\theta^q} f(s, x(s)) d\theta ds \right] dt.
$$

Now we can invert the last Laplace transform to get

$$
x(t) = q \int_0^\infty \psi_q(\theta) Q\left(\frac{t^q}{\theta^q}\right) \frac{t^{q-1}}{\theta^q} x_0 d\theta
$$

$$+ q \int_0^t \int_0^\infty \psi_q(\theta) Q\left(\frac{(t-s)^q}{\theta^q}\right) \frac{(t-s)^{q-1}}{\theta^q} f(s, x(s)) d\theta ds$$

$$= q t^{q-1} \int_0^\infty \theta \Psi_q(\theta) Q(t^q \theta) x_0 d\theta$$

$$+ q \int_0^t \int_0^\infty \theta(t-s)^{q-1} \Psi_q(\theta) Q((t-s)^q \theta) f(s, x(s)) d\theta ds$$

$$= t^{q-1} P_q(t) x_0 + \int_0^t (t-s)^{q-1} P_q(t-s) f(s, x(s)) ds.$$

The proof is completed. □

Due to Lemma 2.2, we give the following definition of the mild solution of (2.1).

Definition 2.1. *By the mild solution of Cauchy problem (2.1), we mean a function $x \in C(J', X)$ satisfying*

$$x(t) = t^{q-1} P_q(t) x_0 + \int_0^t (t-s)^{q-1} P_q(t-s) f(s, x(s)) ds, \quad \text{for } t \in (0, a].$$

According to Definitions 1.1, 1.2, and 1.3, it is suitable to rewrite problem (2.2) in the equivalent integral equation

$$x(t) = x_0 + \frac{1}{\Gamma(q)} \int_0^t (t-s)^{q-1} \big(Ax(s) + f(s, x(s)) \big) ds, \quad t \in [0, a], \quad (2.10)$$

provided that the integral in (2.10) exists.

Before giving the definition of mild solution of (2.2), we prove the following lemma.

Lemma 2.3. *If*

$$x(t) = x_0 + \frac{1}{\Gamma(q)} \int_0^t (t-s)^{q-1} \big(Ax(s) + f(s, x(s)) \big) ds, \quad \text{for } t \geq 0 \quad (2.11)$$

holds, then we have

$$x(t) = S_q(t) x_0 + \int_0^t (t-s)^{q-1} P_q(t-s) f(s, x(s)) ds, \quad \text{for } t \geq 0.$$

Proof. Let $\lambda > 0$. Applying Laplace transform

$$\nu(\lambda) = \int_0^\infty e^{-\lambda s} x(s) ds \text{ and } \omega(\lambda) = \int_0^\infty e^{-\lambda s} f(s, x(s)) ds, \quad \lambda > 0,$$

to (2.10), we have

$$\begin{aligned}
\nu(\lambda) &= \frac{1}{\lambda} x_0 + \frac{1}{\lambda^q} A\nu(\lambda) + \frac{1}{\lambda^q} \omega(\lambda) \\
&= \lambda^{q-1}(\lambda^q I - A)^{-1} x_0 + (\lambda^q I - A)^{-1} \omega(\lambda) \\
&= \lambda^{q-1} \int_0^\infty e^{-\lambda^q s} Q(s) x_0 ds + \int_0^\infty e^{-\lambda^q s} Q(s) \omega(\lambda) ds,
\end{aligned} \tag{2.12}$$

provided that the integrals in (2.12) exist, where I is the identity operator defined on X.

Using (2.7) and (2.12), we get

$$\begin{aligned}
&\lambda^{q-1} \int_0^\infty e^{-\lambda^q s} Q(s) x_0 ds \\
&= \int_0^\infty q(\lambda t)^{q-1} e^{-(\lambda t)^q} Q(t^q) x_0 dt \\
&= \int_0^\infty -\frac{1}{\lambda} \frac{d}{dt} \left(e^{-(\lambda t)^q} \right) Q(t^q) x_0 dt \\
&= \int_0^\infty \int_0^\infty \theta \psi_q(\theta) e^{-\lambda t \theta} Q(t^q) x_0 d\theta dt \\
&= \int_0^\infty e^{-\lambda t} \left[\int_0^\infty \psi_q(\theta) Q\left(\frac{t^q}{\theta^q} \right) x_0 d\theta \right] dt.
\end{aligned} \tag{2.13}$$

According to (2.9), (2.12), and (2.13), we have

$$\nu(\lambda) = \int_0^\infty e^{-\lambda t} \left[\int_0^\infty \psi_q(\theta) Q\left(\frac{t^q}{\theta^q} \right) x_0 d\theta \right. \\
\left. + q \int_0^t \int_0^\infty \psi_q(\theta) Q\left(\frac{(t-s)^q}{\theta^q} \right) f(s, x(s)) \frac{(t-s)^{q-1}}{\theta^q} d\theta ds \right] dt.$$

Now we can invert the last Laplace transform to get

$$x(t) = \int_0^\infty \psi_q(\theta) Q\left(\frac{t^q}{\theta^q} \right) x_0 d\theta \\
+ q \int_0^t \int_0^\infty \psi_q(\theta) Q\left(\frac{(t-s)^q}{\theta^q} \right) f(s, x(s)) \frac{(t-s)^{q-1}}{\theta^q} d\theta ds$$

$$= \int_0^\infty \Psi_q(\theta) Q(t^q\theta) x_0 d\theta$$

$$+ q \int_0^t \int_0^\infty \theta(t-s)^{q-1} \Psi_q(\theta) Q((t-s)^q\theta) f(s, x(s)) d\theta ds$$

$$= S_q(t) x_0 + \int_0^t (t-s)^{q-1} P_q(t-s) f(s, x(s)) ds.$$

The proof is completed. $\qquad\qquad\qquad\qquad\qquad\qquad\qquad\qquad$ □

Due to Lemma 2.3, we give the following definition of the mild solution of (2.2).

Definition 2.2. *By the mild solution of Cauchy problem (2.2), we mean a function* $x \in C(J, X)$ *satisfying*

$$x(t) = S_q(t) x_0 + \int_0^t (t-s)^{q-1} P_q(t-s) f(s, x(s)) ds, \ \ t \in [0, a].$$

For a positive constant r, let

$$B_r(J) = \{x \in C(J, X) : \|x\| \le r\},$$

where $\|x\| = \sup_{t \in J} |x(t)|$. Observe that $B_r(J)$ is clearly a bounded closed and convex subset in Banach space X. Hence, $(B_r(J), \| \cdot \|)$ is a Banach space.

2.1.3 Equations with Riemann-Liouville Derivative

Define

$$X^{(q)}(J') = \left\{ x \in C(J', X) : \lim_{t \to 0+} t^{1+qp} x(t) \text{ exists and is finite} \right\}.$$

For any $x \in X^{(q)}(J')$, define the norm $\| \cdot \|_q$ as

$$\|x\|_q = \sup_{t \in (0, a]} \left\{ t^{1+qp} |x(t)| \right\}.$$

Then $(X^{(q)}(J'), \| \cdot \|_q)$ is a Banach space.

For $r > 0$, define a closed subset $B_r^{(q)}(J') \subset X^{(q)}(J')$ as follows:

$$B_r^{(q)}(J') = \{x \in X^{(q)}(J') : \|x\|_q \leq r\}.$$

Thus, $B_r^{(q)}(J')$ is a bounded closed and convex subset of $X^{(q)}(J')$.

Let $B(J)$ be the closed ball of the space $C(J, X)$ with radius r and center at 0, that is,

$$B(J) = \{y \in C(J, X) : \|y\| \leq r\}.$$

Thus $B(J)$ is a bounded closed and convex subset of $C(J, X)$.

We introduce the following hypotheses:

(H_0) $Q(t)(t > 0)$ is equicontinuous, i.e., $Q(t)$ is continuous in the uniform operator topology for $t > 0$;

(H_1) for each $t \in J'$, the function $f(t, \cdot) : X \to X$ is continuous and for each $x \in C(J', X)$, the function $f(\cdot, x) : J' \to X$ is strongly measurable;

(H_2) there exists a function $m \in L(J', \mathbb{R}^+)$ such that

$$_0D_t^{qp}m \in C(J', \mathbb{R}^+), \quad \lim_{t \to 0+} t^{1+qp}{}_0D_t^{1+qp}m(t) = 0,$$

and

$$|f(t, x)| \leq m(t) \text{ for all } x \in B_r^{(q)}(J') \text{ and almost all } t \in [0, a];$$

(H_3) there exists a constant $r > 0$ such that

$$C_2\left(|x_0| + \sup_{t \in (0,a]} \left\{t^{1+qp} \int_0^t (t - s)^{-(1+qp)} m(s)ds\right\}\right) \leq r,$$

where C_2 is given by Proposition 2.1.

For any $x \in B_r^q(J')$, we define an operator T as follows:

$$(Tx)(t) = t^{q-1}P_q(t)x_0 + \int_0^t (t-s)^{q-1}P_q(t-s)f(s, x(s))ds, \text{ for } t \in (0, a].$$

It is easy to see that $\lim_{t \to 0+} t^{1+qp}(Tx)(t) = 0$.

For any $y \in B(J)$, set

$$x(t) = t^{-(1+qp)}y(t), \text{ for } t \in (0, a].$$

Then, $x \in B_r^{(q)}(J')$. Define $\mathscr{T}$ as follows

$$(\mathscr{T}y)(t) = \begin{cases} t^{1+qp}(Tx)(t), & \text{if } t \in (0, a], \\ 0, & \text{if } t = 0. \end{cases}$$

Before giving the main results, we first prove the following lemmas.

Lemma 2.4. *Let $A \in \Theta_\omega^p(X)$ with $-1 < p < 0$ and $0 < \omega < \frac{\pi}{2}$. Assume that (H_0)-(H_3) hold, then the operator $\mathscr{T} : B(J) \to B(J)$ is equicontinuous provided $x_0 \in D(A^\beta)$ with $\beta > 1 + p$.*

Proof. For any $y \in B(J)$, and $t_1 = 0, 0 < t_2 \le a$, we get

$$|(\mathscr{T}y)(t_2) - (\mathscr{T}y)(0)|$$

$$= \left| t_2^{q(1+p)} P_q(t_2)x_0 + t_2^{1+qp} \int_0^{t_2} (t_2 - s)^{q-1} P_q(t_2 - s)f(s, x(s))ds \right|$$

$$\le \left| t_2^{q(1+p)} P_q(t_2)x_0 \right| + \left| t_2^{1+qp} \int_0^{t_2} (t_2 - s)^{q-1} P_q(t_2 - s)f(s, x(s))ds \right|$$

$$\le \left| t_2^{q(1+p)} P_q(t_2)x_0 \right| + C_2\, t_2^{1+qp} \int_0^{t_2} (t_2 - s)^{-(1+qp)} m(s)ds$$

$$\to 0, \quad \text{as } t_2 \to 0.$$

For $0 < t_1 < t_2 \le a$, we have

$$|(\mathscr{T}y)(t_2) - (\mathscr{T}y)(t_1)|$$

$$\le \left| t_2^{q(1+p)} P_q(t_2)x_0 - t_1^{q(1+p)} P_q(t_1)x_0 \right|$$

$$+ \left| t_2^{1+qp} \int_0^{t_2} (t_2 - s)^{q-1} P_q(t_2 - s)f(s, x(s))ds \right.$$

$$\left. - t_1^{1+qp} \int_0^{t_1} (t_1 - s)^{q-1} P_q(t_1 - s)f(s, x(s))ds \right|$$

$$\le \left| t_2^{q(1+p)} P_q(t_2)x_0 - t_1^{q(1+p)} P_q(t_1)x_0 \right|$$

$$+ \left| \int_{t_1}^{t_2} t_2^{1+qp}(t_2 - s)^{q-1} P_q(t_2 - s)f(s, x(s))ds \right|$$

$$+ \left| \int_0^{t_1} t_2^{1+qp}(t_2 - s)^{q-1} P_q(t_2 - s)f(s, x(s))ds \right.$$

$$- \int_0^{t_1} t_1^{1+qp}(t_1 - s)^{q-1}P_q(t_2 - s)f(s, x(s))ds \bigg|$$

$$+ \left| \int_0^{t_1} t_1^{1+qp}(t_1 - s)^{q-1}P_q(t_2 - s)f(s, x(s))ds \right.$$

$$- \int_0^{t_1} t_1^{1+qp}(t_1 - s)^{q-1}P_q(t_1 - s)f(s, x(s))ds \bigg|$$

$$\leq \left| t_2^{q(1+p)}P_q(t_2)x_0 - t_1^{q(1+p)}P_q(t_1)x_0 \right|$$

$$+ \left| \int_{t_1}^{t_2} t_2^{1+qp}(t_2 - s)^{q-1}P_q(t_2 - s)f(s, x(s))ds \right|$$

$$+ \left| \int_0^{t_1} \left(t_2^{1+qp}(t_2 - s)^{q-1} - t_1^{1+qp}(t_1 - s)^{q-1} \right)P_q(t_2 - s)f(s, x(s))ds \right|$$

$$+ \left| \int_0^{t_1} t_1^{1+qp}(t_1 - s)^{q-1} \left(P_q(t_2 - s) - P_q(t_1 - s) \right)f(s, x(s))ds \right|$$

$$=: I_0 + I_1 + I_2 + I_3.$$

By Proposition 2.2, it is easy to see that $\lim_{t_2 \to t_1} I_0 = 0$. Since

$$I_1 \leq C_2\, t_2^{1+qp} \int_{t_1}^{t_2} (t_2 - s)^{-(1+qp)}m(s)ds$$

$$\leq C_2 \left| t_2^{1+qp} \int_0^{t_2} (t_2 - s)^{-(1+qp)}m(s)ds \right.$$

$$- t_1^{1+qp} \int_0^{t_1} (t_1 - s)^{-(1+qp)}m(s)ds \bigg|$$

$$+ C_2 \int_0^{t_1} \left(t_1^{1+qp}(t_1 - s)^{-(1+qp)} - t_2^{1+qp}(t_2 - s)^{-(1+qp)} \right)m(s)ds,$$

noting that $\int_0^{t_1} t_1^{1+qp}(t_1 - s)^{-(1+qp)}m(s)ds$ exists ($s \in (0, t_1]$), then by Theorem 1.1, we have

$$\lim_{t_2 \to t_1} \int_0^{t_1} \left(t_1^{1+qp}(t_1 - s)^{-(1+qp)} - t_2^{1+qp}(t_2 - s)^{-(1+qp)} \right)m(s)ds = 0.$$

Thus, by $_0D_t^{qp}\, m \in C(J', \mathbb{R}^+)$, one can deduce that $\lim_{t_2 \to t_1} I_1 = 0$. Since

$$I_2 \leq C_2 \int_0^{t_1} \left| t_2^{1+qp}(t_2 - s)^{q-1} - t_1^{1+qp}(t_1 - s)^{q-1} \right| (t_2 - s)^{-q(1+p)}m(s)ds,$$

noting that

$$\left|t_2^{1+qp}(t_2-s)^{q-1} - t_1^{1+qp}(t_1-s)^{q-1}\right|(t_2-s)^{-q(1+p)}m(s)$$
$$\leq\left(t_2^{1+qp}(t_2-s)^{-(1+qp)} + t_1^{1+qp}(t_1-s)^{q-1}(t_2-s)^{-q(1+p)}\right)m(s)$$
$$\leq\left(t_2^{1+qp}(t_2-s)^{-(1+qp)} + t_1^{1+qp}(t_1-s)^{-(1+qp)}\right)m(s)$$
$$\leq 2t_1^{1+qp}(t_1-s)^{-(1+qp)}m(s),$$

and $\int_0^{t_1} t_1^{1+qp}(t_1-s)^{-(1+qp)}m(s)ds$ exists $(s \in (0,t_1])$, then by Theorem 1.1, we have

$$\int_0^{t_1}\left|t_2^{1+qp}(t_2-s)^{q-1} - t_1^{1+qp}(t_1-s)^{q-1}\right|(t_2-s)^{-q(1+p)}m(s)ds \to 0,$$

as $t_2 \to t_1$. Thus one can deduce that $\lim_{t_2 \to t_1} I_2 = 0$.

For $\varepsilon > 0$ be enough small, we have

$$\begin{aligned}
I_3 &\leq \int_0^{t_1-\varepsilon} t_1^{1+qp}(t_1-s)^{q-1} \\
&\quad \times \|P_q(t_2-s) - P_q(t_1-s)\|_{\mathscr{L}(X)}\, |f(s,x(s))|ds \\
&\quad + \int_{t_1-\varepsilon}^{t_1} t_1^{1+qp}(t_1-s)^{q-1} \\
&\quad \times \|P_q(t_2-s) - P_q(t_1-s)\|_{\mathscr{L}(X)}\, |f(s,x(s))|ds \\
&\leq t_1^{1+qp} \int_0^{t_1} (t_1-s)^{q-1}m(s)ds \\
&\quad \times \sup_{s\in[0,t_1-\varepsilon]} \|P_q(t_2-s) - P_q(t_1-s)\|_{\mathscr{L}(X)} \\
&\quad + C_2 \int_{t_1-\varepsilon}^{t_1} t_1^{1+qp}(t_1-s)^{q-1} \\
&\quad \times \left((t_2-s)^{-q(1+p)} + (t_1-s)^{-q(1+p)}\right)m(s)ds \\
&\leq t_1^{1+q+2qp} \int_0^{t_1} (t_1-s)^{-(1+qp)}m(s)ds \\
&\quad \times \sup_{s\in[0,t_1-\varepsilon]} \|P_q(t_2-s) - P_q(t_1-s)\|_{\mathscr{L}(X)} \\
&\quad + 2C_2 \int_{t_1-\varepsilon}^{t_1} t_1^{1+qp}(t_1-s)^{-(1+qp)}m(s)ds.
\end{aligned}$$

By (H_0) and $\lim_{t_2 \to t_1} I_1 = 0$, it is easy to see I_3 tends to zero independently of $y \in B(J)$ as $t_2 \to t_1$, $\varepsilon \to 0$. Therefore, $|(\mathscr{T}y)(t_2) - (\mathscr{T}y)(t_1)|$ tends to zero independently of $y \in B(J)$ as $t_2 \to t_1$, which means that $\{\mathscr{T}y : y \in B(J)\}$ is equicontinuous. $\square$

Lemma 2.5. *Let $A \in \Theta_\omega^p(X)$ with $-1 < p < 0$ and $0 < \omega < \frac{\pi}{2}$. Assume that (H_1)-(H_3) hold, then the operator $\mathscr{T} : B(J) \to B(J)$ is bounded and continuous provided $x_0 \in D(A^\beta)$ with $\beta > 1 + p$.*

Proof. Claim 1. $\mathscr{T}$ maps $B(J)$ into $B(J)$.

For any $y \in B(J)$, let $x(t) = t^{-(1+qp)} y(t)$. Then $x \in B_r^{(q)}(J')$.

For $t \in [0, a]$, by (H_1)-(H_3), we have

$$|(\mathscr{T}y)(t)| \leq |P_q(t)x_0| + t^{1+qp} \left| \int_0^t (t-s)^{-(1+qp)} P_q(t-s) f(s, x(s)) ds \right|$$

$$\leq C_2 |x_0| + C_2 t^{1+qp} \int_0^t (t-s)^{-(1+qp)} |f(s, x(s))| ds$$

$$\leq C_2 \left(|x_0| + \sup_{t \in [0,a]} \left\{ t^{1+qp} \int_0^t (t-s)^{-(1+qp)} m(s) ds \right\} \right)$$

$$\leq r.$$

Hence, $\|\mathscr{T}y\| \leq r$, for any $y \in B(J)$.
Claim 2. $\mathscr{T}$ is continuous in $B(J)$.

For any $y_m, y \in B(J)$, $m = 1, 2, ...$, with $\lim_{m \to \infty} y_m = y$, we have

$$\lim_{m \to \infty} y_m(t) = y(t) \quad \text{and} \quad \lim_{m \to \infty} t^{-(1+qp)} y_m(t) = t^{-(1+qp)} y(t),$$

for $t \in (0, a]$. Then by (H_1), we have

$$f(t, x_m(t)) = f(t, t^{-(1+qp)} y_m(t))$$
$$\to f(t, t^{-(1+qp)} y(t)) = f(t, x(t)), \quad \text{as } m \to \infty,$$

where $x_m(t) = t^{-(1+qp)} y_m(t)$ and $x(t) = t^{-(1+qp)} y(t)$.

On the one hand, using (H_2), we get for each $t \in J'$,

$$(t-s)^{-(1+qp)} |f(s, x_m(s)) - f(s, x(s))| \leq (t-s)^{-(1+qp)} 2m(s),$$

for a.e. $s \in [0, t]$. On the other hand, the function $s \to (t - s)^{-(1+qp)} 2m(s)$ is integrable for $s \in [0, t]$ and $t \in J$. By Theorem 1.1, we get

$$\int_0^t (t - s)^{-(1+qp)} |f(s, x_m(s)) - f(s, x(s))| ds \to 0, \quad \text{as } m \to 0.$$

For $t \in [0, a]$,

$$\begin{aligned}
&|(\mathscr{T} y_m)(t) - (\mathscr{T} y)(t)| \\
=& |t^{1+qp} (T x_m(t) - T x(t))| \\
\leq& t^{1+qp} \left| \int_0^t (t - s)^{-(1+qp)} P_q(t - s) (f(s, x_m(s)) - f(s, x(s))) ds \right| \\
\leq& C_2 t^{1+qp} \int_0^t (t - s)^{-(1+qp)} |f(s, x_m(s)) - f(s, x(s))| ds \\
\to& 0, \quad \text{as } m \to \infty.
\end{aligned}$$

Therefore, $\mathscr{T} y_m \to \mathscr{T} y$ pointwise on J as $m \to \infty$. By Lemma 2.4, $\mathscr{T} y_m \to \mathscr{T} y$ uniformly on J as $m \to \infty$, and so $\mathscr{T}$ is continuous. $\square$

Lemma 2.6. *Assume that $\{Q(t)\}_{t \geq 0}$ is compact. Then $P_q(t)$ is also compact for every $t > 0$.*

Theorem 2.1. *Let $A \in \Theta_\omega^p(X)$ with $-1 < p < 0$ and $0 < \omega < \frac{\pi}{2}$. Assume that $\{Q(t)\}_{t \geq 0}$ is compact. Furthermore, assume that conditions (H_1)-(H_3) are satisfied. Then Cauchy problem (2.1) has at least one mild solution in $B_r^{(q)}(J')$ for every $x_0 \in D(A^\beta)$ with $\beta > 1 + p$.*

Proof. According to Property 1.12, $\{Q(t)\}_{t \geq 0}$ is equicontinuous, which implies that (H_0) is satisfied. Then, by Lemmas 2.4 and 2.5, we know that $\mathscr{T} : B(J) \to B(J)$ is bounded, continuous, and equicontinuous.

Next, we will show that for any $t \in [0, a]$, $V(t) = \{(\mathscr{T} y)(t), y \in B(J)\}$ is relatively compact in X.

Obviously, $V(0)$ is relatively compact in X. Let $t \in (0, a]$ be fixed. For $\forall \varepsilon \in (0, t)$ and $\forall \delta > 0$, define an operator $\mathscr{T}_{\varepsilon, \delta}$ on $B(J)$ by the formula

$$(\mathscr{T}_{\varepsilon, \delta} y)(t) = t^{q(1+p)} P_q(t) x_0 + q t^{1+qp} \int_0^{t-\varepsilon} \int_\delta^\infty \theta (t - s)^{q-1} \Psi_q(\theta)$$

$$\times\, Q((t-s)^q\theta)f(s,x(s))d\theta ds$$

$$=t^{q(1+p)}P_q(t)x_0 + qt^{1+qp}Q(\varepsilon^q\delta)\int_0^{t-\varepsilon}\int_\delta^\infty \theta(t-s)^{q-1}\Psi_q(\theta)$$

$$\times\, Q((t-s)^q\theta - \varepsilon^q\delta)f(s,x(s))d\theta ds,$$

where $x(t) = t^{-(1+qp)}y(t)$, $t \in (0,a]$. Then from the compactness of $P_q(t)$ $(t > 0)$ and $Q(\varepsilon^q\delta)$ $(\varepsilon^q\delta > 0)$, we obtain that the set $V_{\varepsilon,\delta}(t) = \{(\mathscr{T}_{\varepsilon,\delta}y)(t) : y \in B(J)\}$ is relatively compact in X for $\forall\, \varepsilon \in (0,t)$ and $\forall\, \delta > 0$. Moreover, for every $y \in B(J)$, we have

$$|(\mathscr{T}y)(t) - (\mathscr{T}_{\varepsilon,\delta}y)(t)|$$

$$\leq \left| qt^{1+qp}\int_0^t\int_0^\delta \theta(t-s)^{q-1}\Psi_q(\theta)Q((t-s)^q\theta)f(s,x(s))d\theta ds \right|$$

$$+ \left| qt^{1+qp}\int_{t-\varepsilon}^t\int_\delta^\infty \theta(t-s)^{q-1}\Psi_q(\theta)Q((t-s)^q\theta)f(s,x(s))d\theta ds \right|$$

$$\leq qC_0t^{1+qp}\int_0^t(t-s)^{-(1+qp)}m(s)ds\int_0^\delta \theta^{-p}\Psi_q(\theta)d\theta$$

$$+ qC_0t^{1+qp}\int_{t-\varepsilon}^t(t-s)^{-(1+qp)}m(s)ds\int_0^\infty \theta^{-p}\Psi_q(\theta)d\theta$$

$$\leq qC_0t^{1+qp}\int_0^t(t-s)^{-(1+qp)}m(s)ds\int_0^\delta \theta^{-p}\Psi_q(\theta)d\theta$$

$$+ \frac{qC_0\Gamma(1-p)}{\Gamma(1-qp)}t^{1+qp}\int_{t-\varepsilon}^t(t-s)^{-(1+qp)}m(s)ds$$

$$\to 0, \quad \text{as } \varepsilon \to 0,\ \delta \to 0.$$

Therefore, there are relatively compact sets arbitrarily close to the set $V(t)$, $t > 0$. Hence, the set $V(t)$, $t > 0$ is also relatively compact in X. Therefore, $\{\mathscr{T}y : y \in B(J)\}$ is relatively compact by Lemma 1.2. Thus, the continuity of $\mathscr{T}$ and relatively compactness of $\{\mathscr{T}y : y \in B(J)\}$ imply that $\mathscr{T}$ is a completely continuous operator. Schauder fixed point theorem shows that $\mathscr{T}$ has a fixed point in $y^* \in B(J)$. Let $x^*(t) = t^{q-1}y^*(t)$. Then x^* is a mild solution of (2.1). □

Condition (H_2) can be replaced by the following condition:

$(H_2)'$ there exists a constant $q_1 \in (0, q)$ and a function $m \in L^{-\frac{1}{q_1 p}}(J, \mathbb{R}^+)$ such that

$$|f(t, x(t))| \leq m(t) \text{ for all } x \in B_r^{(q)}(J') \text{ and a.e. } t \in (0, a].$$

Corollary 2.1. *Let* $A \in \Theta_\omega^p(X)$ *with* $-1 < p < 0$ *and* $0 < \omega < \frac{\pi}{2}$. *Let all the assumptions in Theorem 2.1 be given except* (H_2). *Assume that* $(H_2)'$ *holds. Then Cauchy problem (2.1) has at least one mild solution in* $B_r^{(q)}(J')$ *for every* $x_0 \in D(A^\beta)$ *with* $\beta > 1 + p$.

Proof. In fact, if $(H_2)'$ holds, by using Hölder inequality, for any $t_1, t_2 \in J'$ and $t_1 < t_2$, we obtain

$$|_0D_t^{qp}m(t_2) - {_0}D_t^{qp}m(t_1)|$$

$$= \frac{1}{\Gamma(-qp)}\left| \int_0^{t_1} \left((t_2 - s)^{-(1+qp)} - (t_1 - s)^{-(1+qp)} \right) m(s)ds \right.$$

$$\left. + \int_{t_1}^{t_2} (t_2 - s)^{-(1+qp)} m(s)ds \right|$$

$$\leq \frac{1}{\Gamma(-qp)} \left(\int_0^{t_1} \left((t_1 - s)^{-(1+qp)} - (t_2 - s)^{-(1+qp)} \right)^{\frac{1}{1+q_1 p}} ds \right)^{1+q_1 p}$$

$$\times \left(\int_0^{t_1} (m(s))^{-\frac{1}{q_1 p}} ds \right)^{-q_1 p}$$

$$+ \frac{1}{\Gamma(-qp)} \left(\int_{t_1}^{t_2} \left((t_2 - s)^{-(1+qp)} \right)^{\frac{1}{1+q_1 p}} ds \right)^{1+q_1 p}$$

$$\times \left(\int_{t_1}^{t_2} (m(s))^{-\frac{1}{q_1 p}} ds \right)^{-q_1 p}$$

$$\leq \frac{\|m\|_{L^{-\frac{1}{q_1 p}}J}}{\Gamma(-qp)} \left(\int_0^{t_1} \left((t_1 - s)^{\frac{-(1+qp)}{1+q_1 p}} - (t_2 - s)^{\frac{-(1+qp)}{1+q_1 p}} \right) ds \right)^{1+q_1 p}$$

$$+ \frac{\|m\|_{L^{-\frac{1}{q_1 p}}J}}{\Gamma(-qp)} \left(\int_{t_1}^{t_2} (t_2 - s)^{\frac{-(1+qp)}{1+q_1 p}} ds \right)^{1+q_1 p}$$

$$\leq \frac{\|m\|_{L^{-\frac{1}{q_1 p}}J}}{\Gamma(-qp)} \left(\frac{1 + q_1 p}{(q_1 - q)p} \right)^{1+q_1 p}$$

$$\times \left(t_1^{\frac{(q_1 - q)p}{1+q_1 p}} + (t_2 - t_1)^{\frac{(q_1 - q)p}{1+q_1 p}} - t_2^{\frac{(q_1 - q)p}{1+q_1 p}} \right)^{1+q_1 p}$$

$$+ \frac{\|m\|_{L^{-\frac{1}{q_1 p}} J}}{\Gamma(-qp)} \left(\frac{1 + q_1 p}{(q_1 - q)p} \right)^{1+q_1 p} \left((t_2 - t_1)^{\frac{(q_1 - q)p}{1 + q_1 p}} \right)^{1+q_1 p}$$

$$\leq \frac{2\|m\|_{L^{-\frac{1}{q_1 p}} J}}{\Gamma(-qp)} \left(\frac{1 + q_1 p}{(q_1 - q)p} \right)^{1+q_1 p} (t_2 - t_1)^{(q_1 - q)p} \qquad (2.14)$$

$$\to 0, \quad \text{as } t_2 \to t_1,$$

where

$$\|m\|_{L^{-\frac{1}{q_1 p}} J} = \left(\int_0^a (m(t))^{-\frac{1}{q_1 p}} dt \right)^{-q_1 p}.$$

On the other hand,

$$t^{1+qp} \int_0^t (t - s)^{-(1+qp)} m(s) ds$$

$$\leq t^{1+qp} \left(\int_0^t (t - s)^{\frac{-(1+qp)}{1+q_1 p}} ds \right)^{1+q_1 p} \left(\int_0^t (m(s))^{-\frac{1}{q_1 p}} ds \right)^{-q_1 p} \qquad (2.15)$$

$$\leq \left(\frac{1 + q_1 p}{(q_1 - q)p} \right)^{1+q_1 p} t^{1+q_1 p} \|m\|_{L^{-\frac{1}{q_1 p}} J}$$

$$\to 0, \quad \text{as } t \to 0.$$

Thus, (2.14) and (2.15) mean that $_0 D_t^{qp} m \in C(J', \mathbb{R}^+)$, and

$$\lim_{t \to 0+} t^{1+qp} {_0 D_t^{qp}} m(t) = 0.$$

Hence, (H_2) holds. By Theorem 2.1, Cauchy problem (2.1) has at least one mild solution in $B_r^{(q)}(J')$. $\qquad \square$

Let Ω be a bounded domain in $\mathbb{R}^N (N \geq 0)$ with boundary $\partial\Omega$ of class C^4. Let $X = C^\lambda(\Omega)$, $\lambda \in (0, 1)$. Set

$$A = \Delta, \quad D(A) = \{u \in C^{2+\lambda}(\Omega) : u(0) = 0 \text{ on } \partial\Omega\}.$$

It follows from Example 2.3 in Mainardi et al. [175] that there exist $\varsigma, \varepsilon > 0$, such that

$$A + \varsigma \in \Theta_{\frac{\pi}{2} - \varepsilon}^{\frac{\lambda}{2} - 1} (C^\lambda(\Omega)).$$

Example 2.1. Consider the following fractional initial boundary value problem

$$\begin{cases} _0D_t^q x(t, z) = \Delta x(t, z) + f(t, x(t)), & \text{a.e. } t \in [0, a], z \in \Omega, \\ u|_{\partial\Omega} = 0, \\ _0D_t^{q-1} x(0, z) = x_0(z), & z \in \Omega \end{cases} \tag{2.16}$$

in the space X, where $q = \frac{1}{2}$, $f(t, x(t)) = t^{-\frac{1}{3}} \sin x(t)$.

Cauchy problem (2.16) can be written abstractly as

$$\begin{cases} _0D_t^{\frac{1}{2}} x(t) = Ax(t) + t^{-\frac{1}{3}} \sin x(t), & \text{a.e. } t \in (0, a], \\ _0D_t^{\frac{1}{2}} x(0) = x_0. \end{cases}$$

We choose

$$m(t) = t^{-\frac{1}{3}} \text{ and } r = C_2|x_0| + \frac{C_2 a^{\frac{2}{3}} \Gamma(-\frac{p}{2})\Gamma(\frac{2}{3})}{\Gamma(-\frac{p}{2} + \frac{2}{3})}.$$

Then, conditions (H_1)-(H_3) are satisfied. According to Theorem 2.1, Cauchy problem (2.16) has a mild solution on $B_r^{(\frac{1}{2})}((0, a])$.

In the following, we consider the case that $Q(t)$ is noncompact.

If $Q(t)$ is noncompact, we give an assumption as follows:

(H_4) there exists a constant $\ell > 0$ such that for any bounded $D \subset X$,

$$\alpha(F(D)) \leq \ell\alpha(D).$$

Theorem 2.2. Let $A \in \Theta_\omega^p(X)$ with $-1 < p < 0$ and $0 < \omega < \frac{\pi}{2}$. Assume that (H_0)-(H_4) hold. Then Cauchy problem (2.1) has at least one mild solution in $B_r^{(q)}(J')$ for every $x_0 \in D(A^\beta)$ with $\beta > 1 + p$.

Proof. By Lemmas 2.4 and 2.5, we have that $\mathscr{T} : B(J) \to B(J)$ is bounded, continuous, and equicontinuous. Next, we will show that $\mathscr{T}$ is compact in a subset of $B(J)$.

For each bounded subset $B_0 \subset B(J)$, set

$$\mathscr{T}^1(B_0) = \mathscr{T}(B_0), \quad \mathscr{T}^n(B_0) = \mathscr{T}\left(\overline{co}(\mathscr{T}^{n-1}(B_0))\right), \quad n = 2, 3, \dots.$$

From Properties 1.16 and 1.17, for any $\varepsilon > 0$, there is a sequence $\{y_n^{(1)}\}_{n=1}^{\infty} \subset B_0$ such that

$$
\begin{aligned}
&\alpha(\mathscr{T}^1(B_0(t))) \\
&= \alpha(\mathscr{T}(B_0(t))) \\
&\leq 2\alpha \left(t^{1+qp} \int_0^t (t-s)^{q-1} P_q(t-s) F(\{s^{-(1+qp)} y_n^{(1)}(s)\}_{n=1}^{\infty}) ds \right) + \varepsilon \\
&\leq 4C_2 t^{1+qp} \int_0^t (t-s)^{-(1+qp)} \alpha \left(F(\{s^{-(1+qp)} y_n^{(1)}(s)\}_{n=1}^{\infty}) \right) ds + \varepsilon \\
&\leq 4C_2 \ell t^{1+qp} \alpha(\alpha(B_0)) \int_0^t (t-s)^{-(1+qp)} s^{-(1+qp)} ds + \varepsilon \\
&\leq \frac{4C_2 \ell \Gamma^2(-qp) t^{-qp}}{\Gamma(-2qp)} \alpha(B_0) + \varepsilon.
\end{aligned}
$$

Since $\varepsilon > 0$ is arbitrary, we have

$$
\alpha(\mathscr{T}^1(B_0(t))) \leq \frac{4C_2 \ell \Gamma^2(-qp) t^{-qp}}{\Gamma(-2qp)} \alpha(B_0).
$$

From Properties 1.16 and 1.17, for any $\varepsilon > 0$, there is a sequence $\{y_n^{(2)}\}_{n=1}^{\infty} \subset \overline{\text{co}}(\mathscr{T}^1(B_0))$ such that

$$
\begin{aligned}
&\alpha(\mathscr{T}^2(B_0(t))) \\
&= \alpha(\mathscr{T}(\overline{\text{co}}(\mathscr{T}^1(B_0(t))))) \\
&\leq 2\alpha \left(t^{1+qp} \int_0^t (t-s)^{q-1} P_q(t-s) F(\{s^{-(1+qp)} y_n^{(2)}\}_{n=1}^{\infty}) ds \right) + \varepsilon \\
&\leq 4C_2 t^{1+qp} \int_0^t (t-s)^{-(1+qp)} \alpha \left(F(\{s^{-(1+qp)} y_n^{(2)}(s)\}_{n=1}^{\infty}) \right) ds + \varepsilon \\
&\leq 4C_2 \ell t^{1+qp} \int_0^t (t-s)^{-(1+qp)} \alpha \left(\{s^{-(1+qp)} y_n^{(2)}(s)\}_{n=1}^{\infty} \right) ds + \varepsilon \\
&\leq 4C_2 \ell t^{1+qp} \int_0^t (t-s)^{-(1+qp)} s^{-(1+qp)} \alpha \left(\{y_n^{(2)}(s)\}_{n=1}^{\infty} \right) ds + \varepsilon \\
&\leq \frac{(4C_2\ell)^2 t^{1+qp} \Gamma^2(-qp) t^{-qp}}{\Gamma(-2qp)} \alpha(B_0) \int_0^t (t-s)^{-(1+qp)} s^{-(1+2qp)} ds + \varepsilon \\
&= \frac{(4C_2\ell)^2 \Gamma^3(-qp)}{\Gamma(-3qp)} t^{-2qp} \alpha(B_0) + \varepsilon.
\end{aligned}
$$

It can be shown, by mathematical induction, that for every $\bar{n} \in \mathbb{N}^+$,

$$\alpha(\mathscr{T}^{\bar{n}}(B_0(t))) \leq \frac{(4C_2\ell)^{\bar{n}}\Gamma^{\bar{n}+1}(-qp)}{\Gamma(-(\bar{n}+1)qp)} t^{-\bar{n}qp} \alpha(B_0).$$

Since

$$\lim_{\bar{n}\to\infty} \frac{(4C_2\ell a^{-qp})^{\bar{n}}\Gamma^{\bar{n}+1}(-qp)}{\Gamma(-(\bar{n}+1)qp)} = 0,$$

there exists a positive integer $\hat{n}$ such that

$$\frac{(4C_2\ell)^{\hat{n}}\Gamma^{\hat{n}+1}(-qp)}{\Gamma(-(\hat{n}+1)qp)} t^{-\hat{n}qp} \leq \frac{(4C_2\ell a^{-qp})^{\hat{n}}\Gamma^{\hat{n}+1}(-qp)}{\Gamma(-(\hat{n}+1)qp)} = k < 1.$$

Then

$$\alpha(\mathscr{T}^{\hat{n}}(B_0(t))) \leq k\alpha(B_0).$$

We know from Property 1.14, $\mathscr{T}^{\hat{n}}(B_0(t))$ is bounded and equicontinuous. Then, from Property 1.15, we have

$$\alpha(\mathscr{T}^{\hat{n}}(B_0)) = \max_{t\in[0,a]} \alpha(\mathscr{T}^{\hat{n}}(B_0(t))).$$

Hence,

$$\alpha(\mathscr{T}^{\hat{n}}(B_0)) \leq k\alpha(B_0).$$

Let

$$D_0 = B(J), \ D_1 = \overline{\text{co}}(\mathscr{T}^{\hat{n}}(D)), ..., \ D_n = \overline{\text{co}}(\mathscr{T}^{\hat{n}}(D_{n-1})), \ n = 2, 3, ...$$

Then, we can get

(i) $D_0 \supset D_1 \supset D_2 \supset \cdots \supset D_{n-1} \supset D_n \supset \cdots$;
(ii) $\lim_{n\to\infty} \alpha(D_n) = 0$.

Then $\hat{D} = \bigcap_{n=0}^{\infty} D_n$ is a nonempty, compact, and convex subset in $B(J)$.

We will prove $\mathscr{T}(\hat{D}) \subset \hat{D}$. First, we show

$$\mathscr{T}(D_n) \subset D_n, \ n = 0, 1, 2, ... \tag{2.17}$$

From $\mathscr{T}^1(D_0) = \mathscr{T}(D_0) \subset D_0$, we know $\overline{\text{co}}(\mathscr{T}^1(D_0)) \subset D_0$. Therefore

$$\mathscr{T}^2(D_0) = \mathscr{T}(\overline{\text{co}}(\mathscr{T}^1(D_0))) \subset \mathscr{T}(D_0) = \mathscr{T}^1(D_0),$$
$$\mathscr{T}^3(D_0) = \mathscr{T}(\overline{\text{co}}(\mathscr{T}^2(D_0))) \subset \mathscr{T}(\overline{\text{co}}(\mathscr{T}^1(D_0))) = \mathscr{T}^2(D_0),$$

$$\cdots$$

$$\mathscr{T}^{\hat{n}}(D_0) = \mathscr{T}(\overline{\text{co}}(\mathscr{T}^{\hat{n}-1}(D_0))) \subset \mathscr{T}(\overline{\text{co}}(\mathscr{T}^{\hat{n}-2}(D_0))) = \mathscr{T}^{\hat{n}-1}(D_0).$$

Hence, $D_1 = \overline{\text{co}}(\mathscr{T}^{\hat{n}}(D_0)) \subset \overline{\text{co}}(\mathscr{T}^{\hat{n}-1}(D_0))$, so

$$\mathscr{T}(D_1) \subset \mathscr{T}(\overline{\text{co}}(\mathscr{T}^{\hat{n}-1}(D_0))) = \mathscr{T}^{\hat{n}}(D_0) \subset \overline{\text{co}}(\mathscr{T}^{\hat{n}}(D_0)) = D_1.$$

Employing the same method, we can prove $\mathscr{T}(D_n) \subset D_n (n = 0, 1, 2, \ldots)$. By (2.17), we get $\mathscr{T}(\hat{D}) \subset \bigcap_{n=0}^{\infty} \mathscr{T}(D_n) \subset \bigcap_{n=0}^{\infty} D_n = \hat{D}$. Then $\mathscr{T}(\hat{D})$ is compact. Hence, Schauder fixed point theorem shows that $\mathscr{T}$ has a fixed point $y^* \in B(J)$. Let $x^*(t) = t^{q-1}y^*(t)$. Then x^* is a mild solution of (2.1). $\qquad \square$

Corollary 2.2. *Let $A \in \Theta_\omega^p(X)$ with $-1 < p < 0$ and $0 < \omega < \frac{\pi}{2}$. Let all the assumptions in Theorem 2.2 be given except condition (H_2). Assume that $(H_2)'$ holds. Then Cauchy problem (2.1) has at least one mild solution in $B_r^{(q)}(J')$ for every $x_0 \in D(A^\beta)$ with $\beta > 1 + p$.*

The proof of Corollary 2.2 is similar to that of Corollary 2.1, it is thus omitted.

In the following, we also give the existence and uniqueness result which is based on Banach contraction principle. We will need the following assumption:

(H_5) there exists a constant $k > 0$ such that for any $x, y \in B_r^{(q)}(J')$, and $t \in (0, a]$ we have

$$|f(t, x(t)) - f(t, y(t))| < k\|x - y\|_q.$$

Theorem 2.3. *Let $A \in \Theta_\omega^p(X)$ with $-1 < p < 0$ and $0 < \omega < \frac{\pi}{2}$. If assumptions (H_2), (H_3), and (H_5) hold. Then, for every $x_0 \in D(A^\beta)$ with $\beta > 1 + p$, Cauchy problem (2.1) has a unique mild solution in $B_r^{(q)}(J')$*

provided

$$-\frac{C_2 ak}{qp} < 1. \tag{2.18}$$

Proof. By the proof of Lemma 2.5, we know that $\mathscr{T}$ is an operator from $B(J)$ into itself.

For any $y_1, y_2 \in B(J)$, let $x_1(t) = t^{-(1+qp)} y_1(t)$ and $x_2(t) = t^{-(1+qp)} y_2(t)$, $t \in (0, a]$, we have

$$
\begin{aligned}
&|(\mathscr{T} y_1)(t) - (\mathscr{T} y_2)(t)| \\
=& t^{1+qp} |(T x_1)(t) - (T x_2)(t)| \\
=& t^{1+qp} \left| \int_0^t (t-s)^{q-1} P_q(t-s)[f(s, x_1(s)) - f(s, x_2(s))] ds \right| \\
\leq& C_2\, t^{1+qp} \int_0^t (t-s)^{-(1+qp)} |f(s, x_1(s)) - f(s, x_2(s))| ds \\
\leq& -\frac{C_2 ak}{qp} \|x_1 - x_2\|_q \\
=& -\frac{C_2 ak}{qp} \|y_1 - y_2\|,
\end{aligned}
$$

which implies

$$|\mathscr{T} y_1 - \mathscr{T} y_2| \leq -\frac{C_2 ak}{qp} \|y_1 - y_2\|.$$

This means that $\mathscr{T}$ is a contraction according to (2.18). By applying Banach contraction principle, we know that $\mathscr{T}$ has a unique fixed point $y^* \in B(J)$. Let $x^*(t) = t^{q-1} y^*(t)$. Then x^* is a mild solution of (2.1). $\qquad\square$

Corollary 2.3. *Let $A \in \Theta_\omega^p(X)$ with $-1 < p < 0$ and $0 < \omega < \frac{\pi}{2}$. Let all the assumptions in Theorem 2.3 be given except condition (H_2). Assume that $(H_2)'$ holds. Then Cauchy problem (2.1) has a unique mild solution in $B_r^{(q)}(J')$ for every $x_0 \in D(A^\beta)$ with $\beta > 1 + p$.*

The proof of Corollary 2.3 is similar to that of Corollary 2.1, it is thus omitted.

2.1.4 Equations with Caputo Derivative

For a positive constant r, let

$$B_r(J) = \{x \in C(J, X) : \|x\| \leq r\}.$$

Observe that $B_r(J)$ is clearly a bounded closed and convex subset of $C(J, X)$.

We introduce the following hypotheses:

$(\overline{H}_1)$ for each $t \in J$, the function $f(t, \cdot) : X \to X$ is continuous and for each $x \in C(J, X)$, the function $f(\cdot, x) : J \to X$ is strongly measurable;

$(\overline{H}_2)$ there exists a function $m \in L(J, \mathbb{R}^+)$ such that

$$_0D_t^{qp}m \in C(J, \mathbb{R}^+), \quad \lim_{t \to 0+} {_0D_t^{qp}}m(t) = 0$$

and

$$|f(t, x)| \leq m(t) \text{ for all } x \in B_r(J) \text{ and a.e. } t \in [0, a];$$

$(\overline{H}_3)$ there exists a constant $r > 0$ such that

$$\sup_{t \in [0,a]} \left\{ |S_q(t)x_0| + C_2 \int_0^t (t - s)^{-(1+qp)}m(s)ds \right\} \leq r.$$

For any $x \in B_r(J)$, we define an operator T as follows:

$$(Tx)(t) = S_q(t)x_0 + \int_0^t (t - s)^{q-1}P_q(t - s)f(s, x(s))ds, \text{ for } t \in [0, a].$$

Before giving the main results, we first prove the following lemmas.

Lemma 2.7. Let $A \in \Theta_\omega^p(X)$ with $-1 < p < 0$ and $0 < \omega < \frac{\pi}{2}$. Assume that (H_0) and $(\overline{H}_1)$-$(\overline{H}_3)$ hold, then the operator $T : B_r(J) \to B_r(J)$ is equicontinuous for every $x_0 \in D(A^\beta)$ with $\beta > 1 + p$.

Proof. For any $x \in B_r(J)$, for the case $t_1 = 0, 0 < t_2 \leq a$, we get

$$|(Tx)(t_2) - (Tx)(0)|$$

$$\leq |S_q(t_2)x_0 - x_0| + \left| \int_0^{t_2} (t_2 - s)^{q-1}P_q(t_2 - s)f(s, x(s))ds \right|$$

$$\leq \left| S_q(t_2)x_0 - x_0 \right| + C_2 \int_0^{t_2} (t_2 - s)^{-(1+qp)} m(s) ds$$

$$\to 0, \quad \text{as } t_2 \to 0.$$

For $0 < t_1 < t_2 \leq a$, we have

$$\left| (Tx)(t_2) - (Tx)(t_1) \right|$$

$$\leq \left| S_q(t_2)x_0 - S_q(t_1)x_0 \right| + \left| \int_0^{t_2} (t_2 - s)^{q-1} P_q(t_2 - s) f(s, x(s)) ds \right.$$

$$\left. - \int_0^{t_1} (t_1 - s)^{q-1} P_q(t_1 - s) f(s, x(s)) ds \right|$$

$$\leq \left| S_q(t_2)x_0 - S_q(t_1)x_0 \right| + \left| \int_{t_1}^{t_2} (t_2 - s)^{q-1} P_q(t_2 - s) f(s, x(s)) ds \right|$$

$$+ \left| \int_0^{t_1} (t_2 - s)^{q-1} P_q(t_2 - s) f(s, x(s)) ds \right.$$

$$\left. - \int_0^{t_1} (t_1 - s)^{q-1} P_q(t_2 - s) f(s, x(s)) ds \right|$$

$$+ \left| \int_0^{t_1} (t_1 - s)^{q-1} P_q(t_2 - s) f(s, x(s)) ds \right.$$

$$\left. - \int_0^{t_1} (t_1 - s)^{q-1} P_q(t_1 - s) f(s, x(s)) ds \right|$$

$$\leq \left| S_q(t_2)x_0 - S_q(t_1)x_0 \right| + C_2 \int_{t_1}^{t_2} (t_2 - s)^{-(1+qp)} m(s) ds$$

$$+ C_2 \int_0^{t_1} \left((t_1 - s)^{q-1} - (t_2 - s)^{q-1} \right) (t_2 - s)^{-q(1+p)} m(s) ds$$

$$+ \int_0^{t_1} (t_1 - s)^{q-1} \| P_q(t_2 - s) - P_q(t_1 - s) \|_{\mathscr{L}(X)} m(s) ds$$

$$\leq \left| S_q(t_2)x_0 - S_q(t_1)x_0 \right| + C_2 \left| \int_0^{t_2} (t_2 - s)^{-(1+qp)} m(s) ds \right.$$

$$\left. - \int_0^{t_1} (t_1 - s)^{-(1+qp)} m(s) ds \right|$$

$$+ 2C_2 \int_0^{t_1} \left((t_1 - s)^{-(1+qp)} - (t_2 - s)^{-(1+qp)} \right) m(s) ds$$

$$+ \int_0^{t_1} (t_1 - s)^{q-1} \| P_q(t_2 - s) - P_q(t_1 - s) \|_{\mathscr{L}(X)} m(s) ds$$

$$= : I_0 + I_1 + I_2 + I_3.$$

By Proposition 2.2, it is easy to see that $\lim_{t_2 \to t_1} I_0 = 0$. One can deduce that $\lim_{t_2 \to t_1} I_1 = 0$, since $_0 D_t^{qp} m \in C(J, \mathbb{R}^+)$.

Noting that

$$\left((t_1 - s)^{-(1+qp)} - (t_2 - s)^{-(1+qp)} \right) m(s) \leq (t_1 - s)^{-(1+qp)} m(s),$$

and $\int_0^{t_1} (t_1 - s)^{-(1+qp)} m(s) ds$ exists ($s \in (0, t_1]$), then by Theorem 1.1, we have

$$\int_0^{t_1} \left((t_1 - s)^{-(1+qp)} - (t_2 - s)^{-(1+qp)} \right) m(s) ds \to 0, \quad \text{as } t_2 \to t_1.$$

Thus one can deduce that $\lim_{t_2 \to t_1} I_2 = 0$.

For $\varepsilon > 0$ be enough small, we have

$$
\begin{aligned}
I_3 &= \int_0^{t_1-\varepsilon} (t_1 - s)^{q-1} \|P_q(t_2 - s) - P_q(t_1 - s)\|_{\mathscr{L}(X)} \, m(s) ds \\
&\quad + \int_{t_1-\varepsilon}^{t_1} (t_1 - s)^{q-1} \|P_q(t_2 - s) - P_q(t_1 - s)\|_{\mathscr{L}(X)} \, m(s) ds \\
&\leq \int_0^{t_1-\varepsilon} (t_1 - s)^{q-1} m(s) ds \\
&\quad \times \sup_{s \in [0, t_1-\varepsilon]} \|P_q(t_2 - s) - P_q(t_1 - s)\|_{\mathscr{L}(X)} \\
&\quad + 2C_2 \int_{t_1-\varepsilon}^{t_1} (t_1 - s)^{-(1+qp)} m(s) ds \\
&\leq t_1^{q(1+p)} \int_0^{t_1} (t_1 - s)^{-(1+qp)} m(s) ds \\
&\quad \times \sup_{s \in [0, t_1-\varepsilon]} \|P_q(t_2 - s) - P_q(t_1 - s)\|_{\mathscr{L}(X)} \\
&\quad + 2C_2 \left| \int_0^{t_1} (t_1 - s)^{-(1+qp)} m(s) ds \right. \\
&\quad \left. - \int_0^{t_1-\varepsilon} (t_1 - \varepsilon - s)^{-(1+qp)} m(s) ds \right| \\
&\quad + 2C_2 \int_0^{t_1-\varepsilon} \left((t_1 - \varepsilon - s)^{-(1+qp)} - (t_1 - s)^{-(1+qp)} \right) m(s) ds
\end{aligned}
$$

$$= : I_{31} + I_{32} + I_{33}.$$

By (H_0), it is easy to see that $I_{31} \to 0$ as $t_2 \to t_1$. Similar to the proofs that I_1, I_2 tend to zero, we get $I_{32} \to 0$ and $I_{33} \to 0$ as $\varepsilon \to 0$. Thus, I_3 tends to zero independently of $y \in B(J)$ as $t_2 \to t_1$, $\varepsilon \to 0$. Therefore, $|(Tx)(t_2) - (Tx)(t_1)|$ tends to zero independently of $y \in B(J)$ as $t_2 \to t_1$, which means that $\{Tx, \ y \in B(J)\}$ is equicontinuous. $\qquad \square$

Lemma 2.8. *Let $A \in \Theta^p_\omega(X)$ with $-1 < p < 0$ and $0 < \omega < \frac{\pi}{2}$. Assume that $(\overline{H}_1)$-$(\overline{H}_3)$ hold, then the operator $T : B_r(J) \to B_r(J)$ is bounded and continuous for every $x_0 \in D(A^\beta)$ with $\beta > 1 + p$.*

Proof. For any $x \in B_r(J)$, we have

$$|(Tx)(t)| \leq |S_q(t)x_0| + \left| \int_{0^-}^t (t - s)^{q-1} P_q(t - s) f(s, x(s)) ds \right|$$

$$\leq |S_q(t)x_0| + C_2 \int_0^t (t - s)^{-(1+qp)} m(s) ds.$$

Due to $(\overline{H}_3)$, we have

$$\|Tx\| \leq r, \quad \text{for each } x \in B_r(J),$$

which implies that $T : B_r(J) \to B_r(J)$ is bounded.

Next, we will show that T is continuous. For any $x_m, x \in B_r(J)$, $m = 1, 2, \ldots$, with $\lim_{m \to \infty} \|x_m - x\| = 0$, by condition $(\overline{H}_1)$, we have

$$\lim_{m \to \infty} f(s, x_m(s)) = f(s, x(s)).$$

On the one hand, using $(\overline{H}_2)$, we get for each $t \in J$,

$$(t - s)^{-(1+qp)} |f(s, x_m(s)) - f(s, x(s))|$$
$$\leq (t - s)^{-(1+qp)} 2m(s), \quad \text{a.e. } s \in [0, t].$$

On the other hand, the function $s \to (t - s)^{q-1} 2m(s)$ is integrable for $s \in [0, t]$ and $t \in J$. By Theorem 1.1, we get

$$|(Tx_m)(t) - (Tx)(t)|$$
$$= \left| \int_0^t (t - s)^{q-1} P_q(t - s) \big(f(s, x_m(s)) - f(s, x(s)) \big) ds \right|$$

$$\leq C_2 \int_0^t (t-s)^{-(1+qp)} |f(s, x_m(s)) - f(s, x(s))| ds$$
$$\to 0, \quad \text{as } m \to \infty.$$

Therefore, $T x_m \to T x$ pointwise on J as $m \to \infty$. By Lemma 2.7, $T x_m \to T x$ uniformly on J as $m \to \infty$ and so T is continuous. □

Lemma 2.9. *Assume that $\{Q(t)\}_{t \geq 0}$ is compact. Then $S_q(t)$ and $P_q(t)$ are also compact for every $t > 0$.*

Theorem 2.4. *Let $A \in \Theta_\omega^p(X)$ with $-1 < p < 0$ and $0 < \omega < \frac{\pi}{2}$. Assume that $\{Q(t)\}_{t \geq 0}$ is compact. Furthermore, assume that conditions $(\overline{H}_1)$-$(\overline{H}_3)$ hold. Then Cauchy problem (2.2) has at least one mild solution in $B_r(J)$ for every $x_0 \in D(A^\beta)$ with $\beta > 1 + p$.*

Proof. According to Property 1.12, $\{Q(t)\}_{t \geq 0}$ is equicontinuous, which implies that (H_0) is satisfied. Then, by Lemmas 2.7 and 2.8, we know that $T : B_r(J) \to B_r(J)$ is bounded, continuous, and equicontinuous.

Next, we will show that for any $t \in [0, a]$, $V(t) = \{(Tx)(t) : x \in B_r(J)\}$ is relatively compact in X.

Obviously, $V(0) = \{x_0 : x \in B_r(J)\}$ is relatively compact in X. Let $t \in (0, a]$ be fixed. For $\forall \varepsilon \in (0, t)$ and $\forall \delta > 0$, define an operator $T_{\varepsilon, \delta}$ on $B_r(J)$ by the formula

$$(T_{\varepsilon, \delta} x)(t)$$
$$= S_q(t) x_0 + q \int_0^{t-\varepsilon} \int_\delta^\infty \theta(t-s)^{q-1} \Psi_q(\theta) Q((t-s)^q \theta) f(s, x(s)) d\theta ds$$
$$= S_q(t) x_0 + q Q(\varepsilon^q \delta) \int_0^{t-\varepsilon} \int_\delta^\infty \theta(t-s)^{q-1} \Psi_q(\theta)$$
$$\times Q((t-s)^q \theta - \varepsilon^q \delta) f(s, x(s)) d\theta ds,$$

where $x \in B_r(J)$. Then from the compactness of $S_q(t)$ $(t > 0)$ and $Q(\varepsilon^q \delta)$ $(\varepsilon^q \delta > 0)$, we obtain that the set $V_{\varepsilon, \delta}(t) = \{(T_{\varepsilon, \delta} x)(t) : x \in B_r(J)\}$ is relatively compact in X for $\forall \varepsilon \in (0, t)$ and $\forall \delta > 0$. Moreover, for every $x \in B_r(J)$, we have

$$|(Tx)(t) - (T_{\varepsilon, \delta} x)(t)|$$

$$\leq \left| q \int_0^t \int_0^\delta \theta(t-s)^{q-1} \Psi_q(\theta) Q((t-s)^q \theta) f(s, x(s)) d\theta ds \right|$$

$$+ \left| q \int_{t-\varepsilon}^t \int_\delta^\infty \theta(t-s)^{q-1} \Psi_q(\theta) Q((t-s)^q \theta) f(s, x(s)) d\theta ds \right|$$

$$\leq q C_0 \int_0^t (t-s)^{-(1+qp)} m(s) ds \int_0^\delta \theta^{-p} \Psi_q(\theta) d\theta$$

$$+ q C_0 \int_{t-\varepsilon}^t (t-s)^{-(1+qp)} m(s) ds \int_0^\infty \theta^{-p} \Psi_q(\theta) d\theta$$

$$\leq q C_0 \int_0^t (t-s)^{-(1+qp)} m(s) ds \int_0^\delta \theta^{-p} \Psi_q(\theta) d\theta$$

$$+ \frac{q C_0 \Gamma(1-p)}{\Gamma(1-qp)} \int_{t-\varepsilon}^t (t-s)^{-(1+qp)} m(s) ds$$

$$\to 0, \quad \text{as } \varepsilon \to 0, \ \delta \to 0.$$

Then, there are relatively compact sets arbitrarily close to the set $V(t)$, $t > 0$. Thus, the set $V(t)$, $t > 0$ is also relatively compact in X. Hence, $\{Tx : x \in B_r(J)\}$ is relatively compact by Lemma 1.2. Schauder fixed point theorem shows that T has a fixed point in $B_r(J)$. Therefore, Cauchy problem (2.2) has a mild solution in $B_r(J)$. $\square$

Condition $(\overline{H}_2)$ can be replaced by the following condition:

$(\overline{H}_2)'$ there exists a constant $q_1 \in (0, q)$ and a function $m \in L^{-\frac{1}{q_1 p}}(J, \mathbb{R}^+)$ such that

$$|f(t, x(t))| \leq m(t) \text{ for all } x \in B_r(J), \text{ and a.e. } t \in [0, a].$$

Corollary 2.4. *Let all the assumptions in Theorem 2.4 be given except condition $(\overline{H}_2)$. Furthermore, assume that $(\overline{H}_2)'$ holds. Then Cauchy problem (2.2) has at least one mild solution in $B_r(J)$.*

Proof. In fact, if $(\overline{H}_2)'$ holds, similar to Corollary 2.1, we get $_0D_t^{qp} m \in C(J', \mathbb{R}^+)$. Since

$$\int_0^t (t-s)^{-(1+qp)} m(s) ds$$

$$\leq \left(\int_0^t (t-s)^{\frac{-(1+q_1 p)}{1+q_1 p}} ds \right)^{1+q_1 p} \left(\int_0^t (m(s))^{-\frac{1}{q_1 p}} ds \right)^{-q_1 p}$$

$$\leq \left(\frac{1+q_1 p}{(q_1-q)p} \right)^{1+q_1 p} t^{(q_1-q)p} \|m\|_{L^{-\frac{1}{q_1 p}} J}$$

$$\to 0, \quad \text{as } t \to 0,$$

which implies that $\lim_{t \to 0+} (_0 D_t^{qp} m)(t) = 0$, $(\overline{H}_2)$ holds. By Theorem 2.4, Cauchy problem (2.2) has at least one mild solution in $B_r(J)$. $\square$

In the following, we consider the case that $Q(t)$ is noncompact. We will give the following theorems.

Theorem 2.5. *Let* $A \in \Theta_\omega^p(X)$ *with* $-1 < p < 0$ *and* $0 < \omega < \frac{\pi}{2}$. *Assume that* (H_0), (H_4), *and* $(\overline{H}_1)$-$(\overline{H}_3)$ *hold. Then Cauchy problem (2.2) has at least one mild solution in* $B_r(J)$ *for every* $x_0 \in D(A^\beta)$ *with* $\beta > 1+p$.

Proof. By Lemmas 2.7 and 2.8, we have $T : B_r(J) \to B_r(J)$ is bounded, continuous, and equicontinuous. Next, we will show that T is compact in a subset of $B_r(J)$.

For each bounded subset $B_0 \subset B_r(J)$, set

$$T^1(B_0) = T(B_0), \ T^n(B_0) = T\left(\overline{\text{co}}(T^{n-1}(B_0))\right), \ n = 2, 3, \dots .$$

From Properties 1.16 and 1.17, for any $\varepsilon > 0$, there is a sequence $\{x_n^{(1)}\}_{n=1}^\infty \subset B_0$ such that

$$\alpha(T^1(B_0(t)))$$
$$= \alpha(T(B_0(t)))$$
$$\leq 2\alpha \left(\int_0^t (t-s)^{q-1} P_q(t-s) F(\{x_n^{(1)}(s)\}_{n=1}^\infty) ds \right) + \varepsilon$$
$$\leq 4C_2 \int_0^t (t-s)^{-(1+qp)} \alpha \left(F(\{x_n^{(1)}(s)\}_{n=1}^\infty) \right) ds + \varepsilon$$
$$\leq 4C_2 \ell \alpha(B_0) \int_0^t (t-s)^{-(1+qp)} ds + \varepsilon$$
$$\leq \frac{4C_2 \ell t^{-qp}}{-qp} \alpha(B_0) + \varepsilon.$$

Since $\varepsilon > 0$ is arbitrary, we have

$$\alpha(T^1(B_0(t))) \le \frac{4C_2\ell t^{-qp}}{-qp}\alpha(B_0).$$

From Properties 1.16 and 1.17, for any $\varepsilon > 0$, there is a sequence $\{x_n^{(2)}\}_{n=1}^{\infty} \subset \overline{co}(T^1(B_0))$ such that

$$\alpha(T^2(B_0(t)))$$
$$=\alpha(T(\overline{co}(T^1(B_0(t)))))$$
$$\le 2\alpha\left(\int_0^t (t-s)^{q-1}P_q(t-s)F(\{x_n^{(2)}(s)\}_{n=1}^{\infty})ds\right) + \varepsilon$$
$$\le 4C_2\int_0^t (t-s)^{-(1+qp)}\alpha\left(F(\{x_n^{(2)}(s)\}_{n=1}^{\infty})\right)ds + \varepsilon$$
$$\le 4C_2\ell\int_0^t (t-s)^{-(1+qp)}\alpha\left(\{x_n^{(2)}(s)\}_{n=1}^{\infty}\right)ds + \varepsilon$$
$$\le \frac{(4C_2\ell)^2}{-qp}\alpha(B_0)\int_0^t (t-s)^{-(1+qp)}s^{-qp}ds + \varepsilon$$
$$= \frac{(4C_2\ell)^2\Gamma^2(-qp)}{\Gamma(-2qp+1)}t^{-2qp}\alpha(B_0) + \varepsilon.$$

It can be shown, by mathematical induction, that for every $\bar{n} \in \mathbb{N}^+$,

$$\alpha(T^{\bar{n}}(B_0(t))) \le \frac{(4C_2\ell)^{\bar{n}}\Gamma^{\bar{n}}(-qp)}{\Gamma(-\bar{n}qp+1)}t^{-\bar{n}qp}\alpha(B_0).$$

Since

$$\lim_{\bar{n}\to\infty} \frac{(4C_2\ell a^{-qp})^{\bar{n}}\Gamma^{\bar{n}}(-qp)}{\Gamma(-\bar{n}qp+1)} = 0,$$

there exists a positive integer $\hat{n}$ such that

$$\frac{(4C_2\ell)^{\bar{n}}\Gamma^{\hat{n}}(-qp)}{\Gamma(-\hat{n}qp+1)}t^{-\hat{n}qp} \le \frac{(4C_2\ell a^{-qp})^{\hat{n}}\Gamma^{\hat{n}}(-qp)}{\Gamma(-\hat{n}qp+1)} = k < 1.$$

Then

$$\alpha(T^{\hat{n}}(B_0(t))) \le k\alpha(B_0).$$

From Property 1.14, we know that $T^{\hat{n}}(B_0(t))$ is bounded and equicontinuous. Then, from Property 1.15, we have

$$\alpha(T^{\hat{n}}(B_0)) = \max_{t \in [0,a]} \alpha(T^{\hat{n}}(B_0(t))).$$

Hence,

$$\alpha(T^{\hat{n}}(B_0)) \leq k\alpha(B_0).$$

Let

$$D_0 = B_r(J), \ D_1 = \overline{\text{co}}(T^{\hat{n}}(D)), ..., \ D_n = \overline{\text{co}}(T^{\hat{n}}(D_{n-1})), \ n = 2, 3, ...,$$

and $\hat{D} = \bigcap_{n=0}^{\infty} D_n$. By using the similar method in the proof of Theorem 2.2, we can prove $T(\hat{D}) \subset \hat{D}$. Thus $T(\hat{D})$ is compact. Hence, Schauder fixed point theorem shows that T has a fixed point in $\hat{D}$. Therefore, Cauchy problem (2.2) has a mild solution in $B_r(J)$. $\qquad \square$

Corollary 2.5. *Let $A \in \Theta_\omega^p(X)$ with $-1 < p < 0$ and $0 < \omega < \frac{\pi}{2}$. Let all the assumptions in Theorem 2.5 be given except condition $(\overline{H}_2)$. Furthermore, assume that $(\overline{H}_2)'$ holds. Then Cauchy problem (2.2) has at least one mild solution in $B_r(J)$ for every $x_0 \in D(A^\beta)$ with $\beta > 1 + p$.*

The proof of Corollary 2.5 is similar to that of Corollary 2.4, it is thus omitted.

In the following, we also give the existence and uniqueness result which is based on Banach contraction principle. We will need the following assumption:

$(\overline{H}_4)$ there exists a constant $k > 0$ such that for any $x, y \in B_r(J)$ and $t \in [0, a]$,

$$|f(t, x(t)) - f(t, y(t))| < k\|x - y\|.$$

Theorem 2.6. *If assumptions $(\overline{H}_2)$-$(\overline{H}_4)$ are satisfied, then Cauchy problem (2.2) has a unique mild solution in $B_r(J)$ provided*

$$-\frac{kC_2 a^{-qp}}{qp} < 1. \tag{2.19}$$

Proof. From the proof of Lemma 2.7, we know that T is an operator from $B_r(J)$ into itself.

For any $x, y \in B_r(J)$ and $t \in [0, a]$, we have

$$
\begin{aligned}
&|(Tx)(t) - (Ty)(t)| \\
&= \left| \int_0^t (t-s)^{q-1} P_q(t-s) \big(f(s, x(s)) - f(s, y(s)) \big) ds \right| \\
&\leq \left(C_2 k \int_0^t (t-s)^{-(1+qp)} ds \right) \|x - y\| \\
&= -\frac{C_2 k \, t^{-qp}}{qp} \|x - y\|,
\end{aligned}
$$

which implies

$$
\|Tx - Ty\| \leq -\frac{C_2 k \, t^{-qp}}{qp} \|x - y\|.
$$

This means that T is a contraction according to (2.19). By applying Banach contraction principle, we know that T has a unique fixed point in $B_r(J)$. $\square$

Corollary 2.6. *Let all the assumptions in Theorem 2.6 be given except condition* $(\overline{H}_2)$. *Furthermore, assume that* $(\overline{H}_2)'$ *holds. Then Cauchy problem (2.2) has a unique mild solution in* $B_r(J)$.

The proof of Corollary 2.6 is similar to that of Corollary 2.4; it is thus omitted.

2.2 BOUNDED SOLUTIONS ON REAL AXIS

2.2.1 Introduction

In this section, we study the existence and uniqueness of bounded solutions for the fractional evolution equations in an ordered Banach space X

$$
_{-\infty}D_t^q x(t) + Ax(t) = f(t, x(t)), \quad t \in \mathbb{R}, \tag{2.20}
$$

where $_{-\infty}D_t^q$ is Liouville fractional derivative of order $q \in (0, 1)$ with the lower limit $-\infty$, $-A : D(A) \subset X \to X$ is the infinitesimal generator of a C_0-semigroup $\{T(t)\}_{t \geq 0}$.

Applying Fourier transform, we give the reasonable definition of mild solutions of equation (2.20). The existence and uniqueness results for the corresponding linear fractional evolution equations are established, and the spectral radius of resolvent operator is accurately estimated. Then some sufficient conditions are established for the existence and uniqueness of periodic solutions, S-asymptotically periodic solutions, and other types of bounded solutions when $f : \mathbb{R} \times X \to X$ satisfies some ordered or Lipschitz conditions. The main methods are the monotone iterative technique and Banach contraction principle.

2.2.2 Linear Equations

Consider the following linear fractional evolution equation in a Banach space X

$$_{-\infty}D_t^q x(t) + Ax(t) = h(t), \quad t \in \mathbb{R}, \tag{2.21}$$

where $-A : D(A) \subset X \to X$ is the infinitesimal generator of a C_0-semigroup $\{T(t)\}_{t\geq 0}$, $h : \mathbb{R} \to X$ is a continuous function.

For convenience, we assume the following condition:

(H) $x \in C(\mathbb{R}, X)$, $\int_{-\infty}^t g_{1-q}(t-s)x(s)ds \in C^1(\mathbb{R}, X)$, $x(t) \in D(A)$ for $t \in \mathbb{R}$, $Ax \in L^1(\mathbb{R}, X)$ and x satisfies (2.21), where

$$g_{1-q}(t) = \begin{cases} \dfrac{1}{\Gamma(1-q)} t^{-q}, & t > 0, \\ 0, & t \leq 0. \end{cases}$$

Let

$$V(t) = q \int_0^\infty \theta \Psi_q(\theta) T(t^q \theta) d\theta, \tag{2.22}$$

where $\{T(t)\}_{t\geq 0}$ is a C_0-semigroup. We have the following results.

Lemma 2.10.

(i) *Assume that $\{T(t)\}_{t\geq 0}$ is a uniformly bounded C_0-semigroup. Then for any fixed $t \geq 0$, $V(t)$ is a linear and bounded operator, i.e., for any $x \in X$, we have*

$$|V(t)x| \leq \frac{M}{\Gamma(q)} |x|. \tag{2.23}$$

(ii) *If* $\{T(t)\}_{t\geq 0}$ *is a* C_0-*semigroup, then* $\{V(t)\}_{t\geq 0}$ *is strongly continuous.*
(iii) *If* $\{T(t)\}_{t\geq 0}$ *is a positive* C_0-*semigroup, then* $V(t)$ *is positive for* $t \geq 0$.
(iv) *If* $\{T(t)\}_{t\geq 0}$ *is exponentially stable, then*

$$\|V(t)\|_{\mathscr{L}(X)} \leq M e_q(-\delta t^q), \text{ for } t \geq 0;$$

here, $\mathscr{L}(X)$ *is the space of all bounded linear operators on Banach space* X *with the norm* $\|\cdot\|_{\mathscr{L}(X)}$.

Proof. For the proof of (i) and (ii), we can see [292]. By Property 1.11(iii), we obtain (iii). By (1.15) and Property 1.11(v), we have

$$\|V(t)\|_{\mathscr{L}(X)} = \left\| q \int_0^\infty \theta \Psi_q(\theta) T(t^q \theta) d\theta \right\|_{\mathscr{L}(X)}$$
$$\leq q \int_0^\infty \theta \Psi_q(\theta) M e^{-\delta t^q \theta} d\theta$$
$$= M e_q(-\delta t^q).$$

Then (iv) holds. $\qquad\qquad\qquad\qquad\qquad\qquad\qquad\qquad\qquad\qquad\square$

Lemma 2.11. *Assume that* $-A$ *generates an exponentially stable* C_0-*semigroup* $\{T(t)\}_{t\geq 0}$. *If* $x : \mathbb{R} \to X$ *is a function satisfying equation* (2.21) *and assumption* (H), *then* u *satisfies the following integral equation*

$$x(t) = \int_{-\infty}^t (t-s)^{q-1} V(t-s) h(s) ds, \quad t \in \mathbb{R}.$$

Proof. Denote by $\mathcal{F}f$ Fourier transform of f, that is,

$$(\mathcal{F}f)(\lambda) = \int_{-\infty}^\infty e^{i\lambda t} f(t) dt$$

for $\lambda \in \mathbb{R}$ and $f \in L^1(\mathbb{R}, X)$. Thus $(\mathcal{F}_{-\infty} D_t^q f)(\lambda) = (-i\lambda)^q (\mathcal{F}f)(\lambda)$ (see [134]). By applying Fourier transform to (2.21), we get that $(-i\lambda)^q (\mathcal{F}u)(\lambda) + A(\mathcal{F}u)(\lambda) = (\mathcal{F}h)(\lambda)$ for $\lambda \in \mathbb{R}$. In view of (1.14) and Property 1.11(ii), we have

$$(\mathcal{F}x)(\lambda) = ((-i\lambda)^q I + A)^{-1}(\mathcal{F}h)(\lambda)$$
$$= \int_0^\infty e^{-(-i\lambda)^q t} T(t)(\mathcal{F}h)(\lambda) dt$$

$$= \int_0^\infty \int_{-\infty}^\infty qt^{q-1}e^{-(-i\lambda t)^q}$$
$$\times \, T(t^q)h(s)e^{i\lambda s}\,ds\,dt$$

$$= \int_0^\infty \int_{-\infty}^\infty \int_0^\infty \frac{q^2}{\tau^{2q+1}}t^{q-1}\Psi_q\left(\frac{1}{\tau^q}\right)e^{i\lambda t}T\left(\frac{t^q}{\tau^q}\right)h(s)e^{i\lambda s}\,d\tau\,ds\,dt$$

$$= \int_0^\infty \int_{-\infty}^\infty \int_0^\infty q\tau t^{q-1}\Psi_q(\tau)T(t^q\tau)h(s)e^{i\lambda(t+s)}\,d\tau\,ds\,dt$$

$$= \int_{-\infty}^\infty e^{i\lambda t}\int_{-\infty}^t (t-s)^{q-1}h(s)$$
$$\times \left(\int_0^\infty q\tau\Psi_q(\tau)T((t-s)^q\tau)d\tau\right)ds\,dt$$

$$= \int_{-\infty}^\infty e^{i\lambda t}\left(\int_{-\infty}^t (t-s)^{q-1}V(t-s)h(s)ds\right)dt,$$

where $(-i\lambda)^q I \in \rho(-A)$. By the uniqueness of Fourier transform, we deduce that the assertion of lemma holds. This completes this proof. $\quad\square$

Definition 2.3. *A function* $x : \mathbb{R} \to X$ *is said to be a mild solution of equation* (2.21) *if*

$$x(t) = \int_{-\infty}^t (t-s)^{q-1}V(t-s)h(s)ds, \quad t \in \mathbb{R}.$$

Theorem 2.7. *If* $\{T(t)\}_{t\geq 0}$ *is exponentially stable,* h *belongs to one of the space* $\mathcal{M}(X)$, *and*

$$(Rh)(t) = \int_{-\infty}^t (t-s)^{q-1}V(t-s)h(s)ds, \qquad (2.24)$$

then Rh *belongs to the same space as* h.

Proof. $P_\omega(X)$: If $h \in P_\omega(X)$, then

$$(Rh)(t+\omega) = \int_{-\infty}^{t+\omega} (t+\omega-s)^{q-1}V(t+\omega-s)h(s)ds$$
$$= \int_{-\infty}^t (t-\tau)^{q-1}V(t-\tau)h(\tau+\omega)d\tau$$
$$= \int_{-\infty}^t (t-\tau)^{q-1}V(t-\tau)h(\tau)d\tau = (Rh)(t).$$

Therefore, $Rh \in P_\omega(X)$.

$AP(X)$: By the hypotheses, for any $\epsilon > 0$, we can find a real number $l = l(\epsilon) > 0$ for any interval of length $l(\varepsilon)$, there exists a number $m = m(\varepsilon)$ in this interval such that $\|h(t + m) - h(t)\|_\infty < \epsilon$ for all $t \in \mathbb{R}$. From Property 1.10, Lemma 2.10(iv), and $E_q(0) = 1$, we have

$$\sup_{t \in \mathbb{R}} |(Rh)(t + m) - (Rh)(t)|$$

$$= \sup_{t \in \mathbb{R}} \left| \int_{-\infty}^{t} (t - s)^{q-1} V(t - s) \big(h(s + m) - h(s) \big) ds \right|$$

$$\leq \left| \int_{-\infty}^{t} (t - s)^{q-1} V(t - s) ds \right| \sup_{t \in \mathbb{R}} |h(t + m) - h(m)|$$

$$\leq M\varepsilon \int_{-\infty}^{t} (t - s)^{q-1} e_q(-\delta(t - s)^q) ds$$

$$= \frac{M\varepsilon}{\delta} E_q(-\delta(t - s)^q)|_{-\infty}^{t}$$

$$= \frac{M\varepsilon}{\delta},$$

and therefore, Rh has the same property as h, i.e., it is almost periodic.

$AA_c(X)$: Since $h \in AA_c(X)$, there exist a subsequence $\{s_n\}_{n \in \mathbb{N}}$ and a continuous function $y \in C_b(X)$ such that $h(t + s_n)$ converges to $y(t)$ and $y(t - s_n)$ converges to $h(t)$ uniformly on compact subsets of $\mathbb{R}$. Since

$$(Rh)(t + s_n) = \int_{-\infty}^{t+s_n} (t + s_n - s)^{q-1} V(t + s_n - s) h(s) ds$$

$$= \int_{-\infty}^{t} (t - s)^{q-1} V(t - s) h(s + s_n) ds, \tag{2.25}$$

using Theorem 1.1, we obtain that $(Rh)(t + s_n)$ convergence to $z(t) = \int_{-\infty}^{t} (t - s)^{q-1} V(t - s) y(s) ds$ as $n \to \infty$ for each $t \in \mathbb{R}$.

Furthermore, the preceding convergence is uniform on compact subsets of $\mathbb{R}$. To show this assertion, we take a compact set $K = [-a, a]$. For $\varepsilon > 0$, by Property 1.10 and Lemma 2.10, we choose $L_\varepsilon > 0$ and $N_\varepsilon \in \mathbb{N}$ such that

$$\int_{L_\varepsilon}^{\infty} s^{q-1} \|V(s)\|_{\mathscr{L}(X)} ds \leq -\frac{M}{\delta} E_q(-\delta s^q)|_{L_\varepsilon}^{\infty} = \frac{M}{\delta} E_q(-\delta L_\varepsilon^q) \leq \varepsilon,$$

$$|h(s + s_n) - y(s)| \leq \varepsilon, \ , n \geq N_\varepsilon, \ s \in [-L, L],$$

where $L = L_\varepsilon + a$. For $t \in K$, in view of Property 1.10 and Lemma 2.10, we estimate

$$|(Rh)(t + s_n) - z(t)|$$

$$\leq \int_{-\infty}^t (t - s)^{q-1} \|V(t - s)\|_{\mathscr{L}(X)} |h(s + s_n) - y(s)| ds$$

$$\leq \int_{-\infty}^{-L} (t - s)^{q-1} \|V(t - s)\|_{\mathscr{L}(X)} |h(s + s_n) - y(s)| ds$$

$$+ \int_{-L}^t (t - s)^{q-1} \|V(t - s)\|_{\mathscr{L}(X)} |h(s + s_n) - y(s)| ds$$

$$\leq (\|h\|_\infty + \|y\|_\infty) \int_{t+L}^\infty s^{q-1} \|V(s)\|_{\mathscr{L}(X)} ds$$

$$+ \varepsilon \int_0^\infty s^{q-1} \|V(s)\|_{\mathscr{L}(X)} ds$$

$$\leq \varepsilon (\|h\|_\infty + \|y\|_\infty) + \frac{\varepsilon M}{\delta},$$

which proves that the convergence is independent of $t \in K$. Repeating this argument, one can show that $z(t - s_n)$ converges to $(Rh)(t)$ as $n \to \infty$ uniformly for t on compact subsets of $\mathbb{R}$. This completes the proof in the case of the space $AA_c(X)$.

$AA(X)$: Let $\{s_n'\} \subset \mathbb{R}$ be an arbitrary sequence. Since $h \in AA(X)$, there exists a subsequence $\{s_n\}$ of $\{s_n'\}$ such that

$$\lim_{n \to \infty} h(t + s_n) = y(t), \quad t \in \mathbb{R},$$

and

$$\lim_{n \to \infty} y(t - s_n) = h(t), \quad t \in \mathbb{R}.$$

Note that

$$|(Rh)(t + s_n)| \leq \frac{M}{\delta} \|h\|_\infty,$$

from (2.25), Property 1.10, and Lemma 2.10, we have that $V(t - s)h(s + s_n) \to V(t - s)y(s)$ as $n \to \infty$ for each $s \in \mathbb{R}$ fixed and any $t \geq s$. Then by Theorem 1.1, we obtain that $(Rh)(t + s_n)$ converges to $z(t) = \int_{-\infty}^t (t - s)^{q-1} V(t - s)y(s) ds$ as $n \to \infty$ for each $t \in \mathbb{R}$. Similarly we can show that

$$z(t - s_n) \to (Rh)(t), \text{ as } n \to \infty, \text{ for } t \in \mathbb{R}.$$

$SAP_\omega(X)$: Assume that $h \in SAP_\omega(X)$. For $\forall\, \varepsilon > 0$, there is a positive constant $L_\varepsilon > 0$ such that $|h(t + \omega) - h(t)| < \varepsilon$, for every $t \geq L_\varepsilon$. Under these conditions, in view of Property 1.10 and Lemma 2.10, we have

$$|(Rh)(t + \omega) - (Rh)(t)|$$

$$= \left| \int_{-\infty}^{t+\omega} (t + \omega - s)^{q-1} V(t + \omega - s) h(s) ds \right.$$

$$\left. - \int_{-\infty}^{t} (t - s)^{q-1} V(t - s) h(s) ds \right|$$

$$\leq \int_{-\infty}^{t} \left| (t - \tau)^{q-1} V(t - \tau) \big(h(\tau + \omega) - h(\tau) \big) \right| d\tau$$

$$\leq 2\|h\|_\infty \int_{-\infty}^{L_\varepsilon} (t - \tau)^{q-1} \|V(t - \tau)\|_{\mathscr{L}(X)} d\tau$$

$$+ \varepsilon \int_{L_\varepsilon}^{t} (t - \tau)^{q-1} \|V(t - \tau)\|_{\mathscr{L}(X)} d\tau$$

$$\leq 2\|h\|_\infty M \int_{-\infty}^{L_\varepsilon} (t - \tau)^{q-1} e_q(-\delta(t - \tau)^q) d\tau$$

$$+ \varepsilon M \int_{L_\varepsilon}^{t} (t - \tau)^{q-1} e_q(-\delta(t - \tau)^q) d\tau$$

$$= \frac{2\|h\|_\infty M}{\delta} E_q(-\delta(t - \tau)^q) \Big|_{-\infty}^{L_\varepsilon} + \frac{\varepsilon M}{\delta} E_q(-\delta(t - \tau)^q) \Big|_{L_\varepsilon}^{t}$$

$$= \frac{2\|h\|_\infty M - \varepsilon M}{\delta} E_q(-\delta(t - L_\varepsilon)^q) + \frac{\varepsilon M}{\delta},$$

for $t \geq L_\varepsilon$. It follows that $|(Rh)(t + \omega) - (Rh)(t)| \to 0$ as $t \to \infty$. Thus, $Rh \in SAP_\omega(X)$.

Now we will study the asymptotic behavior of the solutions. Let $w \in C_0(X)$ and $\varepsilon > 0$ be given. There exists $T > 0$ such that $|w(s)| < \varepsilon$ for $|s| > T$; hence, we can get from Property 1.10 and Lemma 2.10 that

$$|(Rw)(t)| \leq \int_{-\infty}^{T} (t - s)^{q-1} \|V(t - s)\|_{\mathscr{L}(X)} |w(s)| ds$$

$$+ \varepsilon \int_T^t (t-s)^{q-1} \|V(t-s)\|_{\mathscr{L}(X)} ds$$

$$\leq \frac{\|w\|_\infty M}{\delta} E_q(-\delta(t-T)^q) + \frac{\varepsilon M}{\delta}\left(1 - E_q(-\delta(t-T)^q)\right),$$

and we conclude that $(Rw)(t) \to 0$ as $t \to \infty$. Then we can infer the conclusion of the theorem for the spaces $AP_\omega(X)$, $AAP(X)$, $AAA_c(X)$, and $AAA(X)$.

Vanishing mean value: Let $w \in P_0(X)$. For $L > 0$ we have

$$\frac{1}{2L} \int_{-L}^L |(Rw)(t)| dt$$

$$\leq \frac{1}{2L} \int_{-L}^L \left(\int_{-\infty}^t (t-s)^{q-1} \|V(t-s)\|_{\mathscr{L}(X)} |w(s)| ds \right) dt$$

$$\leq \frac{1}{2L} \int_{-L}^L \left(\int_0^\infty s^{q-1} \|V(s)\|_{\mathscr{L}(X)} |w(t-s)| ds \right) dt$$

$$= \int_0^\infty s^{q-1} \|V(s)\|_{\mathscr{L}(X)} \left(\frac{1}{2L} \int_{-L}^L |w(t-s)| dt \right) ds.$$

Note that the set $P_0(X)$ is translation invariant. Hence, using Theorem 1.1, we obtain from the above inequality that

$$\frac{1}{2L} \int_{-L}^L |(Rw)(t)| dt \to 0, \text{ as } L \to \infty.$$

We conclude that the spaces $PP_\omega(X)$, $PAP(X)$, $PAA_c(X)$, and $PAA(X)$ have the maximal regularity property under the convolution defined by (2.24). □

Theorem 2.8. *Assume that $h \in \Omega(X) \in \mathcal{M}(X)$, $-A$ generates an exponentially stable C_0-semigroup $\{T(t)\}_{t \geq 0}$. Then the linear fractional evolution equation (2.21) has a unique mild solution $x = Rh \in \Omega(X)$, and*

$$\|Rh\|_\infty \leq \frac{M}{\delta} \|h\|_\infty. \tag{2.26}$$

Proof. In view of Definition 2.3 and Theorem 2.7, Rh is a mild solution of equation (2.21) and $Rh \in \Omega(X)$. By Property 1.10 and Lemma 2.10, we have

$$|(Rh)(t)| \leq \int_{-\infty}^{t} (t-s)^{q-1} |V(t-s)h(s)|\, dsr$$

$$\leq M \|h\|_\infty \int_{-\infty}^{t} (t-s)^{q-1} e_q(-\delta(t-s)^q)\, ds$$

$$= \frac{M}{\delta} \|h\|_\infty E_q(-\delta(t-s)^q)|_{-\infty}^{t}$$

$$= \frac{M}{\delta} \|h\|_\infty .$$

Then we obtain

$$\|Rh\|_\infty \leq \frac{M}{\delta} \|h\|_\infty .$$

$\square$

Remark 2.1. (2.26) *is an optimal estimation. In fact, for* $X = \mathbb{R}$*, the periodic solution of the equation* $_{-\infty}D_t^q x + \gamma x = 1$ *is* $x = \frac{1}{\gamma}$*.*

Corollary 2.7. *Let* $h \in \Omega(X) \in \mathcal{M}(X)$*. Assume that* $-A$ *generates a uniformly bounded* C_0*-semigroup* $\{T(t)\}_{t\geq 0}$*. If* $Re(\lambda) > 0$*, then the linear fractional evolution equation*

$$_{-\infty}D_t^q x(t) + Ax(t) + \lambda x(t) = h(t), \quad t \in \mathbb{R}, \tag{2.27}$$

has a unique mild solution $x = R_\lambda h$*, and*

$$\|R_\lambda h\|_\infty \leq \frac{M}{Re(\lambda)} \|h\|_\infty . \tag{2.28}$$

Proof. $-(A + \lambda I)$ generates a C_0-semigroup $\{S(t)\}_{t\geq 0}$, and $S(t) = e^{-\lambda t}T(t)$. Then $\|S(t)\|_{\mathscr{L}(X)} = e^{-Re(\lambda)t} \|T(t)\|_{\mathscr{L}(X)} \leq M e^{-Re(\lambda)t}$, so $\{S(t)\}_{t\geq 0}$ is exponentially stable for $Re(\lambda) > 0$. The conclusion follows by Theorem 2.8. $\square$

Theorem 2.9. *Let* $h \in \Omega(X) \in \mathcal{M}(X)$*. Assume that* $-A$ *generates an exponentially stable* C_0*-semigroup* $\{T(t)\}_{t\geq 0}$*, that is, the growth bound*

$$\nu_0 = \limsup_{t \to +\infty} \frac{\ln \|T(t)\|_{\mathscr{L}(X)}}{t} < 0.$$

Then the linear fractional evolution equation (2.21) *has a unique mild solution* $x = Rh \in \Omega(X)$, $R : \Omega(X) \to \Omega(X)$ *is a bounded linear operator, and spectral radius* $r(R) \leq \frac{1}{|\nu_0|}$.

Proof. By Theorem 2.8 and Lemma 2.10(i), equation (2.21) has a unique mild solution $u = Rh \in \Omega(X)$, and $R : \Omega(X) \to \Omega(X)$ is a bounded linear operator.

For all $\nu \in (0, |\nu_0|)$, there exists $M_1 \geq 1$ such that

$$\|T(t)\|_{\mathscr{L}(X)} \leq M_1 e^{-\nu t}, \quad t \geq 0.$$

Define a new norm $|\cdot|_0$ in X as

$$|x|_0 = \sup_{t \geq 0} \left| e^{\nu t} T(t) x \right|.$$

Since $|x| \leq |x|_0 \leq M_1 |x|$, then $|\cdot|_0$ is equivalent to $|\cdot|$. We denote the norm of $T(t)$ in $X_0 := (X, |\cdot|_0)$ by $\|T(t)\|_{\mathscr{L}(X_0)}$. Then for $t \geq 0$, we have

$$
\begin{aligned}
|T(t)x|_0 &= \sup_{s \geq 0} |e^{\nu s} T(s) T(t) x| \\
&= e^{-\nu t} \sup_{s \geq 0} |e^{\nu(s+t)} T(s+t) x| \\
&= e^{-\nu t} \sup_{\eta \geq t} |e^{\nu \eta} T(\eta) x| \\
&\leq e^{-\nu t} |x|_0.
\end{aligned}
$$

This implies that $\|T(t)\|_{\mathscr{L}(X_0)} \leq e^{-\nu t}$. In view of Property 1.11(v), it follows that

$$
\begin{aligned}
\|V(t)\|_{\mathscr{L}(X_0)} &= \left\| q \int_0^\infty \theta \Psi_q(\theta) T(t^q \theta) d\theta \right\|_{\mathscr{L}(X_0)} \\
&\leq q \int_0^\infty \theta \Psi_q(\theta) e^{-\nu t^q \theta} d\theta \\
&= e_q(-\nu t^q).
\end{aligned}
\tag{2.29}
$$

In view of $h \in \Omega(X)$, we have that $|h|_\infty = \sup_{t \in \mathbb{R}} |h(t)|_0 < \infty$. By (2.29) and Property 1.10, we have

$$|(Rh)(t)|_0 \leq \left| \int_{-\infty}^t (t-s)^{q-1} V(t-s) h(s) ds \right|_0$$

$$\leq |h|_\infty \int_{-\infty}^t (t-s)^{q-1} e_q(-\nu(t-s)^q) ds$$

$$= \frac{|h|_\infty}{\nu} E_q(-\nu(t-s)^q)|_{-\infty}^t$$

$$= \frac{|h|_\infty}{\nu}, \quad \text{for } t \geq 0.$$

Thus $|Rh|_\infty \leq \frac{|h|_\infty}{\nu}$. Then $\|R\|_{\mathscr{L}(X_0)} \leq \frac{1}{\nu}$, and the spectral radius $r(R) \leq \frac{1}{\nu}$. By the randomicity of $\nu \in (0, |\nu_0|)$, we obtain that $r(R) \leq \frac{1}{|\nu_0|}$. $\qquad \square$

Remark 2.2. *For the applications of Theorem 2.9, it is important to estimate the growth bound of* $\{T(t)\}_{t\geq 0}$. *If* $T(t)$ *is continuous in the uniform operator topology for* $t > 0$, *that is,* $T(t)$ *is equicontinuous, it is well known that* ν_0 *can be obtained by* $\sigma(A)$ *which is the spectrum of* A *(see [226]):*

$$\nu_0 = -\inf\{Re(\lambda) : \lambda \in \sigma(A)\}. \tag{2.30}$$

We know that $T(t)$ *is continuous in the uniform operator topology for* $t > 0$ *if* $T(t)$ *is a compact semigroup (see [199]). Assume that* P *is a regeneration cone, and* $\{T(t)\}_{t\geq 0}$ *is a compact and positive analytic semigroup. Then by the characteristic of positive semigroups (see [151]), for sufficiently large* $\lambda_0 > -\inf\{Re(\lambda) : \lambda \in \sigma(A)\}$, *we have that* $\lambda_0 I + A$ *has positive bounded inverse operator* $(\lambda_0 I + A)^{-1}$. *Since* $\sigma(A) \neq \emptyset$, *the spectral radius*

$$r\left((\lambda_0 I + A)^{-1}\right) = \frac{1}{dist(-\lambda_0, \sigma(A))} > 0.$$

By Krein-Rutmann theorem (see [141]), A *has the first eigenvalue* λ_1, *which has a positive eigenfunction* x_1, *and*

$$\lambda_1 = \inf\{Re(\lambda) : \lambda \in \sigma(A)\}, \tag{2.31}$$

that is, $\nu_0 = -\lambda_1$.

Corollary 2.8. *Assume that* $h \in \Omega(X)$, *X is an ordered Banach space, whose positive cone* P *is a regeneration cone,* $-A$ *generates a compact and positive* C_0-*semigroup* $\{T(t)\}_{t\geq 0}$, *and its first eigenvalue of* A

$$\lambda_1 = \inf\{Re(\lambda) : \lambda \in \sigma(A)\} > 0. \tag{2.32}$$

Then (2.21) *has a unique mild solution* $x = Rh \in \Omega(X)$, $R : \Omega(X) \to \Omega(X)$ *is a positive and bounded linear operator, and the spectral radius* $r(R) = \frac{1}{\lambda_1}$.

Proof. By (2.32), we know that the growth bound of $\{T(t)\}_{t \geq 0}$ is $\nu_0 = -\lambda_1 < 0$, i.e., $\{T(t)\}_{t \geq 0}$ is exponentially stable. By Lemma 2.9, we know that $R : \Omega(X) \to \Omega(X)$ is a bounded linear operator, and the spectral radius $r(R) \leq \frac{1}{\lambda_1}$. On the other hand, since λ_1 is the first eigenvalue of A, it has a positive eigenfunction x_1. In equation (2.21), we set $h(t) = x_1$, then $\frac{x_1}{\lambda_1}$ is the corresponding mild solution. By the definition of the operator R, $R(x_1) = \frac{x_1}{\lambda_1}$, that is, $\frac{1}{\lambda_1}$ is an eigenvalue of R. Then $r(R) \geq \frac{1}{\lambda_1}$. Thus, $r(R) = \frac{1}{\lambda_1}$. $\square$

2.2.3 Nonlinear Equations

Theorem 2.10. *Let X be an ordered Banach space, whose positive cone P is normal with normal constant N. Assume that $-A$ generates a positive C_0-semigroup $\{T(t)\}_{t \geq 0}$, $\Omega(X) \in \mathcal{M}(X)$, $f \in \Omega(\mathbb{R} \times X, X) \in \mathcal{M}(\mathbb{R} \times X, X)$ and satisfies (1.19), $f(t, \theta) \geq \theta$ for $\forall\, t \in \mathbb{R}$, and the following conditions are satisfied:*

(H_1) *for any $\kappa > 0$, there exists $C = C(\kappa) > 0$ such that*

$$f(t, x_2) - f(t, x_1) \geq -C(x_2 - x_1),$$

where $t \in \mathbb{R}$, $\theta \leq x_1 \leq x_2$, $|x_1|$, $|x_2| \leq \kappa$;
(H_2) *there exists $L < -\nu_0$ (ν_0 is the growth bound of $\{T(t)\}_{t \geq 0}$), such that*

$$f(t, x_2) - f(t, x_1) \leq L(x_2 - x_1),$$

where $t \in \mathbb{R}$, $\theta \leq x_1 \leq x_2$.

Then equation (2.20) has a unique positive mild solution $x \in \Omega(X)$.

Proof. Let $h_0(t) = f(t, \theta)$. Then $h_0 \in \Omega(X)$, $h_0 \geq \theta$. We consider the linear equation

$$_{-\infty}D_t^q x(t) + (A - LI)x(t) = h_0(t), \ t \in \mathbb{R}. \tag{2.33}$$

We know that $-(A - LI)$ generates a positive C_0-semigroup $\{e^{Lt}T(t)\}$, whose growth bound $L + \nu_0 < 0$. By Theorem 2.9, the linear equation (2.33) has a unique positive mild solution $w_0 \in \Omega(X)$.

Let $\kappa_0 = N \|w_0\|_\infty + 1$, $C = C(\kappa_0)$ be the corresponding constant in (H_1). We may suppose that $C > \max\{\nu_0, -L\}$ (otherwise substitute $C + |\nu_0| + |L|$ for C, (H_1) is also satisfied). Then we consider the linear equation

$$_{-\infty}D_t^q x(t) + (A + CI)x(t) = h(t), \quad t \in \mathbb{R}. \tag{2.34}$$

$-(A+CI)$ generates a positive C_0-semigroup $T_1(t) = e^{-Ct}T(t)$, whose growth bound $-C + \nu_0 < 0$. By Theorem 2.9, for $h \in \Omega(X)$, the linear equation (2.34) has a unique mild solution $x = Q_1 h$, and $Q_1 : \Omega(X) \to \Omega(X)$ is a positive bounded linear operator, and the spectral radius $r(Q_1) \leq \frac{1}{C - \nu_0}$.

Let $F(x) = f(t, x) + Cx$. Then by Lemma 2.10(ii), Corollary 1.1, and Theorem 2.7, it follows that $Q_1 F : \Omega(X) \to \Omega(X)$ is continuous, $F(\theta) = h_0 \geq \theta$. By (H_1), F is an increasing operator on $[\theta, w_0]$. By (2.34), we have that $Q_1(h_0 + Lw_0 + Cw_0)$ is another mild solution of (2.33). Since the mild solution of (2.33) is unique, we have

$$w_0 = Q_1(h_0 + Lw_0 + Cw_0). \tag{2.35}$$

Let $x_1 = \theta$, $x_2 = w_0(t)$ in (H_2), then

$$f(t, w_0) \leq h_0(t) + Lw_0(t),$$

$$\theta \leq F(\theta) \leq F(w_0) \leq h_0 + Lw_0 + Cw_0. \tag{2.36}$$

Set $v_0 \equiv \theta$ and $u_0 = w_0$. We can construct the sequences

$$v_n = (Q_1 \circ F)(v_{n-1}), u_n = (Q_1 \circ F)(u_{n-1}), \quad n = 1, 2, \dots . \tag{2.37}$$

By (2.35) and (2.36), and the definition and the positivity of Q_1, we obtain

$$Q_1\theta = \theta = v_0 \leq v_1 \leq u_1 \leq u_0.$$

Since $Q_1 \circ F$ is an increasing operator on $[\theta, w_0]$, in view of (2.37) we can show that

$$\theta \leq v_1 \leq \cdots \leq v_n \leq \cdots \leq u_n \leq \cdots \leq u_1 \leq u_0. \tag{2.38}$$

Therefore,

$$\theta \leq u_n - v_n = Q_1(F(u_{n-1}) - F(v_{n-1}))$$

$$=Q_1 \left(f(\cdot, u_{n-1}) - f(\cdot, v_{n-1}) + C(u_{n-1} - v_{n-1}) \right)$$
$$\leq (C + L) Q_1 (u_{n-1} - v_{n-1}).$$

By induction,

$$\theta \leq u_n - v_n \leq (C + L)^n Q_1^n (u_0 - v_0) = (C + L)^n Q_1^n (u_0).$$

In view of the normality of the cone P, we have

$$\|u_n - v_n\|_\infty \leq N(C+L)^n \|Q_1^n (u_0)\|_\infty$$
$$\leq N(C+L)^n \|Q_1^n\|_{\mathscr{L}(X)} \|u_0\|_\infty . \tag{2.39}$$

On the other hand, since $0 < C + L < C - \nu_0$, for some $\varepsilon > 0$ we have that $C + L + \varepsilon < C - \nu_0$. By Gelfand formula, $\lim_{n \to \infty} \|Q_1^n\|_{\mathscr{L}(X)}^{\frac{1}{n}} = r(Q_1) \leq \frac{1}{C - \nu_0}$. Then there exists N_0 such that $\|Q_1^n\|_{\mathscr{L}(X)} \leq \frac{1}{(C+L+\varepsilon)^n}$ for $n \geq N_0$. By (2.39), we obtain

$$\|u_n - v_n\|_\infty \leq N \|u_0\|_\infty \left(\frac{C + L}{C + L + \varepsilon} \right)^n \to 0, \quad \text{as } n \to \infty. \tag{2.40}$$

By (2.38) and (2.40), similarly to the nested interval method, we can prove that there exists a unique $x^* \in \bigcap_{n=1}^{\infty} [v_n, u_n]$ such that

$$\lim_{n \to \infty} v_n = \lim_{n \to \infty} u_n = x^*.$$

By the continuity of the operator $Q_1 \circ F$ and (2.37), we have

$$x^* = (Q_1 \circ F)(x^*).$$

By the definition of Q_1 and (2.38), we know that x^* is a positive mild solution of (2.34) when $h(t) = f(t, x^*(t)) + Cx^*(t)$. Hence, x^* is a positive mild solution of equation (2.20).

Finally, we prove the uniqueness. If x_1 and x_2 are the positive mild solutions of (2.20), set $u_0 = x_i$ $(i = 1, 2)$, then $u_n = (Q_1 \circ F)(x_i) = x_i$ $(i = 1, 2)$. By (2.40), we have

$$\|x_i - v_n\|_\infty \to 0, \quad \text{as } n \to \infty, \ i = 1, 2.$$

Thus, $x_1 = x_2 = \lim_{n \to \infty} v_n$, (2.20) has a unique positive mild solution $x \in \Omega(X)$. $\square$

From Theorem 2.10 and Remark 2.2, we obtain the following results.

Corollary 2.9. *Let X be an ordered Banach space, whose positive cone P is a regeneration cone. Assume that $-A$ generates a compact and positive C_0-semigroup $\{T(t)\}_{t\geq 0}$, $\Omega(X) \in \mathcal{M}(X)$, $f \in \Omega(\mathbb{R} \times X, X) \in \mathcal{M}(\mathbb{R} \times X, X)$ and satisfies (1.19), $f(t, \theta) \geq \theta$ for $\forall t \in \mathbb{R}$, f satisfies (H_1) and the following condition:*

$(H_2)'$ *there exists $L < \lambda_1$ (λ_1 is the first eigenvalue of A) such that*

$$f(t, x_2) - f(t, x_1) \leq L(x_2 - x_1),$$

for any $t \in \mathbb{R}$, $\theta \leq x_1 \leq x_2$.

Then equation (2.20) has a unique positive mild solution $x \in \Omega(X)$.

Remark 2.3. *In Corollary 2.9, since λ_1 is the first eigenvalue of A, condition "$L < \lambda_1$" in $(H_2)'$ cannot be extended to "$L \leq \lambda_1$". Otherwise equation (2.20) does not always have a mild solution. For example, $f(t, x) = \lambda_1 x + h(t)$.*

Theorem 2.11. *Assume that $-A$ generates an exponentially stable C_0-semigroup $\{T(t)\}_{t\geq 0}$, $\Omega(X) \in \mathcal{M}(X)$, $f \in \Omega(\mathbb{R} \times X, X) \in \mathcal{M}(\mathbb{R} \times X, X)$ and satisfies (1.19). If the following condition is satisfied:*

(H_3) *$f(t, x)$ is Lipschitz continuous in x, i.e., there exists a constant $L \geq 0$ such that*

$$|f(t, x_2) - f(t, x_1)| \leq L|x_2 - x_1|, \quad for\ t \in \mathbb{R},\ x_1, x_2 \in X,$$

then for $\delta > ML$ equation (2.20) has a unique mild solution $x \in \Omega(X)$.

Proof. We define the operator Q by

$$(Qx)(t) = \int_{-\infty}^{t} (t - s)^{q-1} V(t - s) f(s, x(s)) ds, \quad t \in \mathbb{R}.$$

Form Lemma 2.10(ii), Corollary 1.1 and Theorem 2.7, it follows that $Q : \Omega(X) \to \Omega(X)$ is continuous.

By (H_3), Property 1.10 and Lemma 2.10, for all $t \in \mathbb{R}$, $x_1, x_2 \in \Omega(X)$, we have

$$|(Qx_2)(t) - (Qx_1)(t)|$$

$$\leq \int_{-\infty}^{t} (t-s)^{q-1} |V(t-s)(f(s,x_2(s)) - f(s,x_1(s)))| ds$$

$$\leq ML \|x_2 - x_1\|_\infty \int_{-\infty}^{t} (t-s)^{q-1} e_q(-\delta(t-s)^q) ds$$

$$= \frac{ML}{\delta} \|x_2 - x_1\|_\infty E_q(-\delta(t-s)^q)|_{-\infty}^{t}$$

$$= \frac{ML}{\delta} \|x_2 - x_1\|_\infty .$$

Thus,

$$\|Qx_2 - Qx_1\|_\infty \leq \frac{ML}{\delta} \|x_2 - x_1\|_\infty .$$

When $\delta > ML$, Q is a contraction in $\Omega(X)$. By Banach contraction principle we have that Q has a unique fixed point in $\Omega(X)$. This completes the proof. □

Corollary 2.10. *Assume that* $-A$ *generates a uniformly bounded C_0-semigroup* $\{T(t)\}_{t\geq 0}$, $\Omega(X) \in \mathcal{M}(X)$, $f \in \Omega(\mathbb{R} \times X, X) \in \mathcal{M}(\mathbb{R} \times X, X)$ *and satisfies* (1.19). *If f satisfies* (H_3), *then equation* (2.20) *has a unique mild solution* $x \in \Omega(X)$ *for* $Re(\lambda) > ML$.

Proof. Similar to the proof in Corollary 2.7, we know that $\{S(t)\}_{t\geq 0}$ is exponentially stable. By Theorem 2.11, equation (2.20) has a unique mild solution for $Re(\lambda) > ML$. □

Theorem 2.12. *Assume that* $-A$ *generates an exponentially stable C_0-semigroup* $\{T(t)\}_{t\geq 0}$, $\Omega(X) \in \mathcal{M}(X)$, $f \in \Omega(\mathbb{R} \times X, X) \in \mathcal{M}(\mathbb{R} \times X, X)$ *and satisfies* (1.19). *Assume that the following conditions are satisfied:*

(H_4) $f(t,x)$ *is locally Lipschitz continuous in x, i.e., for all $r > 0$, there exists $L(r) > 0$ such that*

$$|f(t,x_2) - f(t,x_1)| \leq L(r)|x_2 - x_1|,$$

for $t \in \mathbb{R}$, $|x_1|, |x_2| \leq r$.

Denote $f_0 = f(t,\theta)$, then for $\delta > ML(r) + \frac{M}{r}\|f_0\|_\infty$, equation (2.20) has a unique mild solution in $B(\theta,r) = \{x \in \Omega(X) : \|x\|_\infty < r\}$.

Proof. Let $F(x) = f(t, x)$, we know that the mild solution of equation (2.20) in $B(\theta, r)$ is the fixed point of $R \circ F$ in $B(\theta, r)$.

For any $x \in B(\theta, r)$, by (2.26) and (H_4), we have

$$
\begin{aligned}
\|R(F(x))\|_\infty &\leq \|R(F(\theta))\|_\infty + \|R(F(x) - F(\theta))\|_\infty \\
&\leq \frac{M\|f_0\|_\infty}{\delta} + \frac{M\|F(x) - F(\theta)\|_\infty}{\delta} \\
&\leq \frac{M\|f_0\|_\infty + ML(r)}{\delta} \\
&< r,
\end{aligned}
$$

where $\delta > ML(r) + \frac{M}{r}\|f_0\|_\infty$. By Lemma 2.10(ii), Corollary 1.1, and Theorem 2.7, we obtain that $R \circ F : B(\theta, r) \to B(\theta, r)$ is continuous. On the other hand, for $\forall x_1, x_2 \in B(\theta, r)$, by (H_4) we get

$$
\begin{aligned}
\|R(F(x_2)) - R(F(x_1))\|_\infty &\leq \|R(F(x_2) - F(x_1))\|_\infty \\
&\leq \frac{ML(r)}{\delta}\|x_2 - x_1\|_\infty.
\end{aligned}
$$

For $\delta > ML(r) + \frac{M}{r}\|f_0\|_\infty$, we have $\frac{ML(r)}{\delta} < 1$. Thus Q is a contraction in $\Omega(X)$.

By Banach contraction principle, there exists a unique $\tilde{x} \in B(\theta, r)$ such that $R(F(\tilde{x})) = \tilde{x}$. Therefore, equation (2.20) has a unique mild solution in $B(\theta, r)$. $\qquad\square$

Corollary 2.11. *Assume that $-A$ generates a uniformly bounded C_0-semigroup $\{T(t)\}_{t\geq 0}$, $\Omega(X) \in \mathcal{M}(X)$, $f \in \Omega(\mathbb{R} \times X, X) \in \mathcal{M}(\mathbb{R} \times X, X)$, and satisfies (1.19) and (H_4). Set $f_0 = f(t, \theta)$, then equation (2.20) has a unique mild solution in $B(\theta, r) = \{x \in \Omega(X) : \|x\|_\infty < r\}$ for $Re(\lambda) > ML(r) + \frac{M}{r}\|f_0\|_\infty$.*

Proof. Similar to the proof in Corollary 2.7, we know that $\{S(t)\}_{t\geq 0}$ is exponentially stable. By Theorem 2.12, equation (2.20) has a unique mild solution in $B(\theta, r)$ for $Re(\lambda) > ML(r) + \frac{M}{r}\|f_0\|_\infty$. $\qquad\square$

Example 2.2. Let $X = C_0(\bar{\Omega})$, we briefly discuss the existence of mild solutions of the following fractional parabolic partial differential equation:

$$\begin{cases} -_\infty D_t^{\frac{1}{2}} x - \Delta x = g(\xi, t, x(\xi, t)), & (\xi, t) \in \Omega \times \mathbb{R}^N, \\ x|_{\partial\Omega} = 0, \end{cases} \tag{2.41}$$

where $-_\infty D_t^{\frac{1}{2}}$ is Liouville fractional partial derivative of order $\frac{1}{2}$ with the lower limit $-\infty$, $\Omega \subset \mathbb{R}^N$ is a bounded domain with a sufficiently smooth boundary $\partial\Omega$, Δ is Laplace operator.

Theorem 2.13. *Let* $g(\cdot, t, x(\cdot, t)) \in \Omega(X) \in \mathcal{M}(X)$, *and* $g(\xi, t, 0) \geq 0$. *Assume that* g *has continuous partial derivatives for* x *in any bounded domain, and* $\sup g_x(\xi, t, x) < \lambda_1$, *where* λ_1 *is the first eigenvalue of Laplace operator* $-\Delta$ *under the condition* $x|_{\partial\Omega} = 0$. *Then the fractional partial differential equation* (2.41) *has a unique positive mild solution* $x(\cdot, t) \in \Omega(X)$.

Proof. Let $K = C_0^+(\bar{\Omega}) := \{f \in C(\bar{\Omega}, \mathbb{R}_+) : f|_{\partial\Omega} = 0\}$, then K is a positive cone in X. Define the operator A in X as follows:

$$D(A) = \{x \in X : \Delta x \in X\}, \quad Ax = -\Delta x.$$

In view of [199], $-A$ generates a compact and analytic semigroup $\{T(t)\}_{t\geq 0}$. Thus, equation (2.41) can be formulated as the abstract fractional partial differential equation (2.20), where $f(t, x) = g(\cdot, t, x(\cdot, t))$. By the maximum principle of parabolic equations, $\{T(t)\}_{t\geq 0}$ is a positive C_0-semigroup. It is easy to see that f satisfies (H_1) and $(H_2)'$. By Corollary 2.9, the fractional parabolic partial differential equation (2.41) has a unique positive mild solution $x(\cdot, t) \in \Omega(X)$. □

Example 2.3. Consider the problem

$$\begin{cases} -_\infty D_t^{\frac{1}{2}} x(\xi, t) = \dfrac{\partial^2}{\partial\xi^2} x(\xi, t) + b(t) \sin x(\xi, t), & (\xi, t) \in [0, \pi] \times \mathbb{R}, \\ x(0, t) = x(\pi, t) = 0, \end{cases}$$

$$\tag{2.42}$$

where $-_\infty D_t^{\frac{1}{2}}$ is Liouville fractional partial derivative of order $\frac{1}{2}$ with the lower limit $-\infty$, $b \in C_b(\mathbb{R}, \mathbb{R})$.

Proof. Let $X = L^2([0, \pi], \mathbb{R})$ and define $A = \frac{\partial^2}{\partial \xi^2}$ with domain $D(A) = \{g \in H^2([0, \pi] \times \mathbb{R}, \mathbb{R}) : g(0, t) = g(\pi, t) = 0\}$. Let us consider the nonlinearity $f(t, x) = b(t) \sin x$ for $x \in X$, $t \in \mathbb{R}$, we observe that $t \to f(t, x)$ belongs to $\Omega(X) \in \mathcal{M}(X)$ for each $x \in X$, and

$$\|f(t, x_1) - f(t, x_2)\|^2_{L^2[0, \pi]} \leq \int_0^\pi |b(t)|^2 |\sin x_1(s) - \sin x_2(s)|^2 ds$$
$$\leq \|b\|^2_\infty \|x_1 - x_2\|^2_{L^2[0, \pi]},$$

for $x_1, x_2 \in X$. In consequence, problem (2.42) has a unique mild solution $x(\cdot, t) \in \Omega(X)$ (by Theorem 2.11). $\qquad\square$

2.3 NOTES AND REMARKS

The subject of fractional differential equations is gaining much importance and attention. The so-called fractional differential equations are specified by generalizing the standard integer order derivative to arbitrary order. Due to the effective memory function of fractional derivative, fractional differential equations have been widely used to describe many physical phenomena such as seepage flow in porous media and in fluid dynamic traffic model. For more interesting theory results and scientific applications of fractional differential equations, we cite the monographs of Diethelm [81], Kilbas et al. [134], Hilfer [119], Lakshmikantham et al. [144], Miller and Ross [184], Podlubny [202], Tarasov [221], Zhou [292], and the recent papers [86, 140, 204, 275] and the references therein.

The existence of mild solutions for Cauchy problem of fractional evolution equations has been considered in several recent papers (see, e.g., Agarwal and Ahmad [1], Belmekki and Benchohra [35], Darwish et al. [70], Hernandez et al. [117], Hu et al. [122], Kumar and Sukavanam [142], Li et al. [155], Shu et al. [217], Wang et al. [275], Wang and Zhou [250], Wang et al. [236], and Zhou and Jiao [301]).

In [275], authors studied the Cauchy problem for the linear evolution equation

$$\begin{cases} {}^C_0 D_t^q x(t) + A x(t) = f(t), & t > 0, \\ x(0) = x_0, \end{cases} \tag{2.43}$$

as well as Cauchy problem for the corresponding semilinear fractional evolution equation

$$
\begin{cases}
{}^{C}_{0}D_{t}^{q}x(t) + Ax(t) = f(t, u(t)), & t > 0, \\
x(0) = x_0
\end{cases}
\tag{2.44}
$$

in X, where A is an almost sectorial operator, that is, $A \in \Theta_{\omega}^{p}(X)$ $(-1 < p < 0, 0 < \omega < \frac{\pi}{2})$. They constructed two operator families based on the generalized Mittag-Leffler functions and the resolvent operators associated with A, present deep anatomy on basic properties for these families consisting on the study of the compactness, and prove that, under natural assumptions, reasonable concept of solutions can be given to problems (2.43) and (2.44), which in turn is used to find solutions to the Cauchy problems.

Wang et al. [238] considered Cauchy problem

$$
\begin{cases}
{}^{C}_{0}D_{t}^{q}x(t) = f(t, x(t)), & t \in J = [0, +\infty), \\
x(0) = x_0,
\end{cases}
\tag{2.45}
$$

where ${}^{C}_{0}D_{t}^{q}$ is Caputo fractional derivative of order $q \in (0, 1)$ with the lower limit zero, $f : J \times \mathbb{R} \to \mathbb{R}$ is a nonlinear function. Using the final value theorem of Laplace transform, it is shown that the corresponding nonhomogeneous fractional Cauchy problem does not have nonzero periodic solution. Further, two basic existence and uniqueness results for asymptotically periodic solution of semilinear fractional Cauchy problem (2.45) in an asymptotically periodic function space are obtained.

For integer differential equations, many authors studied the periodic solutions. In fact, they studied the corresponding periodic boundary value problems, see [152]. In paper [189], authors discuss the existence and uniqueness of positive mild solutions of periodic boundary value problems for fractional evolution equations. However, since the fractional derivatives provide the description of memory property, the solution of periodic boundary value problems cannot be periodically extended to the time $t \in \mathbb{R}$. To get the existence results for periodic solutions of fractional evolution equations on $\mathbb{R}$, the key point is to find the reasonable solutions as a whole.

In [203], Ponce studied the existence and uniqueness of bounded solutions for the semilinear fractional integro-differential equation

$$_{-\infty}D_t^q x(t) = Ax(t) + \int_{-\infty}^{t} a(t-s)Ax(s)ds + f(t, x(t)), \quad t \in \mathbb{R}, \quad (2.46)$$

where A is a closed linear operator defined on a Banach space X, $_{-\infty}D_t^q$ is Liouville-Weyl fractional derivative of order $q > 0$ with the lower limit $-\infty$, $a \in L^1(\mathbb{R}^+)$ is a scalar-valued kernel and $f : \mathbb{R} \times X \to X$ satisfies some Lipschitz-type conditions. Assume that A is the generator of an q-resolvent family $\{S_q(t)\}_{t \geq 0}$ which is uniformly integrable. The mild solutions of (2.46) have been given by

$$x(t) = \int_{-\infty}^{t} S_q(t-s)f(s, x(s))ds, \quad t \in \mathbb{R}.$$

In [204], Ponce studied Hölder continuous solutions of (2.46) in which $_{-\infty}D_t^q$ is replaced by Caputo fractional derivative $_{-\infty}^{C}\mathbf{D}_t^q$ (see Definition 1.6), and $f(t, x(t)) = h(t)$, $h \in C^q(\mathbb{R}, X)$. When $A = \rho I$, $\rho \in \mathbb{R} \setminus \{0\}$, the unique solution of (2.46) has been explicitly given by

$$x(t) = \int_{-\infty}^{t} (t-s)^{q-1} e_q(\rho(t-s)^q)h(s)ds, \quad t \in \mathbb{R},$$

where e_q denotes Mittag-Leffler function.

Lizama and Ponce [164] studied the existence and uniqueness of bounded solutions for equation (2.46) in the case $q = 1$, and $a(t - s) = ae^{-\beta(t-s)}$. In [162], Lizama and N'Guérékata studied the structure of bounded solutions for equation (2.46) in the case $q = 1$.

In addition, for integer order integral/differential equations, the existence of bounded solutions (periodic solutions, asymptotically periodic solutions, S-asymptotically periodic solutions, pseudo-periodic solutions, almost periodic solutions, etc.) has been investigated by many authors, see, e.g., [152, 162–164, 203, 238].

The results in Section 2.1 are taken from Zhang and Zhou [289]. The results in Section 2.2 are adopted from Mu, Zhou, and Huang [190].

CHAPTER 3

Fractional Evolution Inclusions with Hille-Yosida Operators

Abstract   This chapter deals with fractional evolution inclusions involving a nondensely defined closed linear operator satisfying the Hille-Yosida condition and source term of multivalued type in Banach spaces. First, a definition of integral solutions for fractional differential inclusions is given. Then the topological structure of solution sets is investigated. It is shown that the solution set is nonempty, compact, and, moreover, an R_δ-set. An example is given to illustrate the feasibility of the abstract results. The problem of controllability of these inclusions and topological structure of solution sets are considered too.

Keywords   Fractional evolution inclusions, Integral solutions, Hille-Yosida conditions, Weak topology approach, Existence, Compact semigroup, Non-compact semigroup, Topological structure of solution sets, R_δ-set, Controllability.

3.1 EXISTENCE OF INTEGRAL SOLUTIONS

3.1.1 Introduction

In this chapter, we consider the following nonlinear fractional evolution inclusion

$$\begin{cases} {}^{C}_{0}D^{q}_{t}x(t) \in Ax(t) + F(t, x(t)), & t \in [0, b], \\ x(0) = x_0, \end{cases} \tag{3.1}$$

where ${}^{C}_{0}D^{q}_{t}$ is Caputo fractional derivative operator of order $q \in (0, 1]$, the state $x(\cdot)$ takes value in a Banach space X with norm $|\cdot|$, F is a multivalued map defined on a subset of $[0, b] \times X$, $A : D(A) \subset X \to X$ is a nondensely defined closed linear operator satisfying the Hille-Yosida condition.

In Subsection 3.1.2, by using Laplace transform and the Wright function, we give an equivalent form and the definition of an integral solution of nonhomogeneous Cauchy problem. Subsection 3.1.3 gives the concept of

Fractional Evolution Equations and Inclusions: Analysis and Control. http://dx.doi.org/10.1016/B978-0-12-804277-9.50003-1

an integral solution for fractional differential inclusion with Hille-Yosida operators (3.1). In Subsection 3.1.4, we use weak topology approach to obtain the existence of solutions, avoiding hypotheses of compactness on the semigroup generated by the linear part and any conditions on the multi-valued nonlinearity expressed in terms of measures of noncompactness.

3.1.2 Nonhomogeneous Cauchy Problem

In the following, we assume X is reflexive and introduce the following hypothesis:

(H_A) The linear operator $A : D(A) \subset X \to X$ satisfies the Hille-Yosida condition, i.e., there exist two constants $\omega \in \mathbb{R}$ and $\overline{M} > 0$ such that $(\omega, +\infty) \subset \rho(A)$ and

$$\|(\lambda I - A)^{-k}\|_{\mathscr{L}(X)} \leq \frac{\overline{M}}{(\lambda - \omega)^k}$$

for all $\lambda > \omega, k \geq 1$.

Let $X_0 = \overline{D(A)}$ and A_0 be the part of A in X_0 defined by

$$A_0 x = Ax \text{ on } D(A_0) = \{x \in D(A) : Ax \in X_0\}.$$

Lemma 3.1. *[199] The part A_0 of A generates a C_0-semigroup $\{T(t)\}_{t \geq 0}$ on X_0.*

Furthermore, we assume that

(H_T) $\{T(t)\}_{t \geq 0}$ is uniformly bounded, i.e., there exists $M > 1$ such that $\sup_{t \in [0,+\infty)} \|T(t)\|_{\mathscr{L}(X)} < M$.

In this subsection, by using Laplace transform and the Wright function, we will give the integral solution for the following Cauchy problem:

$$\begin{cases} {}^C_0 D^q_t x(t) = Ax(t) + f(t), & t \in (0, b], \\ x(0) = x_0, \end{cases} \tag{3.2}$$

where $f \in C([0, b], X)$ and $x_0 \in X_0$ are given.

For convenience, we set

$$I^q_{0+} := {}_0 D^{-q}_t.$$

Definition 3.1. *A function $x(t)$ is said to be an integral solution of (3.2) if the following conditions hold:*

(i) $x : [0, b] \to X$ *is continuous;*
(ii) $I_{0+}^q x(t) \in D(A)$ *for $t \in [0, b]$;*
(iii) $x(t) = x_0 + AI_{0+}^q x(t) + I_{0+}^q f(t)$ *for $t \in [0, b]$.* (3.3)

Remark 3.1. *If $x(t)$ is an integral solution of (3.2), then $x(t) \in X_0$ for $t \in [0, b]$. In fact, by $I_{0+}^q x(t) \in D(A)$, we have $I_{0+}^1 x(t) = I_{0+}^{1-q} I_{0+}^q x(t) \in D(A)$ for $t \in [0, b]$. Then $x(t) = \lim_{h \to 0+} \frac{1}{h} \int_t^{t+h} x(s)ds \in X_0$ for $t \in [0, b]$.*

Consider the integral solution of the following equation:

$$\begin{cases} {}_0^C D_t^q x(t) = A_0 x(t) + f(t), & t \in (0, b], \\ x(0) = x_0. \end{cases} \tag{3.4}$$

By Definition 3.1, we can obtain an integral solution of (3.4) given by

$$x(t) = x_0 + A_0 I_{0+}^q x(t) + I_{0+}^q f(t) \tag{3.5}$$

for $x_0 \in X_0$ and $t \geq 0$.

Lemma 3.2. *If f takes values in X_0, then the integral equation (3.5) can be solved by*

$$x(t) = S_q(t)x_0 + \int_0^t K_q(t - s)f(s)ds, \ t \in [0, b], \tag{3.6}$$

where

$$S_q(t) = I_{0+}^{1-q} K_q(t), \quad K_q(t) = t^{q-1} P_q(t),$$
$$P_q(t) = \int_0^\infty q\theta \Psi_q(\theta) T(t^q \theta) d\theta,$$

and $\Psi_q(\theta)$ is the Wright function (see Definition 1.8).

Proof. Let $\xi > 0$. Applying the Laplace transform and using the similar method in the proof of Lemma 2.2, we have

$$\chi(\xi) = \xi^{q-1} \int_0^\infty e^{-\xi t} t^{q-1} P_q(t) x_0 dt$$

$$+ \int_0^\infty e^{-\xi t} \left(\int_0^t (t-s)^{q-1} P_q(t-s) f(s) ds \right) dt.$$

Since the inverse Laplace transform of ξ^{q-1} is

$$\mathcal{L}^{-1}(\xi^{q-1}) = \frac{t^{-q}}{\Gamma(1-q)}.$$

Thus, for $t \in [0, b]$ we obtain

$$\begin{aligned}
x(t) &= \left(\mathcal{L}^{-1}(\xi^{q-1}) * K_q(t) \right) x_0 + \int_0^t K_q(t-s) f(s) ds \\
&= \left(I_{0+}^{1-q} K_q(t) \right) x_0 + \int_0^t K_q(t-s) f(s) ds \\
&= S_q(t) x_0 + \int_0^t K_q(t-s) f(s) ds.
\end{aligned}$$

This completes the proof. $\square$

Lemma 3.3. *Assume that* (H_T) *holds. Then* $P_q(t)$ *is continuous in the uniform operator topology for* $t > 0$.

Proof. For any $t > 0$, $h > 0$ and $x \in X_0$, we have

$$\begin{aligned}
|T(t+h)x - T(t)x| &\leq \|T(t)\|_{\mathscr{L}(X)} |T(h)x - x| \\
&\leq M|T(h)x - x| \\
&\to 0,
\end{aligned}$$

as $h \to 0$, which implies that $T(t)$ is continuous in the uniform operator topology for $t > 0$.

Then, for any $t > 0$, $h > 0$, and $x \in X_0$, we have

$$|P_q(t+h)x - P_q(t)x| = \left| \int_0^\infty q\theta \Psi_q(\theta) \left(T((t+h)^q \theta) - T(t^q \theta) \right) x d\theta \right|.$$

Since

$$\begin{aligned}
\left| \int_0^\infty q\theta \Psi_q(\theta) \left(T((t+h)^q \theta) - T(t^q \theta) \right) x d\theta \right| &\leq 2M \int_0^\infty q\theta \Psi_q(\theta) d\theta |x| \\
&= \frac{2M}{\Gamma(q)} |x|,
\end{aligned}$$

by Theorem 1.1, we have

$$|P_q(t+h)x - P_q(t)x| \to 0 \text{ independently of } t, \text{ as } h \to 0.$$

Thus, $P_q(t)$ is continuous in the uniform operator topology for $t > 0$.   □

Lemma 3.4. *[302] Assume that* (H_T) *holds. Then, for any fixed* $t > 0$, $\{K_q(t)\}_{t>0}$ *and* $\{S_q(t)\}_{t>0}$ *are linear operators, and for any* $x \in X_0$,

$$|K_q(t)x| \le \frac{Mt^{q-1}}{\Gamma(q)}|x| \quad \text{and} \quad |S_q(t)x| \le M|x|.$$

Lemma 3.5. *[302] Assume that* (H_T) *holds. Then* $\{K_q(t)\}_{t>0}$ *and* $\{S_q(t)\}_{t>0}$ *are strongly continuous, which means that, for any* $x \in X_0$ *and* $0 < t' < t'' \le b$, *we have*

$$|K_q(t')x - K_q(t'')x| \to 0 \quad \text{and} \quad |S_q(t')x - S_q(t'')x| \to 0, \text{ as } t'' \to t'.$$

Assume for a moment that f takes values in X_0. Then (3.6) can be written as

$$x(t) = S_q(t)x_0 + \int_0^t K_q(t-s) \lim_{\lambda \to +\infty} \mathcal{J}_\lambda f(s)ds \qquad (3.7)$$

or

$$x(t) = S_q(t)x_0 + \lim_{\lambda \to +\infty} \int_0^t K_q(t-s)\mathcal{J}_\lambda f(s)ds, \qquad (3.8)$$

where $\mathcal{J}_\lambda = \lambda(\lambda I - A)^{-1}$, and $\lim_{\lambda \to +\infty} \mathcal{J}_\lambda x = x$ for $x \in X_0$. Also from Hille-Yosida condition it is easy to see that $\lim_{\lambda \to +\infty} |\mathcal{J}_\lambda x| \le \overline{M}|x|$. When f takes values in X, but not in X_0, the limit in (3.8) exists (as we will prove). But the limit in (3.7) will no longer exist.

Lemma 3.6. *A solution of integral equation (3.1) with values in* X_0 *is represented by (3.8).*

Proof. Let

$$x_\lambda(t) = \mathcal{J}_\lambda x(t), \ f_\lambda(t) = \mathcal{J}_\lambda f(t) \text{ and } x_\lambda = \mathcal{J}_\lambda x_0.$$

By applying $\mathcal{J}_\lambda$ to (3.1), we have

$$x_\lambda(t) = x_\lambda + A_0 I_{0+}^q x_\lambda(t) + I_{0+}^q f_\lambda(t).$$

Hence, by Lemma 3.2, we obtain

$$x_\lambda(t) = S_q(t)x_\lambda + \int_0^t K_q(t-s)f_\lambda(s)ds.$$

Since $x(t), x_0 \in X_0$, we have

$$x_\lambda(t) \to x(t), x_\lambda \to x_0 \text{ and } S_q(t)x_\lambda \to S_q(t)x_0, \text{ as } \lambda \to +\infty.$$

Thus, (3.8) holds. This completes the proof.   □

Let

$$\Phi_q^0(t)x = \int_0^t K_q(t-s)x\,ds = \int_0^t K_q(s)x\,ds$$

for $x \in X_0$ and $t \geq 0$. Then, we define

$$\Phi_q(t) = (\lambda I - A)\Phi_q^0(t)(\lambda I - A)^{-1} \tag{3.9}$$

for $\lambda > \omega$, which extends $\Phi_q^0(t)$ from X_0 to X. The definition is independent of λ due to the resolvent identity. Since $\Phi_q(t)$ maps X into X_0, we have

$$\Phi_q(t)x = \lim_{\lambda \to +\infty} \mathcal{J}_\lambda \Phi_q(t)x = \lim_{\lambda \to +\infty} \Phi_q^0(t)\mathcal{J}_\lambda x, \text{ for } x \in X.$$

Remark 3.2. *For $x \in X$ and $t \geq 0$, $\Phi_q(t)$ in (3.9) defines a bounded linear operator and*

$$\Phi_q(t)x = \lim_{\lambda \to +\infty} \int_0^t K_q(t-s)\mathcal{J}_\lambda x\,ds = \lim_{\lambda \to +\infty} \int_0^t K_q(s)\mathcal{J}_\lambda x\,ds. \tag{3.10}$$

Lemma 3.7. ${}_0^C D_t^q \Phi_q^0(t)x = S_q(t)x$ *and* $S_q(t)x = A\Phi_q^0(t)x + x$, *for* $x \in X_0$ *and* $t \geq 0$.

Proof. It is easily proved by the definitions of $S_q(t)$ and $\Phi_q^0(t)$.   □

Lemma 3.8.

(i) *For $x \in X$ and $t \geq 0$, $I_{0+}^q \Phi_q(t) \in D(A)$ and*

$$\Phi_q(t)x = A\big(I_{0+}^q \Phi_q(t)x\big) + \frac{t^q}{\Gamma(1+q)}x. \qquad (3.11)$$

(ii) *For $x \in D(A)$,*

$$\Phi_q(t)Ax + x = S_q(t)x.$$

Proof. (i) For $x \in X$ and $t \geq 0$, let

$$
\begin{aligned}
V(t) =& \lambda I_{0+}^q \Phi_q^0(t)(\lambda I - A)^{-1}x \\
& + \frac{t^q}{\Gamma(1+q)}(\lambda I - A)^{-1}x - \Phi_q^0(t)(\lambda I - A)^{-1}x,
\end{aligned} \qquad (3.12)
$$

and it is obvious that $V(0) = 0$. By Lemma 3.7, we have

$$
\begin{aligned}
{}_0^C D_t^q V(t) =& \lambda \Phi_q^0(t)(\lambda I - A)^{-1}x + (\lambda I - A)^{-1}x \\
& - {}_0^C D_t^q \Phi_q^0(t)(\lambda I - A)^{-1}x \\
=& \lambda \Phi_q^0(t)(\lambda I - A)^{-1}x + (\lambda I - A)^{-1}x \\
& - S_q(t)(\lambda I - A)^{-1}x \\
=& \lambda \Phi_q^0(t)(\lambda I - A)^{-1}x + (\lambda I - A)^{-1}x \\
& - A\Phi_q^0(t)(\lambda I - A)^{-1}x - (\lambda I - A)^{-1}x \\
=& \lambda \Phi_q^0(t)(\lambda I - A)^{-1}x - A\Phi_q^0(t)(\lambda I - A)^{-1}x \\
=& (\lambda I - A)\Phi_q^0(t)(\lambda I - A)^{-1}x \\
=& \Phi_q(t)x.
\end{aligned}
$$

Then

$$V(t) = I_{0+}^q \Phi_q(t)x + V(0) = I_{0+}^q \Phi_q(t)x. \qquad (3.13)$$

By (3.9), (3.12), and (3.13), we have

$$(\lambda I - A)V(t) = (\lambda I - A)I_{0+}^q \Phi_q(t)x = \lambda I_{0+}^q \Phi_q(t)x + \frac{t^q}{\Gamma(1+q)}x - \Phi_q(t)x.$$

Thus,

$$\Phi_q(t)x = A\big(I_{0+}^q \Phi_q(t)x\big) + \frac{t^q}{\Gamma(1+q)}x.$$

(ii) For $x \in D(A)$, by Lemma 3.7, we obtain

$$
\begin{aligned}
\Phi_q(t)Ax &= \lim_{\lambda \to +\infty} \int_0^t K_q(s) \mathcal{J}_\lambda Ax \, ds \\
&= \lim_{\lambda \to +\infty} A_0 \int_0^t K_q(s) \mathcal{J}_\lambda x \, ds \\
&= A_0 \Phi_q^0(t)x \\
&= S_q(t)x - x.
\end{aligned}
$$

This completes the proof.   □

The following theorem gives an equivalent form of (3.2).

Theorem 3.1. $x(t)$ *is an integral solution of (3.2) if and only if*

$$
x(t) = S_q(t)x_0 + \lim_{\lambda \to +\infty} \int_0^t K_q(t-s) \mathcal{J}_\lambda f(s) \, ds \tag{3.14}
$$

for $t \in [0, b]$ and $x_0 \in X_0$.

Proof. By Lemma 3.6, we only need to prove that (3.14) is an integral solution of (3.2). Actually it is sufficient to prove the theorem for $x_0 = 0$, because it is easily proved for the special case $f = 0$. Next, we complete the proof by splitting two steps.

Step 1. Assume that f is continuously differentiable, then for $t \in [0, b]$,

$$
\begin{aligned}
x_\lambda(t) &= \int_0^t K_q(t-s) \mathcal{J}_\lambda f(s) \, ds \\
&= \int_0^t K_q(t-s) \mathcal{J}_\lambda \left(f(0) + \int_0^s f'(r) \, dr \right) ds \\
&= \int_0^t K_q(t-s) \mathcal{J}_\lambda f(0) \, ds + \int_0^t K_q(t-s) \mathcal{J}_\lambda \left(\int_0^s f'(r) \, dr \right) ds \\
&= \Phi_q^0(t) \mathcal{J}_\lambda f(0) + \int_0^t \Phi_q^0(t-r) \mathcal{J}_\lambda f'(r) \, dr.
\end{aligned}
$$

By Lemma 3.8, for $t \in [0, b]$, we have

$$
x(t) = \lim_{\lambda \to +\infty} x_\lambda(t) = \Phi_q(t) f(0) + \int_0^t \Phi_q(t-r) f'(r) \, dr
$$

$$= A\big(I_{0+}^q \Phi_q(t) f(0)\big) + \frac{t^q}{\Gamma(1+q)} f(0)$$

$$+ \int_0^t \left(A\big(I_{0+}^q \Phi_q(t-r)\big) + \frac{(t-r)^q}{\Gamma(1+q)} \right) f'(r) dr$$

$$= A\left(I_{0+}^q \Phi_q(t) f(0) + \int_0^t I_{0+}^q \Phi_q(t-r) f'(r) dr \right)$$

$$+ \frac{t^q}{\Gamma(1+q)} f(0) + \frac{1}{\Gamma(1+q)} \int_0^t (t-r)^q f'(r) dr$$

$$= A\left[I_{0+}^q \Phi_q(t) f(0) + I_{0+}^q \left(\int_0^t \Phi_q(t-r) f'(r) dr \right) \right]$$

$$+ \frac{t^q}{\Gamma(1+q)} f(0) + \frac{1}{\Gamma(1+q)} \int_0^t (t-r)^q f'(r) dr$$

$$= A\big(I_{0+}^q x(t)\big) + I_{0+}^q f(t).$$

Step 2. Approximation by continuously differentiable functions.

We approximate f by continuously differentiable functions f_n such that

$$\sup_{t \in [0,b]} |f(t) - f_n(t)| \to 0, \text{ as } n \to \infty.$$

Let

$$x_n(t) = \lim_{\lambda \to \infty} \int_0^t K_q(s) \mathcal{J}_\lambda f_n(s) ds.$$

Then, we have

$$x_n(t) = A\big(I_{0+}^q x_n(t)\big) + I_{0+}^q f_n(t). \tag{3.15}$$

Hence,

$$|x_n(t) - x_m(t)| = \left| \lim_{\lambda \to \infty} \int_0^t K_q(s) \mathcal{J}_\lambda \big(f_n(s) - f_m(s)\big) ds \right|$$

$$\leq \frac{M\overline{M}}{\Gamma(q)} \int_0^t (t-s)^{q-1} |f_n(s) - f_m(s)| ds$$

$$\leq \frac{M\overline{M}b^q}{\Gamma(q)} \|f_n - f_m\|,$$

which implies that $\{x_n\}$ is a Cauchy sequence and its limit exists. Denote the limit by $x(t)$. Taking limit on the two sides of (3.15), we obtain

$$x(t) = A\big(I_{0+}^q x(t)\big) + I_{0+}^q f(t), \quad \text{for } t \in [0, b].$$

Therefore, (3.14) is an integral solution of (3.2). This completes the proof. □

Remark 3.3.

(i) *The integral solution (3.14) is equal to*

$$x(t) = S_q(t)x_0 + \frac{d}{dt}\int_0^t \Phi_q(t - s)f(s)ds.$$

This is easily seen by integrating the last term in (3.14) and Remark 3.2.
(ii) $(\lambda^q I - A)^{-1}x = \lambda \int_0^\infty e^{-\lambda t}\Phi_q(t)x\,ds$ *for $x \in X$ and $\lambda^q > \omega$.*
In fact, by taking Laplace transform to two sides of (3.11), we obtain

$$\begin{aligned}
\mathfrak{L}[\Phi_q(t)x] &= A\mathfrak{L}[I_{0+}^q\Phi_q(t)x] + \mathfrak{L}\left[\frac{t^q}{\Gamma(1+q)}x\right] \\
&= \lambda^{-q}A\mathfrak{L}[\Phi_q(t)x] + \lambda^{-q-1}x \\
&= \lambda^{-1}(\lambda^q I - A)^{-1}x.
\end{aligned}$$

(iii) *We can say that A generates the operator $\{\Phi_q(t)\}_{t\geq 0}$. When $q = 1$, $\{\Phi_q(t)\}_{t\geq 0}$ degenerates into $\{S(t)\}_{t\geq 0}$, which is an integrated semi-group generated by A in [223].*
(iv) *Theorem 3.1 also holds if $f \in L^{\frac{1}{q_1}}([0, b], X)$, $q_1 \in (0, q)$.*

Let $c \in [0, b)$. Consider the singular integral equation of the form

$$x(t) = \phi(t) + \lim_{\lambda\to+\infty}\int_c^t K_q(t - s)\mathcal{J}_\lambda g(s, x(s))ds, \quad \text{for } t \in [c, b]. \quad (3.16)$$

Similar to the proof of Lemma 3.2 in [280], we can get the following lemma.

Lemma 3.9. *Let $q_1 \in (0, q)$, and $g(\cdot, x)$ be $L^{\frac{1}{q_1}}$-integrable for every $x \in X_0$. Assume that $\{T(t)\}_{t\geq 0}$ is equicontinuous and for all $t \in [c, b]$, $g(t, \cdot)$ is compact or $\{T(t)\}_{t\geq 0}$ is compact. Suppose in addition that*

(i) *for any compact subset $K \subset X_0$, there exist $\delta > 0$ and $L_K \in$ $L^{\frac{1}{q_1}}([c,b], \mathbb{R}^+)$ such that*

$$|g_n(t, x_1) - g_n(t, x_2)| \le L_K(t)|x_1 - x_2|,$$

for a.e. $t \in [c,b]$ and each $x_1, x_2 \in B_\delta(K) := K + B_\delta(0)$;
(ii) *there exists $\gamma(t) \in L^{\frac{1}{q_1}}([c,b], \mathbb{R}^+)$ such that $|g(t,x)| \le \gamma(t)(c' + |x|)$ for a.e. $t \in [c,b]$ and every $x \in X_0$, where c' is arbitrary, but fixed.*

Then integral equation (3.16) admits a unique solution for every $\phi(t) \in C([c,b], X_0)$. Moreover, the solution of (3.16) depends continuously on ϕ.

3.1.3 Integral Solutions of (3.1)

We assume that the multivalued nonlinearity $F : [0,b] \times X \to P_{cl,cv}(X)$ satisfies:

(H_1) $F(\cdot, x) : [0,b] \multimap X$ has a measurable selection for every $x \in X$, i.e., there exists a measurable function $f : [0,b] \to X$ such that $f(t) \in F(t,x)$ for a.e. $t \in [0,b]$;
(H_2) $F(t, \cdot) : X \multimap X$ is weakly sequentially closed for a.e. $t \in [0,b]$, i.e., it has a weakly sequentially closed graph;
(H_3) $F(t, \cdot)$ is weakly u.s.c. for a.e. $t \in [0,b]$;
(H_4) there exists a constant $q_1 \in (0,q)$ and for every $r > 0$, there exists a function $\mu_r \in L^{\frac{1}{q_1}}([0,b], \mathbb{R}^+)$ such that for each $x \in X$, $|x| \le r$

$$|F(t,x)| = \sup\{|f| : f \in F(t,x)\} \le \mu_r(t) \text{ for a.e. } t \in [0,b];$$

(H_5) there exists a function $\alpha(t) \in L^{\frac{1}{q_1}}([0,b], \mathbb{R}^+)$ such that

$$|F(t,x)| \le \alpha(t)(1+|x|) \text{ for a.e. } t \in [0,b] \text{ and } x \in X.$$

Given $x \in C([0,b], X)$, let us denote

$$Sel_F(x) = \left\{f \in L^{\frac{1}{q_1}}([0,b], X) : f(t) \in F(t, x(t)), \text{ for a.e. } t \in [0,b]\right\}.$$

The set $Sel_F(x)$ is always nonempty as the following Lemmas 3.10 and 3.11 show.

Lemma 3.10. *Assume that the multivalued map F satisfies conditions (H_1), (H_2), and (H_4). Then the set $Sel_F(x)$ is nonempty for any $x \in C([0, b], X)$.*

Proof. Let $x \in C([0, b], X)$, by the uniform continuity of x, there exists a sequence $\{x_n\}$ of step functions, $x_n : [0, b] \to X$ such that

$$\sup_{t \in [0,b]} |x_n(t) - x(t)| \to 0, \quad \text{as } n \to \infty. \tag{3.17}$$

Hence, by (H_1), there exists a sequence of functions $\{f_n\}$ such that $f_n(t) \in F(t, x_n(t))$ for a.e. $t \in [0, b]$ and $f_n : [0, b] \to X$ is measurable for any $n \in \mathbb{N}^+$. From (3.17), there exists a bounded set $E \subset X$ such that $x_n(t)$, $x(t) \in E$, for any $t \in [0, b]$ and $n \in \mathbb{N}^+$, and by (H_4) there exists $\mu_n \in L^{\frac{1}{q_1}}([0, b], \mathbb{R})$ such that

$$|f_n(t)| \leq |F(t, x_n(t))| \leq \mu_n(t), \quad \forall\, n \in \mathbb{N}^+, \quad \text{and a.e. } t \in [0, b].$$

Hence, $\{f_n\} \subset L^{\frac{1}{q_1}}([0, b], X)$ is bounded and uniformly integrable and $\{f_n(t)\}$ is bounded in X for a.e. $t \in [0, b]$. According to the reflexivity of the space X and Lemma 1.4, we have the existence of a subsequence, denoted as the sequence, such that

$$f_n \rightharpoonup g \in L^{\frac{1}{q_1}}([0, b], X).$$

By Mazur lemma, we obtain a sequence

$$\tilde{f}_n = \sum_{i=0}^{k_n} \lambda_{n,i} f_{n+i}, \quad \lambda_{n,i} \geq 0, \quad \sum_{i=0}^{k_n} \lambda_{n,i} = 1$$

such that $\tilde{f}_n \to g$ in $L^{\frac{1}{q_1}}([0, b], X)$ and, up to subsequence, $\tilde{f}_n(t) \to g(t)$ for all $t \in [0, b]$. By (H_4), the multivalued map $F(t, \cdot)$ is locally weakly compact for a.e. $t \in [0, b]$, i.e., for a.e. $t \in [0, b]$ and every $x \in X$, there is a neighborhood V of x such that the restriction of $F(t, \cdot)$ to V is weakly compact. Hence, by (H_2) and the locally weak compactness, we easily get that $F(t, \cdot) : X_w \multimap X_w$ is u.s.c. for a.e. $t \in [0, b]$. Thus, $F(t, \cdot) : X \multimap X_w$ is u.s.c. for a.e. $t \in [0, b]$.

To conclude we only need to prove that $g(t) \in F(t, x(t))$ for a.e. $t \in [0, b]$. Indeed, let N_0 be a set with Lebesgue measure zero such that $F(t, \cdot) :$

$X \multimap X_w$ is u.s.c., $f_n(t) \in F(t, x_n(t))$ and $\tilde{f}_n(t) \to g(t)$ for all $t \in [0, b] \setminus N_0$ and $n \in \mathbb{N}^+$.

Fix $t_0 \notin N_0$ and assume, by contradiction, that $g(t_0) \notin F(t_0, x(t_0))$. Since $F(t_0, x(t_0))$ is closed and convex, from Hahn-Banach theorem there is a weakly open convex set $V \supset F(t_0, x(t_0))$ satisfying $g(t_0) \notin \overline{V}$. Since $F(t_0, \cdot) : X \multimap X_w$ is u.s.c., we can find a neighborhood U of $x(t_0)$ such that $F(t_0, x) \subset V$ for all $x \in U$. The convergence $x_n(t_0) \rightharpoonup x(t_0)$ as $n \to \infty$ then implies the existence of $n_0 \in \mathbb{N}^+$ such that $x_n(t_0) \in U$ for all $n > n_0$. Therefore, $f_n(t_0) \in F(t_0, x_n(t_0)) \subset V$ for all $n > n_0$. Since V is convex, we also have that $\tilde{f}_n(t_0) \in V$ for all $n > n_0$ and, by the convergence, we arrive at the contradictory conclusion that $g(t_0) \in \overline{V}$. We obtain that $g(t) \in F(t, x(t))$ for a.e. $t \in [0, b]$. $\qquad\square$

Lemma 3.11. *Let conditions (H_1), (H_3), and (H_5) be satisfied. Then $Sel_F : C([0, b], X) \to P(L^{\frac{1}{q_1}}([0, b], X))$ is weakly u.s.c. with nonempty, convex, and weakly compact values.*

Definition 3.2. *A continuous function $x : [0, b] \to X$ is said to be an integral solution of fractional evolution inclusion (3.1) if $x(0) = x_0$ and there exists $f(t) \in Sel_F(x)(t)$ satisfying the following integral equation:*

$$x(t) = S_q(t)x_0 + \lim_{\lambda \to +\infty} \int_0^t K_q(t - s)\mathcal{J}_\lambda f(s)ds.$$

Remark 3.4. *For any $x \in C([0, b], X_0)$, now define solution multioperator $\mathcal{F} : C([0, b], X_0) \to P(C([0, b], X_0))$ as follows:*

$$\mathcal{F} = S_q(t)x_0 + \mathcal{W} \circ Sel_F,$$

where $\mathcal{W}(f) = \lim_{\lambda \to +\infty} \int_0^t K_q(t - s)\mathcal{J}_\lambda f(s)ds$. It is easy to verify that the fixed points of the multioperator $\mathcal{F}$ are integral solutions of (3.1).

3.1.4 Existence

We study the existence of solutions for fractional evolution inclusion (3.1).

Fix $n \in \mathbb{N}^+$, consider Q_n the closed ball of radius n in $C([0, b], X_0)$ centered at the origin, and denote by $\mathcal{F}_n = \mathcal{F}|_{Q_n} : Q_n \to C([0, b], X_0)$ the restriction of the multioperator $\mathcal{F}$ on the set Q_n. We describe some properties of $\mathcal{F}_n$.

Lemma 3.12. *The multioperator $\mathcal{F}_n$ has a weakly sequentially closed graph.*

Proof. Let $\{x_m\} \subset Q_n$ and $\{y_m\} \subset C([0,b], X_0)$ satisfy $y_m \in \mathcal{F}_n(x_m)$ for all m and $x_m \rightharpoonup x$, $y_m \rightharpoonup y$ in $C([0,b], X_0)$. We will prove that $y \in \mathcal{F}_n(x)$.

Since $x_m \in Q_n$ for all m and $x_m(t) \rightharpoonup x(t)$ for every $t \in [0,b]$, it follows that $|x(t)| \leq \liminf_{m\to\infty} |x_m(t)| \leq n$ for all t (see [47]). The fact that $y_m \in \mathcal{F}(x_m)$ means that there exists a sequence $\{f_m\}$, $f_m \in Sel_F(x_m)$, such that for every $t \in [0,b]$,

$$y_m(t) = S_q(t)x_0 + \lim_{\lambda \to +\infty} \int_0^t K_q(t-s)\mathcal{J}_\lambda f_m(s)ds.$$

According to (H_4), we observe that $|f_m(t)| \leq \mu_n(t)$ for a.e. t and every m, i.e., $\{f_m\}$ is bounded and uniformly integrable and $\{f_m(t)\}$ is bounded in X for a.e. $t \in [0,b]$. Hence, by the reflexivity of the space X and Lemma 1.4, we have the existence of a subsequence, denoted as the sequence, and a function g such that $f_m \rightharpoonup g$ in $L^{\frac{1}{q_1}}([0,b], X)$.

Moreover, we have that $\mathcal{W}(f_m) \rightharpoonup \mathcal{W}(g)$. Indeed, let $x' : X \to \mathbb{R}$ be a linear continuous operator. We first prove that the operator $\mathcal{W}$ is continuous.

For any $y_m, y \in L^{\frac{1}{q_1}}([0,b], X)$ and $y_m \to y$ ($m \to \infty$), using (H_4), we get for each $t \in [0,b]$,

$$(t-s)^{q-1}\|\mathcal{J}_\lambda\|_{\mathscr{L}(X)}|y_m(s)-y(s)| \leq 2\overline{M}(t-s)^{q-1}\mu_n(s), \quad \text{for a.e. } s \in [0,t).$$

On the other hand,

$$\int_0^t (t-s)^{q-1}\mu_n(s)ds \leq \frac{b^{(1+a)(1-q_1)}}{(1+a)^{1-q_1}}\|\mu_n\|_{L^{\frac{1}{q_1}}[0,b]},$$

where $a = \frac{q-1}{1-q_1} \in (-1,0)$, which means that $\int_0^t (t-s)^{q-1}\mu_n(s)ds$ is integrable for $t \in [0,b]$.

By Theorem 1.1, we have

$$\int_0^t (t-s)^{q-1}\|\mathcal{J}_\lambda\|_{\mathscr{L}(X)}|y_m(s) - y(s)|ds \to 0, \quad \text{as } m \to \infty.$$

For $t \in [0, b]$,

$$\|\mathcal{W}(y_m) - \mathcal{W}(y)\|$$

$$\leq \left| \lim_{\lambda \to +\infty} \int_0^t K_q(t - s)\mathcal{J}_\lambda(y_m(s) - y(s))ds \right|$$

$$\leq \frac{M\overline{M}}{\Gamma(q)} \int_0^t (t - s)^{q-1}|y_m(s) - y(s)|ds$$

$$\to 0, \quad \text{as } m \to \infty.$$

Hence, the operator $\mathcal{W}$ is continuous. Clearly, $\mathcal{W}$ is linear; thus, we have that the operator $g \to x' \circ \mathcal{W}(g)$ is a linear and continuous operator from $L^{\frac{1}{q_1}}([0, b], X)$ to $\mathbb{R}$ for all $t \in [0, b]$. Then, from the definition of weak convergence, we have that for every $t \in [0, b]$,

$$x' \circ \mathcal{W}(f_m) \to x' \circ \mathcal{W}(g).$$

Thus

$$y_m(t) \rightharpoonup S_q(t)x_0 + \lim_{\lambda \to +\infty} \int_0^t K_q(t - s)\mathcal{J}_\lambda g(s)ds = y^*(t), \quad \forall\, t \in [0, b],$$

which implies, for the uniqueness of the weak limit in X_0, that $y^*(t) = y(t)$ for all $t \in [0, b]$.

By using the similar method in Lemma 3.10, we can prove that $g(t) \in F(t, x(t))$ for a.e. $t \in [a, b]$. $\qquad\square$

Lemma 3.13. *The multioperator $\mathcal{F}_n$ is weakly compact.*

Proof. We first prove that $\mathcal{F}_n(Q_n)$ is relatively weakly sequentially compact.

Let $\{x_m\} \subset Q_n$ and $\{y_m\} \subset C([0, b], X_0)$ satisfy $y_m \in \mathcal{F}_n(x_m)$ for all m. By the definition of the multioperator $\mathcal{F}_n$, there exists a sequence $\{f_m\}$, $f_m \in Sel_F(x_m)$, such that

$$y_m(t) = S_q(t)x_0 + \lim_{\lambda \to +\infty} \int_0^t K_q(t - s)\mathcal{J}_\lambda f_m(s)ds, \quad \forall\, t \in [0, b].$$

Further, as the reason for Lemma 3.12, we have that there exist a subsequence, denoted as the sequence, and a function g such that $f_m \rightharpoonup g$

in $L^{\frac{1}{q_1}}([0, b], X)$. Therefore,

$$y_m(t) \rightharpoonup l(t) = S_q(t)x_0 + \lim_{\lambda \to +\infty} \int_0^t K_q(t-s)\mathcal{J}_\lambda g(s)ds, \quad \forall\, t \in [0, b].$$

Furthermore, by the weak convergence of $\{f_m\}$ and (H_A), we have that for all $m \in \mathbb{N}^+$ and $t \in [0, b]$

$$|y_m(t)| \leq M|x_0| + \frac{M\overline{M}b^{(1+a)(1-q_1)}}{\Gamma(q)(1+a)^{1-q_1}}\|\mu_n\|_{L^{\frac{1}{q_1}}[0,b]}.$$

Recalling the weak convergence of $C([0, b], X)$, it is easy to prove that $y_m \rightharpoonup l$ in $C([0, b], X_0)$. Thus $\mathcal{F}_n(Q_n)$ is relatively weakly sequentially compact, hence relatively weakly compact by Theorem 1.3. $\qquad \square$

Lemma 3.14. *The multioperator $\mathcal{F}_n$ has convex and weakly compact values.*

Proof. Fix $x \in Q_n$, since F is convex valued, from the linearity of the integral, and the operators $S_q(t)$ and $K_q(t)$, it follows that the set $\mathcal{F}_n(x)$ is convex. The weak compactness of $\mathcal{F}_n(x)$ follows by Lemmas 3.12 and 3.13. $\qquad \square$

Now we state the main results of this subsection.

Theorem 3.2. *Assume that (H_A), (H_T), (H_1), and (H_2) hold. Further, suppose that*

$(H_4)'$ *there exists a sequence of functions $\{\omega_n\} \subset L^{\frac{1}{q_1}}([0, b], \mathbb{R}^+)$ such that*

$$\sup_{|x|\leq n} |F(t, x)| \leq \omega_n(t)$$

for a.e. $t \in [0, b]$, $n \in \mathbb{N}^+$ with

$$\liminf_{n\to\infty} \frac{1}{n}\left(\int_0^b |\omega_n(s)|^{\frac{1}{q_1}} ds\right)^{q_1} = 0. \tag{3.18}$$

Then (3.1) has at least one integral solution.

Proof. We show that there exists $n \in \mathbb{N}^+$ such that the operator $\mathcal{F}_n$ maps the ball Q_n into itself.

Assume, on the contrary, that there exist sequences $\{z_n\}$, $\{y_n\}$ such that $z_n \in Q_n$, $y_n \in \mathcal{F}_n(z_n)$, and $y_n \notin Q_n$, $\forall\, n \in \mathbb{N}^+$. Then there exists a sequence $\{f_n\} \subset L^{\frac{1}{q_1}}([0,b], X)$, $f_n(s) \in F(s, z_n(s))$, $\forall\, n \in \mathbb{N}^+$, and a.e. $s \in [0,b]$ such that

$$y_n(t) = S_q(t)x_0 + \lim_{\lambda \to +\infty} \int_0^t K_q(t-s) \mathcal{J}_\lambda f_n(s)\,ds, \quad \forall\, t \in [0,b].$$

As the reason for Lemma 3.13, we have

$$1 < \frac{\|y_n\|}{n} \leq \frac{M|x_0|}{n} + \frac{M\overline{M}b^{(1+a)(1-q_1)}}{n\Gamma(q)(1+a)^{1-q_1}} \left(\int_0^b |\omega_n(\eta)|^{\frac{1}{q_1}}\,d\eta \right)^{q_1}, \quad n \in \mathbb{N}^+,$$

which contradicts (3.18).

Now, choose $n \in \mathbb{N}^+$ such that $\mathcal{F}_n(Q_n) \subseteq Q_n$. By Lemma 3.13, the set $V_n = \overline{\mathcal{F}_n(Q_n)}^w$ is weakly compact. Let now $W_n = \overline{co}(V_n)$, where $\overline{co}(V_n)$ denotes the closed convex hull of V_n. By Theorem 1.4, W_n is a weakly compact set. Moreover, from the fact that $\mathcal{F}_n(Q_n) \subset Q_n$ and that Q_n is a convex closed set we have that $W_n \subset Q_n$, and hence

$$\mathcal{F}_n(W_n) = \mathcal{F}_n(\overline{co}(\mathcal{F}_n(Q_n))) \subseteq \mathcal{F}_n(Q_n) \subseteq \overline{\mathcal{F}_n(Q_n)}^w = V_n \subset W_n.$$

In view of Lemma 3.12, $\mathcal{F}_n$ has a weakly sequentially closed graph. Thus from Theorem 1.6, inclusion (3.1) has a solution. The proof is completed. $\qquad\square$

Remark 3.5. *Suppose, for example, that there exist $\alpha \in L^{\frac{1}{q_1}}([0,b], \mathbb{R}^+)$ and a nondecreasing function $\varrho : [0,+\infty) \to [0,+\infty)$ such that $|F(t,x)| \leq \alpha(t)\varrho(|x|)$ for a.e. $t \in [0,b]$ and every $x \in X$. Then condition (3.18) is equivalent to*

$$\liminf_{n \to \infty} \frac{\varrho(n)}{n} = 0.$$

Theorem 3.3. *Assume that (H_A), (H_T), (H_1), (H_2), and (H_5) hold. If*

$$\frac{M\overline{M}b^{(1+a)(1-q_1)}}{\Gamma(q)(1+a)^{1-q_1}} \|\alpha\|_{L^{\frac{1}{q_1}}[0,b]} < 1, \tag{3.19}$$

then the fractional evolution inclusion (3.1) *has at least one integral solution.*

Proof. As the reason for Theorem 3.2, and assume that there exist $\{z_n\}$, $\{y_n\}$ such that $z_n \in Q_n$, $y_n \in \mathcal{F}_n(z_n)$, and $y_n \notin Q_n$, $\forall\, n \in \mathbb{N}^+$, we would get

$$n < \|y_n\| \leq M|x_0|$$
$$+ \frac{M\overline{M}b^{(1+a)(1-q_1)}}{\Gamma(q)(1+a)^{1-q_1}} \left(\int_0^b |\alpha(\eta)|^{\frac{1}{q_1}} (1 + |z_n(\eta)|)^{\frac{1}{q_1}} \, d\eta \right)^{q_1}$$
$$\leq M|x_0| + \frac{M\overline{M}b^{(1+a)(1-q_1)}}{\Gamma(q)(1+a)^{1-q_1}} (1 + n)\|\alpha\|_{L^{\frac{1}{q_1}}[0,b]}, \quad n \in \mathbb{N}^+,$$

which contradicts (3.19).

The conclusion then follows by Theorem 1.6, like Theorem 3.2. $\qquad\square$

Furthermore, we also consider superlinear growth condition, as next theorem shows.

Theorem 3.4. *Assume that* (H_A), (H_T), (H_1), *and* (H_2) *hold. Further, suppose that*

$(H_4)''$ *there exists* $\alpha \in L^{\frac{1}{q_1}}([0, b], \mathbb{R}^+)$ *and a nondecreasing function* ϱ : $[0, +\infty) \to [0, +\infty)$ *such that*

$$|F(t, x)| \leq \alpha(t)\varrho(|x|), \text{ for a.e. } t \in [0, b], \forall\, x \in X.$$

Furthermore, there exists $R > 0$ *such that*

$$\frac{R}{M|x_0| + \frac{M\overline{M}b^{(1+a)(1-q_1)}}{\Gamma(q)(1+a)^{1-q_1}}\|\alpha\|_{L^{\frac{1}{q_1}}[0,b]}\varrho(R)} > 1. \tag{3.20}$$

Then inclusion (3.1) *has at least one integral solution.*

Proof. It suffices to prove that the operator $\mathcal{F}$ maps the ball Q_R into itself. In fact, given any $z \in Q_R$ and $y \in \mathcal{F}(z)$, it holds

$$\|y_n\| \leq M|x_0| + \frac{M\overline{M}b^{(1+a)(1-q_1)}}{\Gamma(q)(1+a)^{1-q_1}} \left(\int_0^b |\alpha(s)|^{\frac{1}{q_1}} (\varrho(|z(s)|))^{\frac{1}{q_1}} \, ds \right)^{q_1}$$

$$\leq M|x_0| + \frac{M\overline{M}b^{(1+a)(1-q_1)}}{\Gamma(q)(1+a)^{1-q_1}} \|\alpha\|_{L^{\frac{1}{q_1}}[0,b]} \varrho(R) < R.$$

The conclusion then follows by Theorem 1.6, like Theorem 3.2.   □

3.2 TOPOLOGICAL STRUCTURE OF SOLUTION SETS

3.2.1 Introduction

In this section, we study the topological structure of solution sets for (3.1) in cases that $T(t)$ is compact and noncompact, respectively. Subsection 3.2.2 is devoted to proving that the solution set for fractional evolution inclusion (3.1) is a nonempty compact R_δ-set in the case that the semigroup is compact. An example is given to illustrate the obtained result. Subsection 3.2.3 provides the existence of integral solutions for fractional evolution inclusion (3.1) in the case that the semigroup is noncompact, then proceed to study the R_δ-structure of solution set of (3.1). In Subsection 3.2.4, the controllability for the fractional control system is investigated.

Lemma 3.15. *[280] Let hypotheses (H_1), (H_3), and (H_5) be satisfied. Then there exists a sequence $F_n : [0, b] \times X \to P_{cl,cv}(X)$ such that*

(i) $F(t, x) \subset F_{n+1}(t, x) \subset F_n(t, x) \subset \overline{co}(F(t, B_{3^{1-n}}(x)))$, $n \geq 1$, *for each* $t \in [0, b]$ *and* $x \in X$;

(ii) $|F_n(t, x)| \leq \alpha(t)(2 + |x|)$, $n \geq 1$, *for a.e.* $t \in [0, b]$ *and each* $x \in X$;

(iii) *there exists* $E \subset [0, b]$ *with* $mes(E) = 0$ *such that for each* $x^* \in X^*$, $\varepsilon > 0$ *and* $x \in X$, *there exists* $N > 0$ *such that for all* $n \geq N$,

$$x^*(F_n(t, x)) \subset x^*(F(t, x)) + (-\varepsilon, \varepsilon);$$

(iv) $F_n(t, \cdot) : X \to P_{cl,cv}(X)$ *is continuous for a.e.* $t \in [0, b]$ *with respect to Hausdorff metric for each* $n \geq 1$;

(v) *for each* $n \geq 1$, *there exists a selection* $g_n : [0, b] \times X \to X$ *of* F_n *such that* $g_n(\cdot, x)$ *is measurable for each* $x \in X$ *and for any compact subset* $\mathscr{D} \subset X$, *there exist constants* $C_V > 0$ *and* $\delta > 0$ *for which the estimate*

$$|g_n(t, x_1) - g_n(t, x_2)| \leq C_V \alpha(t)|x_1 - x_2|$$

holds for a.e. $t \in [0, b]$ *and each* $x_1, x_2 \in V$ *with* $V = \mathscr{D} + B_\delta(0)$;

(vi) F_n *verifies condition (H_3) with F_n instead of F for each $n \geq 1$, provided that X is reflexive.*

3.2.2 Compact Semigroup Case

Let us also present the following approximation result, the proof of which is similar to the proof in [279].

Lemma 3.16. *Suppose X is a Banach space and $\{T(t)\}_{t \geq 0}$ is compact in X_0. If the two sequences $\{f_m\} \subset L^{\frac{1}{q_1}}([0,b], X)$ and $\{x_m\} \subset C([0,b], X_0)$, where x_m is an integral solution of Cauchy problem*

$$\begin{cases} {}^C_0 D^q_t x_m(t) = Ax_m(t) + f_m(t), & t \in (0,b], \\ x_m(0) = x_0, \end{cases}$$

$f_m \rightharpoonup f$ *in* $L^{\frac{1}{q_1}}([0,b], X)$ *and* $x_m \to x$ *in* $C([0,b], X_0)$, *then x is an integral solution of the limit problem*

$$\begin{cases} {}^C_0 D^q_t x(t) = Ax(t) + f(t), & t \in (0,b], \\ x(0) = x_0. \end{cases}$$

Theorem 3.5. *Let conditions (H_A), (H_1), (H_3), and (H_5) be satisfied. Suppose in addition that $\{T(t)\}_{t \geq 0}$ is compact in X_0. Then the solution set of inclusion (3.1) is a nonempty compact subset of $C([0,b], X_0)$ for each initial value $x_0 \in X_0$.*

Proof. Set

$$\mathcal{M}_0 = \{x \in C([0,b], X_0) : |x(t)| \leq \psi(t), \ t \in [0,b]\},$$

where $\psi(t)$ is the solution of the integral equation

$$\psi(t) = a_1 + a_2 \int_0^t (t-s)^{q-1} \alpha(s) \psi(s) ds, \ t \in [0,b],$$

in which a_1 and a_2 are defined as

$$a_1 = M|x_0| + \frac{M\overline{M}b^{(1+a)(1-q_1)}}{\Gamma(q)(1+a)^{1-q_1}} \|\alpha\|_{L^{\frac{1}{q_1}}[0,b]}, \quad a_2 = \frac{M\overline{M}}{\Gamma(q)}.$$

It is clear that $\mathcal{M}_0$ is a closed and convex subset of $C([0, b], X_0)$. We first show that $\mathcal{F}(\mathcal{M}_0) \subset \mathcal{M}_0$. Indeed, taking $x \in \mathcal{M}_0$ and $y(t) \in \mathcal{F}(x)$, we have

$$
|y(t)| \leq |S_q(t)x_0| + \left| \lim_{\lambda \to +\infty} \int_0^t K_q(t-s)\mathcal{J}_\lambda f(s)ds \right|
$$

$$
\leq M|x_0| + \lim_{\lambda \to +\infty} \int_0^t |K_q(t-s)| \|\mathcal{J}_\lambda\|_{\mathscr{L}(X)} \alpha(s)(1 + |x(s)|)ds
$$

$$
\leq M|x_0| + \frac{M\overline{M}}{\Gamma(q)} \int_0^t (t-s)^{q-1} \alpha(s)(1 + |x(s)|)ds
$$

$$
\leq a_1 + a_2 \int_0^t (t-s)^{q-1} \alpha(s)\psi(s)ds.
$$

Thus $y \in \mathcal{M}_0$. Set $\widetilde{\mathcal{M}} = \overline{co}\mathcal{F}(\mathcal{M}_0)$. It is clear that $\widetilde{\mathcal{M}}$ is a closed, bounded, and convex set. Moreover, $\mathcal{F}(\widetilde{\mathcal{M}}) \subset \widetilde{\mathcal{M}}$.

We will prove that for each $t \in [0, b], V(t) = \{\mathcal{F}(x)(t), x \in \mathcal{M}_0\}$ is relatively compact in X_0. Obviously, $V(0) = \{x_0\}$ is relatively compact in X_0. Let $t \in [0, b]$ be fixed. For $x \in \mathcal{M}_0$ and $y \in \mathcal{F}(x)$, there exists $f \in Sel_F(x)$ such that

$$
y(t) = S_q(t)x_0 + \lim_{\lambda \to +\infty} \int_0^t K_q(t-s)\mathcal{J}_\lambda f(s)ds.
$$

For arbitrary $\varepsilon \in (0, t)$ and $\delta > 0$, define an operator $J_{\varepsilon,\delta} : V(t) \to X$ by

$$
J_{\varepsilon,\delta}y(t) = S_q(t)x_0 + \lim_{\lambda \to +\infty} T(\varepsilon^q\delta) \int_0^{t-\varepsilon} \int_\delta^\infty q\theta(t-s)^{q-1}M_q(\theta)
$$
$$
\times T((t-s)^q\theta - \varepsilon^q\delta)\mathcal{J}_\lambda f(s)d\theta ds.
$$

Then from the compactness of $\{T(t)\}_{t\geq 0}$, we get that for $t \in [0, b]$, the set $V_{\varepsilon,\delta}(t) = \{J_{\varepsilon,\delta}y(t), y(t) \in V(t)\}$ is relatively compact in X_0 for each $\varepsilon \in (0, t)$ and $\delta > 0$. Moreover, it follows that

$$
|y(t) - J_{\varepsilon,\delta}y(t)|
$$

$$
\leq qM\overline{M} \int_0^t (t-s)^{q-1}|f(s)|ds \int_0^\delta \theta M_q(\theta)d\theta
$$

$$
+ qM\overline{M} \int_{t-\varepsilon}^t (t-s)^{q-1}|f(s)|ds \int_0^\infty \theta M_q(\theta)d\theta.
$$

$$\leq \frac{qM\overline{M}}{(1+a)^{1-q_1}} \left(\frac{\varepsilon^{(1+a)(1-q_1)}}{\Gamma(1+q)} + b^{(1+a)(1-q_1)} \int_0^\delta \theta M_q(\theta) d\theta \right)$$
$$\times (1 + |\psi|) \|\alpha\|_{L^{\frac{1}{q_1}}[0,b]}.$$

Since $\lim_{\delta \to 0} \int_0^\delta M_q(\theta) d\theta = 0$ and $\lim_{\delta \to 0} \int_0^\delta \theta M_q(\theta) d\theta = 0$, we conclude that there are relatively compact sets arbitrarily close to the set $V(t)$. Hence, the set $V(t)$ is also relatively compact in X_0, which yields that the set $\widetilde{\mathcal{M}}(t)$ is relatively compact in X_0 for each $t \in [0, b]$.

Next, we verify that the set $\mathcal{F}(\mathcal{M}_0)$ is equicontinuous on $[0, b]$. Taking $0 < t_1 < t_2 \leq b$ and $\delta > 0$ small enough, for $y(t) \in \mathcal{F}(\mathcal{M}_0)$, we obtain

$$|y(t_2) - y(t_1)|$$

$$\leq \|S_q(t_2) - S_q(t_1)\|_{\mathscr{L}(X)} |x_0| + \left| \lim_{\lambda \to +\infty} \int_{t_1}^{t_2} K_q(t_2 - s) \mathcal{J}_\lambda f(s) ds \right|$$

$$+ \left| \lim_{\lambda \to +\infty} \int_0^{t_1} \left(K_q(t_2 - s) - K_q(t_1 - s) \right) \mathcal{J}_\lambda f(s) ds \right|$$

$$\leq \|S_q(t_2) - S_q(t_1)\|_{\mathscr{L}(X)} |x_0| + \frac{M\overline{M}}{\Gamma(q)} \int_{t_1}^{t_2} (t_2 - s)^{q-1} \alpha(s)(1 + |x(s)|) ds$$

$$+ \overline{M} \int_0^{t_1 - \delta} (t_1 - s)^{q-1} \|P_q(t_2 - s) - P_q(t_1 - s)\|_{\mathscr{L}(X)}$$
$$\times \alpha(s)(1 + |x(s)|) ds$$

$$+ \overline{M} \int_{t_1 - \delta}^{t_1} (t_1 - s)^{q-1} \|P_q(t_2 - s) - P_q(t_1 - s)\|_{\mathscr{L}(X)}$$
$$\times \alpha(s)(1 + |x(s)|) ds$$

$$+ \frac{M\overline{M}}{\Gamma(q)} \int_0^{t_1} |(t_1 - s)^{q-1} - (t_2 - s)^{q-1}| \alpha(s)(1 + |x(s)|) ds$$

$$=: I_1 + I_2 + I_3 + I_4 + I_5,$$

where

$$I_1 = \|S_q(t_2) - S_q(t_1)\|_{\mathscr{L}(X)} |x_0|,$$

$$I_2 = \frac{M\overline{M}}{\Gamma(q)(1+a)^{1-q_1}} (t_2 - t_1)^{(1+a)(1-q_1)} (1 + |\psi|) \|\alpha\|_{L^{\frac{1}{q_1}}[0,b]},$$

$$I_3 = \sup_{s \in [0, t_1 - \delta]} \|P_q(t_2 - s) - P_q(t_1 - s)\|_{\mathscr{L}(X)}$$

$$\times \frac{\overline{M}(t_1^{1+a} - \delta^{1+a})^{1-q_1}}{(1+a)^{1-q_1}}(1 + |\psi|)\|\alpha\|_{L^{\frac{1}{q_1}}[0,b]},$$

$$I_4 = \frac{2M\overline{M}\delta^{(1+a)(1-q_1)}}{\Gamma(q)(1+a)^{1-q_1}}(1 + |\psi|)\|\alpha\|_{L^{\frac{1}{q_1}}[0,b]},$$

$$I_5 = \frac{M\overline{M}}{\Gamma(q)(1+a)^{1-q_1}}\left((t_2 - t_1)^{1+a} + t_1^{1+a} - t_2^{1+a}\right)^{1-q_1}$$
$$\times (1 + |\psi|)\|\alpha\|_{L^{\frac{1}{q_1}}[0,b]}.$$

Therefore, it is not difficult to see that I_i ($i = 1, 2, 3, 4, 5$) tends to zero as $t_2 - t_1 \to 0, \delta \to 0$, which ensures that $\mathcal{F}(\mathcal{M}_0)$ is equicontinuous. Thus $\widetilde{\mathcal{M}}$ is equicontinuous as well.

Thus, an application of Theorem 1.2 justifies that $\widetilde{\mathcal{M}}$ is relatively compact in $C([0, b], X_0)$.

We consider $\mathcal{F} : \widetilde{\mathcal{M}} \to \mathcal{P}(\widetilde{\mathcal{M}})$. In order to apply the fixed point principle given by Theorem 1.13, it remains to show that $\mathcal{F}$ is u.s.c. with contractible values.

By Lemma 1.7, it suffices to show that $\mathcal{F}$ has closed graph. Let $x_n \subset \widetilde{\mathcal{M}}$ with $x_n \to x$ and $y_n \in \mathcal{F}(x_n)$ with $y_n \to y$. We shall prove that $y \in \mathcal{F}(x)$. By the definition of $\mathcal{F}$, there exists $f_n \in Sel_F(x_n)$ such that

$$y_n(t) = S_q(t)x_0 + \mathcal{W}(f_n).$$

We need to prove that there exists $f \in Sel_F(x)$ such that for a.e. $t \in [0, b]$,

$$y(t) = S_q(t)x_0 + \mathcal{W}(f).$$

Therefore, noticing that Sel_F is weakly u.s.c. with weakly compact and convex values due to Lemma 3.11, an application of Lemma 1.5 yields that there exist $f \in Sel_F(x)$ and a subsequence of f_n, still denoted by f_n, such that $f_n \rightharpoonup f$ in $L^{\frac{1}{q_1}}([0, b], X)$. From this and Lemma 3.16, we see that $y(t) = S_q(t)x_0 + \mathcal{W}(f)$ and $y \in \mathcal{F}(x)$. It follows that $\mathcal{F}$ is closed.

After that, we can show that $\mathcal{F}$ has contractible values. Given $x \in \widetilde{\mathcal{M}}$. Fix $f^* \in Sel_F(x)$ and put $y^* = S_q(t)x_0 + \mathcal{W}(f^*)$. Define a function h :

$[0, 1] \times \mathcal{F}(x) \to \mathcal{F}(x)$ as

$$h(\lambda, y)(t) = \begin{cases} y(t), & t \in [0, \lambda b], \\ x(t, \lambda b, y(\lambda b)), & t \in [\lambda b, b], \end{cases}$$

for each $(\lambda, y) \in [0, 1] \times \mathcal{F}(x)$, where

$$x(t, \lambda b, y(\lambda b)) = S_q(t)x_0 + \lim_{\lambda \to +\infty} \int_0^{\lambda b} K_q(t - s)\mathcal{J}_\lambda \tilde{f}(s)ds$$
$$+ \lim_{\lambda \to +\infty} \int_{\lambda b}^t K_q(t - s)\mathcal{J}_\lambda f^*(s)ds;$$

here, $\tilde{f} \in Sel_F(x)$ and $y = S_q(t)x_0 + \mathcal{W}(\tilde{f})$.

It is easy to see that h is well defined. Also, it is clear that

$$h(0, y) = y^*, \quad h(1, y) = y, \text{ on } \mathcal{F}(x).$$

Moreover, it follows readily that h is continuous. Thus, we have proved that $\mathcal{F}(x)$ is contractible.

Now, let $\Theta(x_0)$ denote the set of all integral solutions of (3.1), then $\Theta(x_0)$ is a compact subset of $C([0, b], X_0)$. Indeed, $\Theta(x_0) = \text{Fix}\mathcal{F}$, so $\Theta(x_0) \subset \mathcal{F}\Theta(x_0)$. Assume that $\{y_n\} \subset \Theta(x_0)$, then one can choose $f_n \in Sel_F(y_n)$ such that $y_n = S_q(t)x_0 + \mathcal{W}(f_n)$. By using similar estimates as the above proof of existence for $\{y_n\}$, we obtain that $\{y_n\}$ is relatively compact. The proof is completed. $\square$

Theorem 3.6. *Let all conditions in Theorem 3.5 be satisfied. Then the solution set of inclusion (3.1) is a compact R_δ-set.*

Proof. Consider the following fractional evolution inclusion

$$\begin{cases} {}^C_0D_t^q x(t) \in Ax(t) + F_n(t, x(t)), & t \in [0, b], \\ x(0) = x_0, \end{cases} \tag{3.21}$$

where $n \geq 1$ and the sequence of multivalued functions $\{F_n\}$ with $F_n : [0, b] \times X \to P_{cl,cv}(X)$ is established in Lemma 3.15.

From Lemma 3.15(ii) and (vi), it follows that $\{F_n\}$ verifies conditions (H_3) and (H_5) for each $n \geq 1$. Then from Lemma 3.11 one finds that Sel_{F_n} is weakly u.s.c. with convex and weakly compact values. Moreover, one can see from the above arguments that the solution set of (3.1) is nonempty and compact in $C([0, b], X_0)$ for each $n \geq 1$.

Let $\Theta^n(x_0)$ denote the set of all integral solutions of (3.21). We show that $\Theta^n(x_0)$ is contractible for each $n \geq 1$. To do this, for any $\lambda \in [0, 1]$, let $x \in \Theta^n(x_0)$ and g_n be the selection of $\{F_n\}$, $n \geq 1$. We deal with the existence and uniqueness of solutions to the integral equation

$$
\begin{aligned}
y(t) =& S_q(t)x_0 + \lim_{\lambda \to +\infty} \int_0^{\lambda b} K_q(t - s)\mathcal{J}_\lambda f^x(s)ds \\
& + \lim_{\lambda \to +\infty} \int_{\lambda b}^t K_q(t - s)\mathcal{J}_\lambda g_n(s, y(s))ds,
\end{aligned}
\tag{3.22}
$$

where $f^x \in Sel_{F_n}(x)$. Moreover, it follows from Lemma 3.15(ii) that $|g_n(t, x)| \leq \alpha(t)(2 + |x|)$ for a.e. $t \in [0, b]$. Therefore, noticing Lemma 3.15(v), together with Lemma 3.9, one finds that equation (3.22) has a unique solution on $[\lambda b, b]$, denoted by $y(t, \lambda b, x(\lambda b))$.

We define $\Phi : [0, 1] \times \Theta^n(x_0) \to \Theta^n(x_0)$ by

$$
\Phi(\lambda, x)(t) = \begin{cases} x(t), & t \in [0, \lambda b], \\ y(t, \lambda b, x(\lambda b)), & t \in [\lambda b, b], \end{cases}
$$

for each $(\lambda, x) \in [0, 1] \times \Theta^n(x_0)$.

It is easy to see that Φ is well defined. Also, it is clear that

$$
\Phi(0, x) = y(t, 0, x_0), \quad \Phi(1, x) = x, \text{ on } \Theta^n(x_0).
$$

Moreover, it follows readily that Φ is continuous. Thus, we have proved that $\Theta^n(x_0)$ is contractible for each $n \geq 1$.

Finally, We claim that $\Theta(x_0) = \bigcap_{n \geq 1} \Theta^n(x_0)$. In view of Lemma 3.15(i), it is easy to verify that $\Theta(x_0) \subset \cdots \subset \Theta^n(x_0) \cdots \subset \Theta^2(x_0) \subset \Theta^1(x_0)$, then $\Theta(x_0) \subset \bigcap_{n \geq 1} \Theta^n(x_0)$. To prove the reverse inclusion, we take $x \in \bigcap_{n \geq 1} \Theta^n(x_0)$. Therefore, there exists a sequence $\{g_n\} \subset L^p([0, b], X)$ such that $g_n \in Sel_{F_n}(x)$, $x = S_q(t)x_0 + \mathcal{W}(g_n)$, and for all $n \geq 1$,

$$
|g_n(t, x)| \leq \alpha(t)(2 + |x|), \text{ for } t \in [0, b],
$$

in view of Lemma 3.15(ii). From the fact that X is reflexive, it follows that $\{g_n\}$ is relatively weakly compact in $L^p([0, b], X)$ due to Lemma 1.4.

Thus, there exists a subsequence of $\{g_n\}$, still denoted by $\{g_n\}$, such that g_n converges weakly to f in $L^p([0, b], X)$. An application of Mazur lemma yields that there exists a sequence $\{\tilde{g}_n\} \subset L^p([0, b], X)$ such that $\tilde{g}_n \in \mathrm{co}\{g_k : k \geq n\}$ for each $n \geq 1$ and $\tilde{g}_n \to f$ in $L^p([0, b], X)$ as $n \to \infty$. Denote by $\mathfrak{J}_c$ the set of all $t \in [0, b]$ such that $\tilde{g}_n(t) \to f(t)$ in X and $g_n(t) \in F(t, x_n(t))$ for all $n \geq 1$. Clearly, $[0, b] \setminus \mathfrak{J}_c$ has null measure.

By Lemma 3.15(iii), we have that there exists $J \subset [0, b]$ with mes $(J) = 0$ such that for $t \in ([0, b] \setminus J) \cap \mathfrak{J}_c > 0$, $\varepsilon > 0$, and $x^* \in X^*$,

$$\langle x^*, \tilde{g}_n(t) \rangle \in \mathrm{co}\{\langle x^*, g_k(t) \rangle, k \geq n\} \subset \langle x^*, F_n(t, x(t)) \rangle$$
$$\subset \langle x^*, F(t, x(t)) \rangle + (-\varepsilon, \varepsilon).$$

Therefore, we obtain that $\langle x^*, f(t) \rangle \in \langle x^*, F(t, x) \rangle$ for each $x^* \in X^*$ and $t \in ([0, b] \setminus J) \cap \mathfrak{J}_c$. Since x^* is arbitrary and F has convex and closed values, we conclude that $f(t) \in F(t, x(t))$ for each $t \in ([0, b] \setminus J) \cap \mathfrak{J}_c$, which implies $f \in Sel_F(x)$. Moreover, noticing g_n converges weakly to f in $L^p([0, b], X)$, we deduce, thanks to Lemma 3.16, that $S_q(t)x_0 + \mathcal{W}(f) = x$. This proves that $x \in \Theta(x_0)$, as desired.

Consequently, we conclude that $\Theta(x_0)$ is an R_δ-set, completing this proof. $\qquad\square$

Example 3.1. Consider the following fractional differential inclusions:

$$\begin{cases} {}^C_0 D_t^q z(t, \xi) \in \Delta z(t, \xi) + G(t, z(t, \xi)), & t \in [0, b], \ \xi \in \Omega, \\ z(t, \xi) = 0, & t \in [0, b], \ \xi \in \partial\Omega, \quad (3.23) \\ z(0, \xi) = x_0(\xi), & \xi \in \Omega, \end{cases}$$

where Ω is a bounded open set in $\mathbb{R}^N$ with regular boundary $\partial\Omega$, $x_0 \in L^2(\Omega, \mathbb{R}^N)$.

We choose $X = L^2(\Omega, \mathbb{R}^N)$, and consider the operator $A : D(A) \subset X \to X$ defined by

$$D(A) = \{z \in H^2(\Omega) : \Delta z \in X \text{ and } z = 0 \text{ on } \partial\Omega\},$$
$$Az = \Delta z.$$

We have

$$(0, \infty) \subset \rho(A),$$

$$\|R(\lambda; A)\|_{\mathscr{L}(X)} \leq \frac{1}{\lambda}, \text{ for } \lambda > 0.$$

This implies that operator A satisfies condition (H_A). Moreover, the operator $T(t)$ generated by A_0 is compact in $\overline{D(A)}$ with $M = 1$ (see [199]).

Now, we assume that

$$f_i : [0, b] \times \mathbb{R}^N \to \mathbb{R}^N, \quad i = 1, 2$$

satisfy

(F_1) f_1 is l.s.c. and f_2 is u.s.c.;
(F_2) $f_1(t, z) \leq f_2(t, z)$ for each $(t, z) \in [0, b] \times \mathbb{R}^N$;
(F_3) there exist $\alpha_1, \alpha_2 \in L^\infty([0, b], \mathbb{R})$ such that

$$|f_i(t, z)| \leq \alpha_1(t)|z| + \alpha_2(t), \quad i = 1, 2,$$

for each $(t, z) \in [0, b] \times \mathbb{R}^N$.

Let $G(t, z) = [f_1(t, z), f_2(t, z)]$. From assumptions (F_1)-(F_3), it follows readily that the multivalued function $G(\cdot, \cdot) : [0, b] \times \overline{\Omega} \to 2^{\mathbb{R}^N}$ satisfies (H_1), (H_3), and (H_5).

Then inclusion (3.23) can be reformulated as

$$\begin{cases} {}^C_0 D_t^q x(t) \in Ax(t) + F(t, x(t)), & t \in [0, b], \\ x(0) = x_0, \end{cases}$$

where $x(t)(\xi) = z(t, \xi)$, $F(t, x(t))(\xi) = G(t, z(t, \xi))$.

Thus, all the assumptions in Theorem 3.5 are satisfied; our result can be used to inclusion (3.23).

3.2.3 Noncompact Semigroup Case

We study the fractional evolution inclusion (3.1) under the following assumption:

(H_6) there exists a function $k(s) \in L^{\frac{1}{q_1}}([0, b], \mathbb{R}^+)$ such that

$$\beta(F(t, D)) \leq k(s)\beta(D)$$

for every bounded set D, where β denotes Hausdorff measure of noncompactness.

Lemma 3.17. *The operators* $\mathcal{W}(f)$ *have the following properties:*

(i) *there exists a constant* $C > 0$, *such that*

$$\|\mathcal{W}(f) - \mathcal{W}(g)\|^{\frac{1}{q_1}} \leq C^{\frac{1}{q_1}} \int_0^t |f(s) - g(s)|^{\frac{1}{q_1}} ds, \ \forall \, f, g \in L^{\frac{1}{q_1}}([0, b], X);$$

(ii) *for each compact set* $K \subset X$ *and sequence* $\{f_n\} \subset L^{\frac{1}{q_1}}([0, b], X)$ *such that* $\{f_n\} \subset K$ *for a.e.* $t \in [0, b]$, *the weak convergence* $f_n \rightharpoonup f$ *implies* $\mathcal{W}(f_n) \to \mathcal{W}(f)$.

Proof. (i) By using Hölder inequality we have

$$|\mathcal{W}(f)(t) - \mathcal{W}(g)(t)| \leq \left| \lim_{\lambda \to +\infty} \int_0^t K_q(t-s) \mathcal{J}_\lambda(f(s) - g(s)) ds \right|$$

$$\leq \frac{M\overline{M}}{\Gamma(q)} \int_0^t (t-s)^{q-1} |f(s) - g(s)| ds$$

$$\leq \frac{M\overline{M}b^{(1+a)(1-q_1)}}{\Gamma(q)(1+a)^{1-q_1}} \left(\int_0^t |f(s) - g(s)|^{\frac{1}{q_1}} ds \right)^{q_1}.$$

Then

$$\|\mathcal{W}(f) - \mathcal{W}(g)\|^{\frac{1}{q_1}} \leq C^{\frac{1}{q_1}} \int_0^t |f(s) - g(s)|^{\frac{1}{q_1}} ds, \ \forall \, f, g \in L^{\frac{1}{q_1}}([0, b], X),$$

where

$$C = \frac{M\overline{M}b^{(1+a)(1-q_1)}}{\Gamma(q)(1+a)^{1-q_1}}.$$

(ii) Notice that, without loss of generality, $\{f_n(t)\} \subset K'$ for all $t \in [0, b]$, where $K' = \overline{\text{sp}K}$ is the separable Banach space spanned by the compact set K. Moreover, it is clear that $\{f_n\} \subset K'$ for all $t \in [0, b]$. Then, applying Property 1.19, we obtain

$$\beta(\{\mathcal{W}(f_n)(t)\}) \leq \frac{M\overline{M}}{\Gamma(q)} \int_0^t (t-s)^{q-1} \beta(\{f_n(s)\}) ds = 0.$$

Hence, the sequence $\{\mathcal{W}(f_n)(t)\} \subset X$ is relatively compact for every $t \in [0, b]$.

On the other hand, we have

$$|\mathcal{W}(f)(t_2) - \mathcal{W}(f)(t_1)|$$

$$\leq \left| \lim_{\lambda \to +\infty} \int_{t_1}^{t_2} K_q(t_2 - s)\mathcal{J}_\lambda f(s)ds \right|$$

$$+ \left| \lim_{\lambda \to +\infty} \int_0^{t_1} ((t_2 - s)^{q-1} - (t_1 - s)^{q-1})P_q(t_2 - s)\mathcal{J}_\lambda f(s)ds \right|$$

$$+ \left| \lim_{\lambda \to +\infty} \int_0^{t_1} (t_1 - s)^{q-1}(P_q(t_2 - s) - P_q(t_1 - s))\mathcal{J}_\lambda f(s)ds \right|$$

$$\leq \frac{M\overline{M}}{\Gamma(q)} \int_{t_1}^{t_2} (t - s)^{q-1}\alpha(s)ds$$

$$+ \frac{M\overline{M}}{\Gamma(q)} \int_0^{t_1} ((t_2 - s)^{q-1} - (t_1 - s)^{q-1})\alpha(s)ds$$

$$+ \overline{M} \sup_{s \in [0, t_1 - \delta]} \|P_q(t_2 - s) - P_q(t_1 - s)\|_{\mathscr{L}(X)} \int_0^{t_1 - \delta} (t_1 - s)^{q-1}\alpha(s)ds$$

$$+ \frac{2M\overline{M}}{\Gamma(q)} \int_{t_1 - \delta}^{t_1} (t - s)^{q-1}\alpha(s)ds.$$

Since $\{f_n(t)\} \subset K$ for a.e. $t \in [0, b]$, the right-hand side of this inequality tends to zero as $t_2 \to t_1$ uniformly with respect to n. Hence, $\{\mathcal{W}(f_n)\}$ is an equicontinuous set. Thus from Theorem 1.2, we obtain that the sequence $\{\mathcal{W}(f_n)\} \subset C([0, b], X_0)$ is relatively compact.

Property (i) ensures that $\mathcal{W} : L^{\frac{1}{q_1}}([0, b], X) \to C([0, b], X_0)$ is a bounded linear operator. Then it is continuous with respect to the topology of weak sequential convergence, that is, the weak convergence $f_n \rightharpoonup f$ ensuring $\mathcal{W}(f_n) \rightharpoonup \mathcal{W}(f)$. Taking into account that $\{\mathcal{W}(f_n)\}$ is relatively compact, we arrive at the conclusion that $\mathcal{W}(f_n) \to \mathcal{W}(f)$ strongly in $C([0, b], X_0)$. $\qquad\square$

Lemma 3.18. *[133] Let the sequence $\{f_n\} \subset L^{\frac{1}{q_1}}([0, b], X)$ be $L^{\frac{1}{q_1}}$-integrably bounded:*

$$|f_n(t)| \leq r(t)$$

for all $n = 1, 2, ...,$ and a.e. $t \in [0, b]$, where $r(t) \subset L^{\frac{1}{q_1}}([0, b], \mathbb{R}^+)$. Assume that

$$\beta(f_n(t)) \leq \zeta(t)$$

for a.e. $t \in [0, b]$, where $\zeta \in L^{\frac{1}{q_1}}([0, b], \mathbb{R}^+)$. Then we have

$$\beta(\{\mathcal{W}(f_n)(t)\}) \leq 2C \left(\int_0^t |\zeta(s)|^{\frac{1}{q_1}} ds \right)^{q_1}$$

for any $t \in [0, b]$, where $C \geq 0$ is the constant in Lemma 3.17(i).

With the similar proof of Lemma 3.17, we can get the following lemma.

Lemma 3.19. *Let $\{f_n\}$ be a semicompact sequence in $L^{\frac{1}{q_1}}([0, b], X)$. Then $\{f_n\}$ is weakly compact in $L^{\frac{1}{q_1}}([0, b], X)$, and $\{\mathcal{W}(f_n)\}$ is relatively compact in $C([0, b], X_0)$. Moreover, if $f_n \rightharpoonup f$, then $\mathcal{W}(f_n) \to \mathcal{W}(f)$.*

Theorem 3.7. *Let conditions (H_A), (H_T), (H_1), (H_3), (H_5), and (H_6) be satisfied. Then the solution set of inclusion (3.1) is a nonempty compact subset of $C([0, b], X_0)$ for each initial value $x_0 \in X_0$.*

Proof. For the same $\mathcal{M}_0$, as the reason for Theorem 3.5, we see that $\mathcal{M}_0$ is a closed and convex subset of $C([0, b], X_0)$.

Claim 1. The multioperator $\mathcal{F} = S_q(t)x_0 + \mathcal{W} \circ Sel_F$ has closed graph with compact values.

Let $x_n \subset \mathcal{M}_0$ with $x_n \to x$ and $y_n \in \mathcal{F}(x_n)$ with $y_n \to y$. We shall prove that $y \in \mathcal{F}(x)$. By the definition of $\mathcal{F}$, there exists $f_n \in Sel_F(x_n)$ such that $y_n(t) = S_q(t)x_0 + \mathcal{W}(f_n)$.

We need to prove that there exists $f \in Sel_F(x)$ such that for a.e. $t \in [0, b]$, $y(t) = S_q(t)x_0 + \mathcal{W}(f)$.

In view of (H_5) we have that $\{f_n\}$ is bounded in $L^{\frac{1}{q_1}}([0, b], X)$, one obtains $f_n \rightharpoonup f$ in $L^{\frac{1}{q_1}}([0, b], X)$ (see Lemma 1.4). Since Sel_F is weakly u.s.c. with weakly compact and convex values (see Lemma 3.11), together with Lemma 1.5, we then see $f \in Sel_F(x)$.

We see that $\{f_n\}$ is integrably bounded by (H_5) and the following inequality holds by (H_6):

$$\beta(\{f_n(t)\}) \leq k(t)\beta(\{x_n(t)\}).$$

For the sequence $\{x_n\}$ converges in $C([0, b], X_0)$, thus $\beta(\{f_n(t)\}) = 0$ for a.e. $t \in [0, b]$, then $\{f_n\}$ is a semicompact sequence. By Lemma 3.19, we may assume, without loss of generality, that there exists $f \in Sel_F(x)$ such that

$$f_n \rightharpoonup f \text{ and } y_n(t) = S_q(t)x_0 + \mathcal{W}(f_n) \to S_q(t)x_0 + \mathcal{W}(f) = y(t).$$

It remains to show that, for $x \in \mathcal{M}_0$ and $\{f_n\}$ chosen in $Sel_F(x)$, the sequence $\{\mathcal{W}(f_n)\}$ is relatively compact in $C([0, b], X_0)$. Hypotheses (H_5) and (H_6) imply that $\{f_n\}$ is semicompact. Using Lemma 3.19, we obtain that $\{\mathcal{W}(f_n)\}$ is relatively compact in $C([0, b], X_0)$. Thus $\mathcal{F}(x)$ is relatively compact in $C([0, b], X_0)$. Together with the closeness of $\mathcal{F}$, then $\mathcal{F}(x)$ has compact values.

Claim 2. The multioperator $\mathcal{F}$ is u.s.c.

In view of Lemma 1.7, it suffices to check that $\mathcal{F}$ is a quasicompact multivalued map. Let Q be a compact set. We prove that $\mathcal{F}(Q)$ is a relatively compact subset of $C([0, b], X_0)$. Assume that $\{y_n\} \subset \mathcal{F}(Q)$. Then $y_n = S_q(t)x_0 + \mathcal{W}(f_n)$, where $\{f_n\} \in Sel_F(x_n)$, for a certain sequence $\{x_n\} \subset Q$. Hypotheses (H_5) and (H_6) yield the fact that $\{f_n\}$ is semicompact and then it is a weakly compact sequence in $L^{\frac{1}{q_1}}([0, b], X)$. Similar arguments as the previous proof of closeness imply that $\{\mathcal{W}(f_n)\}$ is relatively compact in $C([0, b], X_0)$. Thus, $\{y_n\}$ converges in $C([0, b], X_0)$, so the multioperator $\mathcal{F}$ is u.s.c.

Claim 3. The multioperator $\mathcal{F}$ is a condensing multioperator.

First, we need an MNC constructed suitably for our problem. For a bounded subset $\Omega \subset \mathcal{M}_0$, let $\text{mod}_C(\Omega)$ be the modulus of equicontinuity of the set of functions Ω given by

$$\text{mod}_C(\Omega) = \lim_{\delta \to 0} \sup_{x \in \Omega} \max_{|t_2 - t_1| < \delta} |x(t_2) - x(t_1)|.$$

Let χ be the real MNC defined on bounded set $D \subset C([0, b], X_0)$ by

$$\chi(D) = \sup_{t \in [0,b]} e^{-Lt} \beta(D(t));$$

here, the constant L is chosen such that

$$l = 2C \sup_{t \in [0,b]} \left(\int_0^t e^{-\frac{L}{q_1}(t-s)} k^{\frac{1}{q_1}}(s) ds \right)^{q_1} < 1,$$

where $k(t)$ is the function from condition (H_6).

Consider the function $\nu(\Omega) = \max_{D \in \Delta(\Omega)} \left(\chi(D), \mathrm{mod}_C(D) \right)$ in the space of $C([0, b], X_0)$, where $\Delta(\Omega)$ is the collection of all countable subsets of Ω.

To show that $\mathcal{F}$ is ν-condensing, let $\Omega \subset \mathcal{M}_0$ be a bounded set in $\mathcal{M}_0$ such that

$$\nu(\Omega) \leq \nu(\mathcal{F}(\Omega)). \tag{3.24}$$

We will show that Ω is relatively compact. Let $\nu(\mathcal{F}(\Omega))$ be achieved on a sequence $\{y_n\} \subset \mathcal{F}(\Omega)$, i.e.,

$$\nu(\{y_n\}) = \left(\chi(\{y_n\}), \mathrm{mod}_C(\{y_n\}) \right).$$

Then

$$y_n = S_q(t)x_0 + \mathcal{W}(f_n), \quad f_n \in Sel_F(x_n),$$

where $\{x_n\} \subset \Omega$. Now inequality (3.24) implies

$$\chi(\{y_n\}) \geq \chi(\{x_n\}). \tag{3.25}$$

It follows from (H_6) that $\beta(\{f_n(t)\}) \leq k(t)\beta(x_n(t))$ for $t \in [0, b]$. Then

$$\beta(\{f_n(t)\}) \leq k(t)e^{Lt} \left(\sup_{s \in [0,t]} e^{-Ls} \beta(x_n(t)) \right)$$

$$\leq k(t)e^{Lt} \chi(\{x_n\}).$$

Now the application of Lemma 3.18 for $\mathcal{W}$ yields

$$e^{-Lt}\beta(\{\mathcal{W}(f_n)(t)\}) \leq 2Ce^{-Lt} \left(\int_0^t e^{\frac{L}{q_1}(s)} k^{\frac{1}{q_1}}(s) ds \right)^{q_1} \chi(\{x_n\})$$

$$\leq 2C\left(\int_0^t e^{-\frac{L}{q_1}(t-s)}k^{\frac{1}{q_1}}(s)ds\right)^{q_1}\chi(\{x_n\}),$$

for any $t \in [0, b]$. Putting this relation together with (3.25), we obtain

$$\chi(\{x_n\}) \leq \chi(\{y_n\}) = \sup_{t\in[0,b]} e^{-Lt}\beta(y_n(t)) \leq l\chi(\{x_n\}).$$

Therefore, $\chi(\{x_n\}) = 0$. This implies that $\beta(x_n(t)) = 0$.

Using (H_5) and (H_6) again, one gets that $\{f_n\}$ is a semicompact sequence. Then, Lemma 3.19 ensures that $\{\mathcal{W}(f_n)\}$ is relatively compact in $C([0, b], X_0)$. This yields that $\{y_n\}$ is relatively compact in $C([0, b], X_0)$. Hence, $\mathrm{mod}_C(\{y_n\}) = 0$. Finally, $\nu(\{y_n\}) = (0, 0)$, and so the map $\mathcal{F}$ is ν-condensing.

From Theorem 1.9, we deduce that the fixed point set $\mathrm{Fix}\mathcal{F}$ is a nonempty compact set. $\qquad\square$

Lemma 3.20. *Under assumptions of Theorem 3.7, there exists a nonempty compact convex subset $\mathcal{M} \subseteq C([0, b], X_0)$ such that*

(i) $x(0) = x_0$, *for all $x \in \mathcal{M}$;*
(ii) $S_q(t)x_0+\lim_{\lambda\to+\infty}\int_0^t K_q(t-s)\mathcal{J}_\lambda\overline{co}F(s, \mathcal{M}(s))ds \subset \mathcal{M}(t)$, *for $t \in [0, b]$, where $\mathcal{M}(t) = \{x(t) : x \in \mathcal{M}\}$.*

Proof. Let us construct the decreasing sequence of closed convex sets $\{\mathcal{M}_n\} \subset C([0, b], X_0)$ by the following inductive process. Let

$$\mathcal{M}_0 = \{x \in C([0, b], X_0) : x(0) = x_0, |x(t)| \leq N, t \in [0, b]\},$$

where $N = a_1 \exp\left(\frac{a_2 b^{(1+a)(1-q_1)}}{(1+a)^{1-q_1}}\|\alpha\|_{L^{\frac{1}{q_1}}[0,b]}\right)$.

Then $\mathcal{M}_n = \overline{\mathcal{N}_n}$, $n \geq 1$, where $\mathcal{N}_n \subset C([0, b], X_0)$ and

$$\mathcal{N}_n = \left\{y \in C([0, b], X_0) : y(t) = S_q(t)x_0 \right.$$
$$\left. + \lim_{\lambda\to+\infty}\int_0^t K_q(t-s)\mathcal{J}_\lambda f(s)ds, \ f \in Sel_{\overline{co}F}(\cdot, \mathcal{M}_{n-1}(\cdot))\right\}.$$

First of all, let us note that all $\mathcal{M}_n$, $n \geq 1$ are nonempty since $\Theta(x_0) \subset \mathcal{M}_n$ for all $n \geq 0$.

As the reason for Theorem 3.5, we known that all sets $\mathcal{M}_n$, $n \geq 1$ are equicontinuous.

Using condition (H_6) we have the following estimation:

$$\beta(\overline{\text{co}}F(s, \mathcal{M}_{n-1}(s))) \leq k(t)e^{Lt}\left(\sup_{s \in [0,t]} e^{-Ls}\beta(\mathcal{M}_{n-1}(s)) \right)$$

$$\leq k(t)e^{Lt}\chi(\mathcal{M}_{n-1}).$$

We have for any $t \in [0, b]$,

$$e^{-Lt}\beta(\mathcal{N}_n(t)) = e^{-Lt}\beta\left(\lim_{\lambda \to +\infty} \int_0^t K_q(t-s)\mathcal{J}_\lambda \overline{\text{co}}F(s, \mathcal{M}_{n-1}(s))ds \right)$$

$$\leq 2Ce^{-Lt}\left(\int_0^t e^{\frac{L}{q_1}(s)}k^{\frac{1}{q_1}}(s)ds \right)^{q_1}\chi(\mathcal{M}_{n-1})$$

$$\leq 2C\left(\int_0^t e^{-\frac{L}{q_1}(t-s)}k^{\frac{1}{q_1}}(s)ds \right)^{q_1}\chi(\mathcal{M}_{n-1}).$$

Therefore, $\chi(\mathcal{N}_n) \leq l\chi(\mathcal{M}_{n-1})$.

Finally, we have $\chi(\mathcal{M}_n) \leq l\chi(\mathcal{M}_{n-1})$ and therefore $\chi(\mathcal{M}_n) \to 0$ $(n \to \infty)$. We obtain a compact set $\mathcal{M} = \bigcap_{n=0}^{\infty} \mathcal{M}_n$, which has the desired properties. $\qquad\square$

In the following, let us note that we may assume, without loss of generality, that F satisfies the following estimation:

$(H_5)'$ $|F(t, x)| \leq \eta(t)$ for every $x \in X$ and a.e. $t \in [0, b]$, where $\eta \in L^{\frac{1}{q_1}}([0, b], \mathbb{R}^+)$.

In fact, let $\|\Theta(x_0)\| \leq N$, B_N be a closed ball in the space X and $\rho : X \to B_N$ be a radial retraction. Then it is easy to see that the multivalued map $\widetilde{F} : [0, b] \times X \to P_{cp,cv}(X)$, defined by $\widetilde{F}(t, u) = F(t, \rho u)$ satisfies conditions (H_1), (H_3), (H_6) (note that ρ is a Lipschitz map), and condition $(H_5)'$ with $\eta(t) = \alpha(t)(1 + N)$. The set $\Theta(x_0)$ coincides with the set of all integral solutions of the problem

$$\begin{cases} {}^C_0D^q_t x(t) \in Ax(t) + \widetilde{F}(t, x(t)), & t \in [0, b], \\ x(0) = x_0. \end{cases}$$

Therefore, in what follows we will suppose that the multivalued map $F : [0, b] \times X \to P_{cl,cv}(X)$ satisfies conditions (H_1), (H_3), (H_6), and $(H_5)'$ instead of (H_5).

Now consider a metric projection $P : [0, b] \times X \to P_{cl,cv}(X)$,

$$P(t, x) = \{y \in \mathcal{M}(t), \|x - y\| = \mathrm{dist}(x, \mathcal{M}(t))\},$$

and a multivalued map $\widehat{F} : [0, b] \times X \to P_{cl,cv}(X)$, defined by

$$\widehat{F}(t, x) = \overline{co}F(t, P(t, x)).$$

From Lemma 5.3.2 in [131], we know the multivalued map P is closed and u.s.c.

Lemma 3.21. *The multivalued map $\widehat{F}$ satisfies conditions (H_1), (H_3), $(H_5)'$, and (H_6).*

Proof. Let $x_n \to x \in X$, $y_n \in \widehat{F}(t, x_n)$, there exists $z_n \in P(t, x_n)$ such that $y_n \in \overline{co}F(t, z_n)$. Noticing that the multivalued map P is closed and u.s.c., Lemma 1.9 yields that $z_n \to z \in P(t, x)$. Since $F(t, \cdot)$ is weakly u.s.c. for a.e. $t \in [0, b]$, we conclude from Lemma 1.5 that there exists a subsequence y_{n_k} of y_n and $y \in \overline{co}F(t, z)$ such that $y_{n_k} \rightharpoonup y$. Therefore, $y \in \overline{co}F(t, P(t, x)) = \widehat{F}(t, x)$, which implies that $\widehat{F}$ satisfies condition (H_3).

For any fixed $x \in X$, the multifunction $P(\cdot, x) : [0, b] \times X \to P_{cl,cv}(X)$ is clearly measurable, and hence it has a measurable selection $p(\cdot)$. Let $x^* \in X^*$ be fixed, $x^*(F(t, \cdot))$ is u.s.c., from Lemma 1.10 it follows that the multifunction $x^*(F(t, p(t)))$ has a measurable selection. Since X is reflexive, the multifunction $F(t, p(t))$ has a measurable selection, which implies the multivalued map $\widehat{F}$ satisfies condition (H_1).

It is clear that $\widehat{F}$ satisfies condition $(H_5)'$. To verify (H_6), for every bounded $D \subset C([0, b], X_0)$, we have

$$\beta(\widehat{F}(t, D)) \leq \beta(F(t, P(t, D)) \leq \mu\beta(P(t, D)) \leq \mu\beta(\mathcal{M}(t)) = 0,$$

so $\widehat{F}(t, \cdot)$ is compact for a.e. $t \in [0, b]$. $\square$

The above result implies that the set $\widehat{\Theta}(x_0)$ of all integral solutions of the problem

$$\begin{cases} {}^{C}_{0}D^{q}_{t}x(t) \in Ax(t) + \widehat{F}(t, u), & t \in [0, b], \\ x(0) = x_0 \end{cases}$$

is nonempty. Moreover, the following statement is valid.

Lemma 3.22. $\widehat{\Theta}(x_0) = \Theta(x_0)$.

Proof. In fact, let $x \in \widehat{\Theta}(x_0)$. Then

$$x(t) \in S_q(t)x_0 + \lim_{\lambda \to +\infty} \int_0^t K_q(t-s)\mathcal{J}_\lambda \widehat{F}(s, x(s))ds$$

$$= S_q(t)x_0 + \lim_{\lambda \to +\infty} \int_0^t K_q(t-s)\mathcal{J}_\lambda \overline{co}F(s, P(t, x(s)))ds$$

$$\subset S_q(t)x_0 + \lim_{\lambda \to +\infty} \int_0^t K_q(t-s)\mathcal{J}_\lambda \overline{co}F(s, \mathcal{M}(s)))ds$$

$$\subset \mathcal{M}(t);$$

hence, $P(t, x(t)) = \{x(t)\}$. Then

$$x(t) = S_q(t)x_0 + \lim_{\lambda \to +\infty} \int_0^t K_q(t-s)\mathcal{J}_\lambda f(s)ds,$$

where $f \in Sel_{\widehat{F}}(x) = Sel_F(x)$, and so $x \in \Theta(x_0)$.

The inclusion $\Theta(x_0) \subset \widehat{\Theta}(x_0)$ easily follows from the observation that $\Theta(x_0) \subset \mathcal{M}$. This completes the proof. $\qquad\square$

Lemma 5.3.4 in [131] yields the following approximation result.

Lemma 3.23. *Let hypotheses* (H_1), (H_3), *and* $(H_5)'$ *be satisfied. Then there exists a sequence* $\{\widehat{F}_n\}$ *with* $\widehat{F}_n : [0, b] \times X \to P_{cl,cv}(X)$ *such that*

(i) $\widehat{F}(t, x) \subset \widehat{F}_{n+1}(t, x) \subset \widehat{F}_n(t, x) \subset \overline{co}(F(t, \mathcal{M}(t)))$, $n \geq 1$, *for each* $t \in [0, b]$ *and* $x \in X$;
(ii) $|\widehat{F}_n(t, x)| \leq \eta(t)$, $n \geq 1$, *for a.e.* $t \in [0, b]$ *and each* $x \in X$;

(iii) *there exists $E \subset [0, b]$ with $mes(E) = 0$ such that for each $x^* \in X^*$,*
$\varepsilon > 0$, *and $x \in X$, there exists $N > 0$ such that for all $n \geq N$,*

$$x^*(\widehat{F}_n(t, x)) \subset x^*(\widehat{F}(t, x)) + (-\varepsilon, \varepsilon);$$

(iv) $\widehat{F}_n(t, \cdot) : X \to P_{cl,cv}(X)$ *is continuous for a.e. $t \in [0, b]$ with respect to Hausdorff metric for each $n \geq 1$;*

(v) *for each $n \geq 1$, there exists a selection $g_n : [0, b] \times X \to X$ of $\widehat{F}_n$ such that $g_n(\cdot, x)$ is measurable and $g_n(t, \cdot)$ is locally Lipschitz.*

Theorem 3.8. *Under the conditions in Theorem 3.7, $\Theta(x_0)$ is a compact R_δ-set.*

Proof. Now we consider the evolution inclusion:

$$\begin{cases} {}^C_0 D^q_t x(t) \in Ax(t) + \widehat{F}_n(t, x), & t \in [0, b], \\ x(0) = x_0. \end{cases} \tag{3.26}$$

Let $\widehat{\Theta}^n(x_0)$ denote the solution set of inclusion (3.26). From Lemma 3.21 it follows that each $\widehat{F}_n$ satisfies conditions (H_1), (H_3), $(H_5)'$, and (H_6); hence, each set $\widehat{\Theta}^n(x_0)$ is nonempty and compact.

We prove that

$$\widehat{\Theta}(x_0) = \bigcap_{n \geq 1} \widehat{\Theta}^n(x_0).$$

It is clear that $\widehat{\Theta}(x_0) \subset \widehat{\Theta}^n(x_0)$ and $\widehat{\Theta}(x_0) \subset \bigcap_{n \geq 1} \widehat{\Theta}^n(x_0)$.

Let $x \in \bigcap_{n \geq 1} \widehat{\Theta}^n(x_0)$, then for each $n \geq 1$, we have

$$x(t) = S_q(t)x_0 + \lim_{\lambda \to +\infty} \int_0^t K_q(t - s)\mathcal{J}_\lambda g_n(s)ds, \quad \text{for } t \in [0, b],$$

where $g_n \in Sel_{\widehat{F}_n}(x)$. From Lemma 3.23(ii), it follows that g_n is relatively weakly compact for a.e. $t \in [0, b]$ due to Lemma 1.4 and, hence, we may assume, up to subsequence, that $g_n \rightharpoonup f \in L^{\frac{1}{q_1}}([0, b], X)$. Similar arguments as the proof in Theorem 3.6, we can know that $f(t) \in \widehat{F}(t, x(t))$ a.e. on $[0, b]$, since $\widehat{F}$ has closed convex values. Therefore,

$$x(t) = S_q(t)x_0 + \lim_{\lambda \to +\infty} \int_0^t K_q(t - s)\mathcal{J}_\lambda f(s)ds, \quad \text{for } t \in [0, b],$$

which means that $x \in \widehat{\Theta}(x_0)$.

We show that $\widehat{\Theta}^n(x_0)$ is contractible for every $n \geq 1$.

To do this, fixed $x \in \widehat{\Theta}^n(x_0)$, we consider the integral equation

$$
\begin{aligned}
y(t) = & S_q(t)x_0 + \lim_{\lambda \to +\infty} \int_0^{\lambda b} K_q(t-s)\mathcal{J}_\lambda f^x(s)ds \\
& + \lim_{\lambda \to +\infty} \int_{\lambda b}^t K_q(t-s)\mathcal{J}_\lambda g_n(s, y(s))ds,
\end{aligned}
\tag{3.27}
$$

where $f^x \in Sel_{\widehat{F}_n}(x)$ and g_n is a measurable locally Lipschitz selection of $\widehat{F}_n$. Moreover, it follows from Lemma 3.23(ii) that $|g_n(t, \cdot)| \leq \eta(t)$ for a.e. $t \in [0, b]$. Since $\widehat{F}(t, \cdot)$ is compact for a.e. $t \in [0, b]$, it is sufficient to note that $g_n(t, \cdot)$ is compact for a.e. $t \in [0, b]$. From Lemma 3.9, equation (3.27) has a unique solution on $[\lambda b, b]$, denoted by $y(t, \lambda b, x(\lambda b))$. Define the same Φ as in Theorem 3.6, then we see that $\widehat{\Theta}^n(x_0)$ is contractible for each $n \geq 1$.

Consequently, we conclude that $\Theta(x_0) = \widehat{\Theta}(x_0)$ is an R_δ-set, which completes this proof.   □

3.2.4 Application to Control Theory

In this subsection, we deal with the controllability for fractional evolution inclusions

$$
\begin{cases}
{}^C_0 D_t^q x(t) \in Ax(t) + Bu(t) + F(t, x(t)), & \text{a.e.} \quad t \in [0, b], \\
x(0) = x_0,
\end{cases}
\tag{3.28}
$$

where $0 < q \leq 1$. We assume that

(H_B) the control function $u(\cdot)$ takes its value in $\overline{U}$, a Banach space of admissible control functions, where $\overline{U} = L^{\frac{1}{q_1}}([0, b], U)$ for $q_1 \in (0, q)$ and U is a Banach space. $\mathscr{L}(U, X)$ is the space of all bounded linear operators from U to X with the norm $\| \cdot \|_{\mathscr{L}(U,X)}$ and $B \in \mathscr{L}(U, X)$ with

$$
\|B\|_{\mathscr{L}(U,X)} = M_2.
$$

Definition 3.3. *A continuous function* $x : [0, b] \to X$ *is said to be an integral solution of inclusion (3.28) if* $x(0) = x_0$ *and there exists* $f(t) \in Sel_F(x)(t)$ *satisfying the following integral equation:*

$$x(t) = S_q(t)(x_0) + \lim_{\lambda \to +\infty} \int_0^t K_q(t - s)\mathcal{J}_\lambda Bu(s)ds$$
$$+ \lim_{\lambda \to +\infty} \int_0^t K_q(t - s)\mathcal{J}_\lambda f(s)ds.$$

We consider the controllability problem for inclusion (3.28), i.e., we study conditions which guarantee the existence of an integral solution to differential inclusion (3.28) satisfying

$$x(b) = x_1, \tag{3.29}$$

where $x_1 \in X_0$ is a given point. A pair (x, u) consisting of an integral solution $x(\cdot)$ to (3.28) satisfying (3.29) and of the corresponding control $u(\cdot) \in L^{\frac{1}{q_1}}([0, b], X)$ is called a solution of the controllability problem.

We assume the standard assumption that the corresponding linear problem (i.e., when $F(t, x) \equiv 0$) has a solution. More precisely, we suppose that

(H_W) the controllability operator $W : \overline{U} \to X$ given by

$$Wu = \lim_{\lambda \to +\infty} \int_0^b K_q(b - s)\mathcal{J}_\lambda Bu(s)ds$$

has a bounded inverse which takes value in $\overline{U}/\ker(W)$ and there exists a positive constant $M_3 > 0$ such that

$$\|W^{-1}\|_{\mathscr{L}(X,\overline{U})} = M_3.$$

Theorem 3.9. *Assume that* (H_A), (H_T), (H_1), (H_2), (H_B), *and* (H_W) *hold. Further, suppose that*

$(H_4)'$ *there exists a sequence of functions* $\{\omega_n\} \subset L^{\frac{1}{q_1}}([0, b], \mathbb{R}^+)$ *such that*

$$\sup_{|x|\leq n} |F(t, x)| \leq \omega_n(t)$$

for a.e. $t \in [0, b]$, $n \in \mathbb{N}^+$ with

$$\liminf_{n \to \infty} \frac{1}{n} \left(\int_0^b |\omega_n(s)|^{\frac{1}{q_1}} ds \right)^{q_1} = 0. \tag{3.30}$$

Then controllability problems (3.28) and (3.29) have a solution.

Proof. Let $q_1 \in (0, q)$. Denote $S_1 : L^{\frac{1}{q_1}}([0, b], X) \to C([0, b], X_0)$ and $S_2 : L^{\frac{1}{q_1}}([0, b], X) \to C([0, b], X_0)$ by the following integral operators:

$$S_1 f(t) = \lim_{\lambda \to +\infty} \int_0^t K_q(t - s)\mathcal{J}_\lambda f(s)ds, \quad \forall t \in [0, b],$$

$$S_2 f(t) = \lim_{\lambda \to +\infty} \int_0^t K_q(t - s)\mathcal{J}_\lambda BW^{-1}$$

$$\times \left(-\lim_{\lambda \to +\infty} \int_0^b K_q(b - s)\mathcal{J}_\lambda f(\eta)d\eta \right)(s)ds, \ \forall t \in [0, b].$$

Then we define the solution multioperator

$$\Gamma : C([0, b], X_0) \to P(C([0, b], X_0))$$

as

$$\Gamma x = \left\{ y(t) \in C([0, b], X_0) : y(t) = S_q(t)x_0 + S_1 f(t) \right.$$

$$+ \lim_{\lambda \to +\infty} \int_0^t K_q(t - s)\mathcal{J}_\lambda BW^{-1}(x_1 - S_q(b)x_0)(s)ds$$

$$\left. + S_2 f(t), \ f \in Sel_F(x) \right\}.$$

Similar to Theorem 3.2, we can see that controllability problems (3.28) and (3.29) have a solution. $\square$

Remark 3.6. *Condition $(H_4)'$ in Theorem 3.9 can be replaced by one of the following conditions:*

$(H_5)'$ *there exists $\alpha \in L^{\frac{1}{q_1}}([0, b], \mathbb{R}^+)$ such that*

$$|F(t, x)| \leq \alpha(t)(1 + |x|), \text{ for a.e. } t \in [0, b], \ \forall x \in X$$

and

$$\frac{M\overline{M}\|\alpha\|_{L^{\frac{1}{q_1}}[0,b]}b^{(1+a)(1-q_1)}}{\Gamma(q)(1+a)^{1-q_1}}\left(1+\frac{M\overline{M}M_2M_3b^{(1+a)(1-q_1)}}{\Gamma(q)(1+a)^{1-q_1}}\right)<1;$$

$(H_5)''$ *there exists* $\alpha \in L^{\frac{1}{q_1}}([0,b],\mathbb{R}^+)$ *and a nondecreasing function* $\varrho :$ $[0,+\infty) \to [0,+\infty)$ *such that*

$$|F(t,x)| \le \alpha(t)\varrho(|x|), \text{ for a.e. } t \in [0,b], \ \forall \, x \in X,$$

and $P > 0$ *such that*

$$\frac{P}{C_1 + C_2\|\alpha\|_{L^{\frac{1}{q_1}}[0,b]}\varrho(P)} > 1,$$

where

$$C_1 = M|x_0| + \frac{M\overline{M}M_2M_3b^{(1+a)(1-q_1)}}{\Gamma(q)(1+a)^{1-q_1}}(|x_1| + M_1|x_0|),$$

$$C_2 = \frac{M\overline{M}b^{(1+a)(1-q_1)}}{\Gamma(q)(1+a)^{1-q_1}}\left(1+\frac{M\overline{M}M_2M_3b^{(1+a)(1-q_1)}}{\Gamma(q)(1+a)^{1-q_1}}\right).$$

Theorem 3.10. *Under assumptions in Theorem 3.5, furthermore, we assume that* (H_B) *and* (H_W) *hold. Then the solution set of problems (3.28) and (3.29) is a nonempty compact* R_δ*-set.*

Theorem 3.11. *Under assumptions in Theorem 3.7, furthermore, we assume that* (H_B) *and* (H_W) *hold. Then the solution set of problems (3.28) and (3.29) is a nonempty compact* R_δ*-set.*

3.3 NOTES AND REMARKS

Fractional evolution inclusion is a kind of important differential inclusions describing the processes behaving in a much more complex way on time, which appear as a generalization of fractional evolution equations (such as time-fractional diffusion equations) through the application of multivalued analysis. Comparing the fractional evolution equations, the researches on the theory of fractional differential inclusions are only on their initial stage of development. It is noted that El-Sayed and Ibrahim initialed the study of fractional differential inclusions in [90] and much interest has developed along this line, see, e.g., [36, 114].

A strong motivation for investigating this class of inclusions comes mainly from two compelling reasons: differential models with the fractional derivative providing an excellent instrument for the description of memory and hereditary properties have recently been proved valuable tools in the modeling of many physical phenomena (see [78, 119, 134, 183]). As in [14, 183], fractional diffusion equations describe anomalous diffusion on fractals (physical objects of fractional dimension, like some amorphous semiconductors or strongly porous materials). In normal diffusion described by, such as the heat equation, the mean square displacement of a diffusive particle behaves like $const \cdot t$ for $t \to \infty$. A typical behavior for anomalous diffusion is $const \cdot t^{\alpha}$ for some $0 < \alpha < 1$. Another of the reasons is that a lot of phenomena investigated in hybrid systems with dry friction, processes of controlled heat transfer, obstacle problems, and others can be described with the help of various differential inclusions, both linear and nonlinear (see [77, 131, 230]). The theory of differential inclusions is highly developed and constitutes an important branch of nonlinear analysis (see, e.g., Bressan and Wang [46], Donchev et al. [82], and Gabor and Quincampoix [100] and the references therein).

In the study of the topological structure of solution sets for integral/differential equations and inclusions, an important aspect is the R_{δ}-property. It is worth mentioning that Aronszajn [16] carried out a systematic study for the topological properties of solution sets of a differential equation defined on a compact interval, where he showed that the solution sets are compact and acyclic, and he in fact specified these continua to be R_{δ}-sets. Since the work of Aronszajn, there have been published, up to now, numerous research papers concerning topological structure of solution sets for differential equations or inclusions of various types, see, e.g., Górniewicz and Pruszko [107], De Blasi and Myjak [72], and Andres and Pavlačková [12] and references therein.

The topological structure of solution sets of differential inclusions on compact intervals has been investigated intensively by many authors, please see De Blasi and Myjak [73], Bothe [44], Deimling [77], Hu and Papageorgiou [121], Staicu [218], and Zhu [311] and references therein. Moreover, one can find results on topological structure of solution sets for differential inclusions defined on non-compact intervals (including infinite intervals) from Andres et al. [11], Bakowska and Gabor [21], Chen et al. [65], Gabor and Grudzka [99], Staicu [219], and Wang et al. [276] and references therein.

However, as far as we know, there have been very few applicable results on the topological structure of solution sets for fractional evolution inclusions. This in fact is the main motivation of Chapter 3. The results in this chapter are taken from Zhou, Gu, Peng, and Zhang [299].

CHAPTER 4

Fractional Control Systems

Abstract In this chapter, we study optimal control, controllability, and topological structure of solution sets for fractional control systems. Section 4.1 concerns the fractional finite time delay evolution systems and optimal controls in infinite dimensional spaces. In Section 4.2, we study optimal feedback controls of a system governed by semilinear fractional evolution equations via a compact semigroup in Banach spaces. Section 4.3 is devoted to the investigation of controllability for a class of Sobolev-type semilinear fractional evolution systems in a separable Banach space. In Section 4.4, we discuss the approximate controllability of Sobolev-type fractional evolution systems with classical nonlocal conditions in Hilbert spaces. Section 4.5 deals with the topological structure of solution sets (compactness and R_δ-property) for control problems of semilinear fractional delay evolution equations.

Keywords Fractional control systems, Sobolev-type evolution systems, Optimal control, Optimal feedback control, Controllability, Approximate controllability, Compactness and R_δ-property, Gronwall inequality, Fixed point theorem, Measure of noncompactness.

4.1 EXISTENCE AND OPTIMAL CONTROL

4.1.1 Introduction

Consider the nonlinear fractional finite time delay evolution system as follows

$$\begin{cases} {}^C_0 D^q_t x(t) = Ax(t) + f(t, x_t, x(t)) + B(t)u(t), & 0 < t \le T, \\ x(t) = \varphi(t), & -r \le t \le 0, \end{cases} \tag{4.1}$$

where ${}^C_0 D^q_t$ denotes Caputo fractional derivative of order $q \in (0, 1)$, A is the generator of a C_0-semigroup $\{T(t)\}_{t \ge 0}$ on a Banach space X, f is X-value function, u takes value from another Banach space Y, B is a linear operator from Y into X, define x_t by $z_t(\theta) = z(t + \theta)$, $\theta \in [-r, 0]$. f, x_t, φ are given and satisfy some conditions that will be specified later.

Fractional Evolution Equations and Inclusions: Analysis and Control. http://dx.doi.org/10.1016/B978-0-12-804277-9.50004-3
139

First, we introduce a suitable mild solution for system (4.1) which is associated with the Wright function and semigroup operator. In Subsection 4.2.2, some sufficient conditions are given for the existence and uniqueness of mild solutions to system (4.1) without compactness condition. The main techniques used here are fractional calculus and singular Gronwall inequality with finite time delay via Banach contraction principle. In Subsection 4.2.3, we discuss the continuous dependence of mild solutions on the initial value, zero arm of time delay and control. In Subsection 4.2.4, we formulate Lagrange problem of fractional finite time delay evolution systems in the Caputo sense and prove existence of fractional optimal controls. At last, an example is given to demonstrate the applicability of the result.

Throughout this section, let X and Y be two Banach spaces, with the norms $|\cdot|$ and $|\cdot|_Y$, respectively. $\mathscr{L}(X, Y)$ denote the space of bounded linear operators from X to Y equipped with the norm $\|\cdot\|_{\mathscr{L}(X,Y)}$. In particular, when $X = Y$, then $\mathscr{L}(X, Y) = \mathscr{L}(X, X) = \mathscr{L}(X)$ and $\|\cdot\|_{\mathscr{L}(X,Y)} = \|\cdot\|_{\mathscr{L}(X,X)} = \|\cdot\|_{\mathscr{L}(X)}$. Suppose $r > 0$, $T > 0$, denote $J = [0, T]$. Denote $M = \sup_{t \in J} \|T(t)\|_{\mathscr{L}(X)}$, which is a finite number. Let $C([-r, a], X)$, $a \geq 0$ be the Banach space of continuous functions from $[-r, a]$ to X with the usual sup-norm. For brevity, we denote $C([-r, a], X)$ simply by $C_{-r,a}$ and its norm by $\|\cdot\|_{-r,a}$. If $a = T$, we denote this space by $C_{-r,T}$ and its norm by $\|\cdot\|_{-r,T}$. If $a = 0$, we denote this space by $C_{-r,0}$ and its norm by $\|\cdot\|_{-r,0}$. Obviously, for any $x \in C_{-r,T}$ and $t \in J$, define $x_t(s) = x(t + s)$ for $-r \leq s \leq 0$, then $x_t \in C_{-r,0}$.

The following results will be used.

Lemma 4.1. *[282] Suppose that $x \in C_{-r,T}$ satisfies the following inequality:*

$$
\begin{cases}
|x(t)| \leq a + b \int_0^t (t - s)^{q-1} \|x_s\|_{-r,0} ds \\
\qquad + c \int_0^t (t - s)^{q-1} |x(s)| ds, \qquad t \in J, \\
x(t) = \varphi(t), \qquad\qquad\qquad\qquad\qquad -r \leq t \leq 0,
\end{cases}
$$

where $\varphi \in C_{-r,0}$ and constants $a, b, c \geq 0$. Then there exists a constant $M^* > 0$ independent of a and φ such that

$$
|x(t)| \leq M^*(a + \|\varphi\|_{-r,0}), \quad \text{for all } t \in J.
$$

Remark 4.1. *From Lemma 4.1, it is obvious that there exists a constant* $\rho = \max\{M^*(a + \|\varphi\|_{-r,0}), \|\varphi\|_{-r,0}\} > 0$ *such that* $\|x\|_{-r,T} \leq \rho$.

Lemma 4.2. *[286] For each* $\psi \in L^p(J, X)$ *with* $1 \leq p < +\infty$,

$$\lim_{h \to 0} \int_0^T |\psi(t + h) - \psi(t)|^p dt = 0;$$

here, $\psi(s) = 0$ *for* $s \notin J$.

4.1.2 Existence and Uniqueness

We make the following assumptions:

(H_1) $f : J \times C_{-r,0} \times X \to X$ satisfies:
  (i) for each $x_t \in C_{-r,0}$, $x \in X$, $t \to f(t, x_t, x(t))$ is measurable;
  (ii) for arbitrary $\xi_1, \xi_2 \in C_{-r,0}, \eta_1, \eta_2 \in X$ satisfying $\|\xi_1\|_{-r,0}, \|\xi_2\|_{-r,0}$, $|\eta_1|, |\eta_2| \leq \rho$, there exists a constant $L_f(\rho) > 0$ such that

$$|f(t, \xi_1, \eta_1) - f(t, \xi_2, \eta_2)| \leq L_f(\rho)(\|\xi_1 - \xi_2\|_{-r,0} + |\eta_1 - \eta_2|),$$

  for all $t \in J$;
  (iii) there exists a constant $a_f > 0$ such that

$$|f(t, \xi, \eta)| \leq a_f(1 + \|\xi\|_{-r,0} + |\eta|), \text{ for all } \xi \in C_{-r,0}, \eta \in X, t \in J.$$

(H_2) Let Y be a reflexive Banach space from which the controls u take the values. The operator $B \in L_\infty(J, \mathscr{L}(Y, X))$, $\|B\|_\infty$ stands for the norm of operator B on Banach space $L_\infty(J, \mathscr{L}(Y, X))$.

(H_3) The multivalued map $U(\cdot) : J \to P(Y)$ has closed, convex, and bounded values. $U(\cdot)$ is graph measurable and $U(\cdot) \subseteq \Omega$, where Ω is a bounded set in Y.

Introduce the admissible set

$$U_{ad} = \{v(\cdot) : J \to Y \text{ strongly measurable}, v(t) \in U(t) \text{ a.e. } t \in J\}.$$

Obviously, $U_{ad} \neq \emptyset$ (see Theorem 2.1 of [286]) and $U_{ad} \subset L^p(J, Y)(1 < p < +\infty)$ is bounded, closed, and convex. It is obvious that $Bu \in L^p(J, X)$ for all $u \in U_{ad}$.

Using the similar method in Subsection 2.1.2, we give the following definition of mild solution for the problem below.

Definition 4.1. *For any* $u \in L^p(J, Y)$, *if there exist* $T = T(u) > 0$ *and* $x \in C([-r, T], X)$ *such that*

$$x(t) = \begin{cases} S_q(t)\varphi(0) + \displaystyle\int_0^t (t-s)^{q-1} P_q(t-s) f(s, x_s, x(s))\, ds \\ \quad + \displaystyle\int_0^t (t-s)^{q-1} P_q(t-s) B(s) u(s)\, ds, \quad 0 \le t \le T, \\ \varphi(t), \hspace{5.5cm} -r \le t \le 0, \end{cases} \quad (4.2)$$

then system (4.1) is called mildly solvable with respect to u *on* $[-r, T]$, *where*

$$S_q(t) = \int_0^\infty \Psi_q(\theta) T(t^q \theta)\, d\theta, \quad P_q(t) = q \int_0^\infty \theta \Psi_q(\theta) T(t^q \theta)\, d\theta,$$

and $\Psi_q(\theta)$ *is the Wright function (see Definition 1.8).*

Lemma 4.3. *[302] The operators* S_q *and* P_q *have the following properties:*

(i) *for any fixed* $t \ge 0$, $S_q(t)$ *and* $P_q(t)$ *are linear and bounded operators, i.e., for any* $x \in X$,

$$|S_q(t)x| \le M|x| \text{ and } |P_q(t)x| \le \frac{M}{\Gamma(q)} |x|;$$

(ii) $\{S_q(t)\}_{t \ge 0}$ *and* $\{P_q(t)\}_{t \ge 0}$ *are strongly continuous.*

In order to discuss the existence of mild solutions of system (4.1), we need the following important a priori estimate.

Lemma 4.4. *Assume that system (4.1) is mildly solvable on* $[-r, T]$ *with respect to* u. *Then there exists a constant* $M_{pr} = M_{pr}(u) > 0$ *such that*

$$|x(t)| \le M_{pr}, \text{ for } t \in J.$$

Proof. If x is a mild solution of system (4.1) with respect to u on $[-r, T]$, then x satisfies (4.2). Using (H_1)(iii), Lemma 4.3(i) and Hölder inequality, we obtain

$$|x(t)| \le |S_q(t)\varphi(0)| + \int_0^t (t-s)^{q-1} |P_q(t-s) f(s, x_s, x(s))|\, ds$$

$$+ \int_0^t (t-s)^{q-1} |P_q(t-s)B(s)u(s)| ds$$

$$\leq M|\varphi(0)| + \frac{M}{\Gamma(q)} \int_0^t (t-s)^{q-1} a_f(1 + \|x_s\|_{-r,0} + |x(s)|) ds$$

$$+ \frac{\|B\|_\infty M}{\Gamma(q)} \int_0^t (t-s)^{q-1} |u(s)|_Y ds$$

$$\leq M|\varphi(0)| + \frac{a_f M T^q}{\Gamma(1+q)} + \frac{\|B\|_\infty M}{\Gamma(q)} \left(\frac{p-1}{pq-1}\right)^{\frac{p-1}{p}} T^{q-\frac{1}{p}} \|u\|_{L^p J}$$

$$+ \frac{a_f M}{\Gamma(q)} \int_0^t (t-s)^{q-1} \|x_s\|_{-r,0} ds$$

$$+ \frac{a_f M}{\Gamma(q)} \int_0^t (t-s)^{q-1} |x(s)| ds.$$

By Lemma 4.1, there exists a constant $M^* > 0$ such that

$$|x(t)| \leq M^*(a + \|\varphi\|_{-r,0}), \text{ for } t \in J,$$

where

$$a = M|\varphi(0)| + \frac{a_f M T^q}{\Gamma(1+q)} + \frac{\|B\|_\infty M}{\Gamma(q)} \left(\frac{p-1}{pq-1}\right)^{\frac{p-1}{p}} T^{q-\frac{1}{p}} \|u\|_{L^p J}.$$

Let

$$M_{pr} = M^*(a + \|\varphi\|_{-r,0}) > 0.$$

Thus, $|x(t)| \leq M_{pr}$, for $t \in J$. $\square$

Theorem 4.1. *Assume that (H_1), (H_2), and (H_3) hold. Then for each $u \in U_{ad}$ and for some p such that $pq > 1$, system (4.1) is mildly solvable on $[-r, T]$ with respect to u and the mild solution is unique.*

Proof. Let $C_{-r,T_1} = C([-r, T_1], X)$ with the usual sup-norm and

$$S(1, T_1) = \Big\{ h \in C_{-r,T_1} : \max_{s \in [0, T_1]} |h(s) - \varphi(0)| \leq 1,$$

$$h(s) = \varphi(s) \text{ for } -r \leq s \leq 0 \Big\}.$$

Then $S(1, T_1) \subseteq C_{-r,T_1}$ is a closed convex subset of C_{-r,T_1}.

According to (H_1)(i) and (H_1)(ii), it is easy to see that $f(t, h_t, h(t))$ is a measurable function on $[0, T_1]$. Let $h \in \mathcal{S}(1, T_1)$, there exists a constant

$$\rho^* = \max\{|\varphi(0)| + 1, \|\varphi\|_{-r,0}\} > 0$$

such that

$$\|h\|_{-r,T_1} \leq \rho^*.$$

Using (H_1)(iii), for $t \in [0, T_1]$, we have

$$|f(t, h_t, h(t))| \leq a_f(1 + \|h_t\|_{-r,0} + |h(t)|) \leq a_f(1 + 2\rho^*) =: K_f.$$

By Lemma 4.3(i), Hölder inequality and (H_1)(iii), we obtain

$$\int_0^t (t - s)^{q-1}|P_q(t - s)f(s, h_s, h(s))|\,ds \leq \frac{MK_f}{\Gamma(1 + q)}T_1^q.$$

Thus, $|(t-s)^{q-1}P_q(t-s)f(s, h_s, h(s))|$ is Lebesgue integrable with respect to $s \in [0, t]$ for all $t \in [0, T_1]$ by Theorem 1.2.

On the other hand, by Lemma 4.3(ii), Hölder inequality and $pq > 1$, we see that

$$\int_0^t (t - s)^{q-1}|P_q(t - s)B(s)u(s)|\,ds$$

$$\leq \frac{\|B\|_\infty M}{\Gamma(q)}\int_0^t (t - s)^{q-1}|u(s)|_Y\,ds$$

$$\leq \frac{\|B\|_\infty M}{\Gamma(q)}\left(\int_0^t (t - s)^{\frac{p}{p-1}(q-1)}\,ds\right)^{\frac{p-1}{p}}\left(\int_0^t |u(s)|_Y^p\,ds\right)^{\frac{1}{p}}$$

$$\leq \frac{\|B\|_\infty M}{\Gamma(q)}\left(\frac{p-1}{pq-1}\right)^{\frac{p-1}{p}}T^{q-\frac{1}{p}}\|u\|_{L^p J}.$$

Thus, $(t - s)^{q-1}P_q(t - s)B(s)u(s)$ is also Bochner integrable with respect to $s \in [0, t]$ for all $t \in [0, T_1]$.

Now we can define $\mathcal{P}: \mathcal{S}(1, T_1) \to C_{-r, T_1}$ as follows:

$$(\mathcal{P}h)(t) = \begin{cases} S_q(t)\varphi(0) + \displaystyle\int_0^t (t - s)^{q-1} P_q(t - s) f(s, h_s, h(s))\, ds \\ \quad + \displaystyle\int_0^t (t - s)^{q-1} P_q(t - s) B(s) u(s)\, ds, & 0 < t \le T_1, \\ \varphi(t), & -r \le t \le 0. \end{cases}$$

By the properties of S_q, P_q, and (H_1), one can verify that $\mathcal{P}$ is a contraction map on $\mathcal{S}(1, T_1)$ with chosen $T_1 > 0$.

For $t \in [0, T_1]$, it is easy to obtain the following inequality:

$$|(\mathcal{P}h)(t) - \varphi(0)|$$

$$\le |S_q(t)\varphi(0) - \varphi(0)| + \int_0^t (t - s)^{q-1} |P_q(t - s) f(s, h_s, h(s))|\, ds$$

$$+ \int_0^t (t - s)^{q-1} |P_q(t - s) B(s) u(s)|\, ds \qquad (4.3)$$

$$\le |S_q(t)\varphi(0) - \varphi(0)| + \frac{M K_f}{\Gamma(1 + q)} t^q$$

$$+ \frac{M \|B\|_\infty \|u\|_{L^p J}}{\Gamma(q)} \left(\frac{p - 1}{pq - 1}\right)^{\frac{p-1}{p}} t^{q - \frac{1}{p}}.$$

Since $\{S_q(t)\}_{t \ge 0}$ is a strongly continuous operator in X, we can choose $\varepsilon = \frac{1}{2}$ such that

$$|S_q(t)\varphi(0) - \varphi(0)| \le \frac{1}{2}. \qquad (4.4)$$

Let

$$T_{11} = \min\left\{\frac{1}{2}, \left(\frac{\Gamma(1 + q)}{2M(K_f + q\|B\|_\infty \|u\|_{L^p J})\left(\frac{p-1}{pq-1}\right)^{\frac{p-1}{p}}}\right)^{\frac{p}{pq-1}}\right\}.$$

Then for all $t \le T_{11}$, we obtain from (4.3) and (4.4) that

$$|(\mathcal{P}h)(t) - \varphi(0)| \le 1.$$

On the other hand,

$$(\mathcal{P}h)(t) = \varphi(t), \quad \text{for} -r \le t \le 0.$$

Hence,

$$\mathcal{P}(\mathcal{S}(1, T_1)) \subseteq \mathcal{S}(1, T_1).$$

Let h_1, $h_2 \in \mathcal{S}(1, T_1)$ and $\|h_1\|_{-r,T_1}, \|h_2\|_{-r,T_1} \le \rho^*$. For $t \in [0, T_1]$, using Lemma 4.3(ii) and (H_1)(iii), we also obtain

$$|(\mathcal{P}h_1)(t) - (\mathcal{P}h_2)(t)|$$

$$\le \int_0^t (t-s)^{q-1}\big|P_q(t-s)\big(f(s, h_{1s}, h_1(s)) - f(s, h_{2s}, h_2(s))\big)\big|ds$$

$$\le \frac{ML_f(\rho^*)}{\Gamma(q)} \int_0^t (t-s)^{q-1}\|h_{1s} - h_{2s}\|_{-r,0}ds$$

$$+ \frac{ML_f(\rho^*)}{\Gamma(q)} \int_0^t (t-s)^{q-1}|h_1(s) - h_2(s)|ds,$$

which implies that

$$|(\mathcal{P}h_1)(t) - (\mathcal{P}h_2)(t)| \le \frac{2ML_f(\rho^*)}{\Gamma(1+q)}t^q\|h_1 - h_2\|_{-r,T_1}.$$

Note that

$$|(\mathcal{P}h_1)(t) - (\mathcal{P}h_2)(t)| = 0, \quad \text{for } t \in [-r, 0].$$

Thus,

$$\|\mathcal{P}h_1 - \mathcal{P}h_2\|_{-r,T_1} \le \frac{2ML_f(\rho^*)}{\Gamma(1+q)}t^q\|h_1 - h_2\|_{-r,T_1}.$$

Let

$$T_{12} = \frac{1}{2}\left(\frac{\Gamma(1+q)}{2ML_f(\rho^*)}\right)^{\frac{1}{q}}, \quad T_1 = \min\{T_{11}, T_{12}\}.$$

Then $\mathcal{P}$ is a contraction map on $\mathcal{S}(1, T_1)$. It follows from the contraction mapping principle that $\mathcal{P}$ has a unique fixed point $h \in \mathcal{S}(1, T_1)$, and h is the unique mild solution of system (4.1) with respect to u on $[-r, T_1]$.

Let

$$T_{21} = T_1 + T_{11}, T_{22} = T_1 + T_{12}, \Delta T = \min\{T_{21} - T_1, T_{12}\} > 0.$$

Similarly, one can verify that system (4.1) has a unique mild solutions on $[-r, \Delta T]$. Repeating the above procedures in each interval $[\Delta T, 2\Delta T]$, $[2\Delta T, 3\Delta T]$, ..., and using the methods of steps, we immediately obtain the global existence of mild solutions for system (4.1). $\qquad\square$

Remark 4.2. *Assume that X and Y are two separable reflexive Banach spaces. If we replace $(H_1)(i)$-(ii) by the conditions that*

$$f : J \times C_{-r,0} \times X \rightarrow X$$

is Hölder continuous with respect to t and for any $\rho > 0$, there exists a constant $L_f(\rho) > 0$ such that

$$|f(t, \xi_1, \eta_1) - f(s, \xi_2, \eta_2)| \leq L_f(\rho)(|t - s|^\gamma + \|\xi_1 - \xi_2\|_{-r,0} + |\eta_1 - \eta_2|),$$

where $\gamma \in (0, 1]$, provided that $\|\xi_1\|_{-r,0}, \|\xi_2\|_{-r,0}, |\eta_1|, |\eta_2| \leq \rho$, condition (H_2) by the condition

$$B \in L_b(L^p(J, Y), L^p(J, X)),$$

and the condition (H_3) by the condition

$$U_{ad} = L^p(J, Y),$$

one can use the same approach to derive the existence of mild solutions.

4.1.3 Continuous Dependence

In this subsection, we show that the mild solution of system (4.1) is on the initial value, zero arm of time delay and control with respect to X, $C_{-r,0}$ and $L^p(J, Y)$.

Theorem 4.2. *Assume that* $\varphi^1(0)$, $\varphi^2(0) \in \Pi \subset X$, *where* Π *is a bounded set. Let*

$$
\begin{cases}
x^1(t, \varphi^1(0), u) = S_q(t)\varphi^1(0) \\
\qquad + \displaystyle\int_0^t (t-s)^{q-1} P_q(t-s) f\left(s, x_s^1, x^1(s)\right) ds \\
\qquad + \displaystyle\int_0^t (t-s)^{q-1} P_q(t-s) B(s) u(s) ds, \qquad 0 \le t \le T, \\
x^1(t) = \varphi^1(t), \qquad\qquad\qquad\qquad\qquad\quad -r \le t \le 0,
\end{cases}
$$

and

$$
\begin{cases}
x^2(t, \varphi^2(0), v) = S_q(t)\varphi^2(0) \\
\qquad + \displaystyle\int_0^t (t-s)^{q-1} P_q(t-s) f\left(s, x_s^2, x^2(s)\right) ds \\
\qquad + \displaystyle\int_0^t (t-s)^{q-1} P_q(t-s) B(s) v(s) ds, \qquad 0 \le t \le T, \\
x^2(t) = \varphi^2(t), \qquad\qquad\qquad\qquad\qquad\quad -r \le t \le 0.
\end{cases}
$$

Then there exists a constant $C^* > 0$ *such that*

$$
\begin{cases}
\left|x^1(t, \varphi^1(0), u) - x^2(t, \varphi^2(0), v)\right| \le C^*(|\varphi^1(0) - \varphi^2(0)| \\
\quad + \|\varphi^1 - \varphi^2\|_{-r,0} + \|u - v\|_{L^p J}), \qquad\qquad t \in J, \\
|x^1(t) - x^2(t)| = |\varphi^1(t) - \varphi^2(t)|, \qquad\qquad -r \le t \le 0,
\end{cases}
$$

where

$$
C^* = \max\left\{ M^* M, M^*, M^* \frac{\|B\|_\infty M}{\Gamma(q)} \left(\frac{p-1}{pq-1}\right)^{\frac{p-1}{p}} T^{q-\frac{1}{p}} \right\} > 0.
$$

Proof. Since $\varphi^1(0)$, $\varphi^2(0) \in \Pi \subset X$ and Π is a bounded set in X, using Lemma 4.4, there exists a constant $\rho > 0$ such that $\|x_s^1\|_{-r,0}$, $\|x_s^2\|_{-r,0}$, $|x^1|$, $|x^2| \le \rho$.

For $t \in J$, by Lemma 4.3, $(H_1)(ii)$, and Hölder inequality, we have

$$
\begin{aligned}
&\left|x^1(t, \varphi^1(0), u) - x^2(t, \varphi^2(0), v)\right| \\
&\le \left|S_q(t)\left(\varphi^1(0) - \varphi^2(0)\right)\right| \\
&\quad + \int_0^t (t-s)^{q-1} \left|P_q(t-s)\left(f\left(s, x_s^1, x^1(s)\right) - f\left(s, x_s^2, x^2(s)\right)\right)\right| ds
\end{aligned}
$$

$$+ \int_0^t (t-s)^{q-1} |P_q(t-s)(B(s)u(s) - B(s)v(s))|\, ds$$

$$\leq M \left| \varphi^1(0) - \varphi^2(0) \right|$$

$$+ \frac{L_f(\rho)M}{\Gamma(q)} \int_0^t (t-s)^{q-1} \left\| x_s^1 - x_s^2 \right\|_{-r,0} ds$$

$$+ \frac{L_f(\rho)M}{\Gamma(q)} \int_0^t (t-s)^{q-1} \left| x^1(s) - x^2(s) \right| ds$$

$$+ \frac{\|B\|_\infty M}{\Gamma(q)} \int_0^t (t-s)^{q-1} |u(s) - v(s)|_Y\, ds$$

$$\leq M \left| \varphi^1(0) - \varphi^2(0) \right|$$

$$+ \frac{\|B\|_\infty M}{\Gamma(q)} \left(\int_0^t (t-s)^{\frac{p}{p-1}(q-1)} ds \right)^{\frac{p-1}{p}} \left(\int_0^t |u(s) - v(s)|_Y^p ds \right)^{\frac{1}{p}}$$

$$+ \frac{L_f(\rho)M}{\Gamma(q)} \int_0^t (t-s)^{q-1} \left\| x_s^1 - x_s^2 \right\|_{-r,0} ds$$

$$+ \frac{L_f(\rho)M}{\Gamma(q)} \int_0^t (t-s)^{q-1} \left| x^1(s) - x^2(s) \right| ds$$

$$\leq M \left| \varphi^1(0) - \varphi^2(0) \right| + \frac{\|B\|_\infty M}{\Gamma(q)} \left(\frac{p-1}{pq-1} \right)^{\frac{p-1}{p}} T^{q-\frac{1}{p}} \|u - v\|_{L^p J}$$

$$+ \frac{L_f(\rho)M}{\Gamma(q)} \int_0^t (t-s)^{q-1} \left\| x_s^1 - x_s^2 \right\|_{-r,0} ds$$

$$+ \frac{L_f(\rho)M}{\Gamma(q)} \int_0^t (t-s)^{q-1} \left| x^1(s) - x^2(s) \right| ds.$$

Using Lemma 4.1 again, we obtain

$$\left| x^1(t, \varphi^1(0), u) - x^2(t, \varphi^2(0), v) \right|$$

$$\leq C^* \left(\left| \varphi^1(0) - \varphi^2(0) \right| + \|\varphi^1 - \varphi^2\|_{-r,0} + \|u - v\|_{L^p J} \right), \text{ for } t \in J,$$

where

$$C^* = \max \left\{ M^* M, M^*, M^* \frac{\|B\|_\infty M}{\Gamma(q)} \left(\frac{p-1}{pq-1} \right)^{\frac{p-1}{p}} T^{q-\frac{1}{p}} \right\} > 0.$$

Note that

$$\left| x^1(t) - x^2(t) \right| \leq \left| \varphi^1(t) - \varphi^2(t) \right|, \text{ for } -r \leq t \leq 0.$$

This completes the proof. □

4.1.4 Optimal Control

In what follows, we consider the fractional optimal control of system (4.1).

We consider the Lagrange problem:

(P) find a control $u^0 \in U_{ad}$ such that

$$J(u^0) \leq J(u), \text{ for all } u \in U_{ad},$$

where

$$J(u) = \int_0^T \mathcal{L}(t, x_t^u, x^u(t), u(t))dt.$$

x^u denotes the mild solution of system (4.1) corresponding to the control $u \in U_{ad}$.

For the existence of solution for problem (P), we introduce the following assumption:

(H_4) (i) the functional

$$\mathcal{L} : J \times C_{-r,0} \times X \times Y \to \mathbb{R} \cup \{\infty\}$$

is Borel measurable;

(ii) $\mathcal{L}(t, \cdot, \cdot, \cdot)$ is sequentially lower semicontinuous on $C_{-r,0} \times X \times Y$ for almost all $t \in J$;

(iii) $\mathcal{L}(t, x, y, \cdot)$ is convex on Y for each $x \in C_{-r,0}$, $y \in X$ and almost all $t \in J$;

(iv) there exist constants $d, e \geq 0$, $j > 0$, φ is nonnegative, and $\varphi \in L^1(J, \mathbb{R})$ such that

$$\mathcal{L}(t, x, y, u) \geq \varphi(t) + d\|x\|_{-r,0} + e|y| + j\|u\|_Y^p.$$

Now we can give the following result on existence of fractional optimal controls for problem (P).

Theorem 4.3. *Under the assumptions in Theorem 4.1 and (H_4), suppose that B is a strongly continuous operator. Then the optimal control problem (P) admits at least one optimal pair, i.e., there exists an admissible control $u^0 \in U_{ad}$ such that*

$$J(u^0) = \int_0^T \mathcal{L}(t, x_t^0, x^0(t), u^0(t))dt \leq J(u), \text{ for } u \in U_{ad}.$$

Proof. If $\inf\{J(u) : u \in U_{ad}\} = +\infty$, there is nothing to prove.

Assume that

$$\inf\{J(u) : u \in U_{ad}\} = \varepsilon < +\infty.$$

Using the assumption (H_4), we have $\varepsilon > -\infty$. By the definition of infimum, there exists a minimizing sequence feasible pair

$$\{(x^m, u^m)\} \subset A_{ad} := \{(x, u) : x \text{ is a mild solution}$$
$$\text{of system (4.1) corresponding to } u \in U_{ad}\},$$

such that $J(x^m, u^m) \to \varepsilon$ as $m \to +\infty$. Since $\{u^m\} \subseteq U_{ad}$, $m = 1, 2, ...,$ $\{u^m\}$ is a bounded subset of the separable reflexive Banach space $L^p(J, Y)$, there exists a subsequence (we denote it by $\{u^m\}$) and $u^0 \in L^p(J, Y)$ such that

$$u^m \rightharpoonup u^0 \text{ in } L^p(J, Y).$$

Since U_{ad} is closed and convex, owing to Marzur lemma, $u^0 \in U_{ad}$.

Assume that $x^m \in C_{-r,T}$ denotes the corresponding sequence of solutions of the integral equation

$$x^m(t) = \begin{cases} S_q(t)\varphi(0) + \displaystyle\int_0^t (t-s)^{q-1} P_q(t-s) f\left(s, x_s^m, x^m(s)\right) ds \\ \qquad + \displaystyle\int_0^t (t-s)^{q-1} P_q(t-s) B(s) u^m(s) ds, \quad 0 \le t \le T, \\ \varphi(t), \qquad\qquad\qquad\qquad\qquad\qquad\qquad\qquad -r \le t \le 0. \end{cases}$$

Owing to Lemmas 4.1 and 4.4, we can verify that there exist a $\rho > 0$ such that

$$\|x^m\|_{-r,T} \le \rho, \text{ where } m = 0, 1, 2, ...,$$

where x^0 denotes the solution corresponding to u^0, i.e.,

$$x^0(t) = \begin{cases} S_q(t)\varphi(0) + \displaystyle\int_0^t (t-s)^{q-1} P_q(t-s) f\left(s, x_s^0, x^0(s)\right) ds \\ \qquad + \displaystyle\int_0^t (t-s)^{q-1} P_q(t-s) B(s) u^0(s) ds, \quad 0 \le t \le T, \\ \varphi(t), \qquad\qquad\qquad\qquad\qquad\qquad\qquad\qquad -r \le t \le 0. \end{cases}$$

Hence, for $t \in J$, by condition (H_1)(ii), Lemma 4.3(i) and Hölder inequality, after an elementary calculation, we have the following inequality:

$$
\begin{aligned}
&|x^m(t) - x^0(t)| \\
&\leq \int_0^t (t-s)^{q-1} \big| P_q(t-s) \big(f(s, x_s^m, x^m(s)) - f(s, x_s^0, x^0(s)) \big) \big| ds \\
&\quad + \int_0^t (t-s)^{q-1} \big| P_q(t-s) \big(B(s)u^m(s) - B(s)u^0(s) \big) \big| ds \\
&\leq \frac{L_f(\rho)M}{\Gamma(q)} \int_0^t (t-s)^{q-1} \| x_s^m - x_s^0 \|_{-r,0} ds \\
&\quad + \frac{L_f(\rho)M}{\Gamma(q)} \int_0^t (t-s)^{q-1} |x^m(s) - x^0(s)| ds \\
&\quad + \frac{M}{\Gamma(q)} \int_0^t (t-s)^{q-1} \big| B(s)u^m(s) - B(s)u^0(s) \big| ds \\
&\leq \frac{L_f(\rho)M}{\Gamma(q)} \int_0^t (t-s)^{q-1} \| x_s^m - x_s^0 \|_{-r,0} ds \\
&\quad + \frac{L_f(\rho)M}{\Gamma(q)} \int_0^t (t-s)^{q-1} |x^m(s) - x^0(s)| ds \\
&\quad + \frac{M}{\Gamma(q)} \left(\frac{p-1}{pq-1} \right)^{\frac{p-1}{p}} T^{q-\frac{1}{p}} \left(\int_0^T |B(s)u^n(s) - B(s)u^0(s)|^p ds \right)^{\frac{1}{p}}.
\end{aligned}
$$

Note that for $-r \leq t \leq 0$, we have

$$x^m(s) - x^0(s) = 0. \tag{4.5}$$

Applying Lemma 4.1 again, we obtain

$$|x^m(t) - x^0(t)| \leq M^* \left(\int_0^T |B(s)u^m(s) - B(s)u^0(s)|^p ds \right)^{\frac{1}{p}}, \tag{4.6}$$

for $t \in J$, where M^* is a constant independent of u, m, and t.

Since B is strongly continuous,

$$\|Bu^m - Bu^0\| \to 0, \quad \text{as } m \to \infty.$$

Applying Lemma 4.2, we have

$$\int_0^T |B(s)u^m(s) - B(s)u^0(s)|^p ds \to 0, \quad \text{as } m \to \infty. \tag{4.7}$$

We see from (4.5)-(4.7) that

$$\|x^m - x^0\|_{-r,T} \to 0, \quad \text{as } m \to \infty,$$

which yields that

$$x^m \to x^0 \text{ in } C_{-r,T}, \quad \text{as } m \to \infty.$$

Note that the assumption (H_4) implies Balder assumptions (see Theorem 2.1 in [28]). Hence, by Balder's theorem we can conclude that

$$(x_t, x, u) \to \int_0^T \mathcal{L}(t, x_t, x(t), u(t)) dt$$

is sequentially lower semicontinuous in the weak topology of $L^p(J, Y) \subset L^1(J, Y)$, and the strong topology of $L^1(J, C_{-r,0} \times X)$. Hence, J is weakly lower semicontinuous on $L^p(J, Y)$, and since by (H_4)(iv), $J > -\infty$, J attains its infimum at $u^0 \in U_{ad}$, i.e.,

$$\begin{aligned}
\varepsilon &= \lim_{m \to \infty} \int_0^T \mathcal{L}\left(t, x_t^m, x^m(t), u^m(t)\right) dt \\
&\geq \int_0^T \mathcal{L}\left(t, x_t^0, x^0(t), u^0(t)\right) dt \\
&= J\left(u^0\right) \\
&\geq \varepsilon.
\end{aligned}$$

The proof is completed. $\qquad\qquad\qquad\qquad\qquad\qquad\qquad\qquad\Box$

Remark 4.3. *Condition (H_3) in Theorem 4.3 can be replaced by the following condition:*

$(H_3)'$ *$\mathcal{U}$ is a weakly compact subset of Y and $t \to U(t)$ is a map with measurable values in $P_{cl,cv}(\mathcal{U})$.*

Theorem 4.4. *Under the assumptions in Theorem 4.3 with (H_3) replaced by $(H_3)'$, let*

$$U_{ad} = \{u(\cdot) : J \to Y \text{ is strongly measurable}, u(t) \in U(t), t \in J\}.$$

Then there exists an optimal control for problem (P).

Proof. The proof is similar to the proof of Theorem 4.3; however, there are two differences. First, the boundedness of the minimizing sequence of controls $\{u^m\}$ follows from $(H_3)'$. Second, under condition $(H_3)'$, we need to apply Theorem1.1 to verify that

$$x^m \to x^0 \text{ in } C_{-r,T}, \text{ as } m \to \infty.$$

We omit it here.   □

Finally, we present an example to illustrate our results.

Example 4.1. Consider the following problem:

$$
\begin{cases}
{}^C_0 D_t^{\frac{5}{6}} x(t,y) = \Delta x(t,y) + f_1(t,y,x(t,y)) \\
\quad + \displaystyle\int_{-r}^{t} h(t-s) f_2(s,y,x(s,y)) ds \\
\quad + \displaystyle\int_{\Omega} K(y,\xi) u(\xi,t) d\xi, & y \in \Omega, \ t \in J, \\
x(t,y) = \varphi(t,y), & y \in \overline{\Omega}, -r \le t \le 0, \\
x(t,y) = 0, & y \in \partial\Omega, \ t \in J,
\end{cases}
\tag{4.8}
$$

where $\Omega \subset \mathbb{R}^3$ is a bounded domain, $\partial\Omega \in C^3$, Δ is Laplace operator,

$$\varphi \in C^{2,1}([-r,0], \overline{\Omega}), \ u \in L^2(J \times \Omega, \mathbb{R}), \ h \in L^1([-r, T+r], \mathbb{R})$$

and

$$K : \overline{\Omega} \times \overline{\Omega} \to \mathbb{R}$$

is continuous.

Assume that

$$f_1 : J \times \overline{\Omega} \times \mathbb{R} \to \mathbb{R}$$

is continuous and there exist constants K_1, $N_1 \ge 0$ such that

$$|f_1(t,y,\varsigma)| \le K_1(1 + |\varsigma|),$$
$$|f_1(t,y,\varsigma) - f_1(t,y,\widetilde{\varsigma})| \le N_1|\varsigma - \widetilde{\varsigma}|.$$

Assume that

$$f_2 : [-r, T] \times \overline{\Omega} \times \mathbb{R} \to \mathbb{R}$$

is continuous and there exist constants $K_2, N_2 \geq 0$ such that

$$|f_2(t, y, \zeta)| \leq K_2(1 + |\zeta|),$$
$$|f_2(t, y, \zeta) - f_2(t, y, \widetilde{\zeta})| \leq N_2|\zeta - \widetilde{\zeta}|.$$

Define

$$X = Y = L^2(J \times \Omega, \mathbb{R}),$$
$$D(A) = H^2(\Omega) \cap H_0^1(\Omega),$$
$$Ax = -\left(\frac{\partial^2 x}{\partial y_1^2} + \frac{\partial^2 x}{\partial y_2^2} + \frac{\partial^2 x}{\partial y_3^2} \right)$$

for $x \in D(A)$. Then A can generate a C_0-semigroup $\{T(t)\}_{t \geq 0}$ on X. The controls are functions $u : Tx(\Omega) \to \mathbb{R}$, such that $u \in L^2(Tx(\Omega), \mathbb{R})$. This claim is that $t \to u(\cdot, t)$ going from J into Y is measurable. We set

$$U(t) = \{u \in Y : |u|_Y \leq \chi\},$$

where $\chi \in L^2(J, \mathbb{R}^+)$. We restrict the admissible controls U_{ad} to be all $u \in L^2(Tx(\Omega), \mathbb{R})$ such that $\|u(\cdot, t)\|_{L^2 Tx(\Omega)} \leq \chi(t)$, almost everywhere.

Define $x(t)(y) = x(t, y)$,

$$B(t)u(t)(y) = \int_\Omega K(y, \xi)u(\xi, t)d\xi,$$

and

$$f(t, x_t, x(t))(y) = F_1(t, x)(y) + \left(\int_{-r}^t h(t - s)F_2(s, x(s))ds \right)(y),$$

where

$$F_1(t, x)(y) = f_1(t, y, x(t, y)) \text{ and } F_2(t, x)(y) = f_2(t, y, x(t, y)).$$

Thus, problem (4.8) can be rewritten as follows:

$$\begin{cases} {}_0^C D_t^q x(t) = Ax(t) + f(t, x_t, x(t)) + B(t)u(t), & t \in J, \\ x(t) = \varphi(t), & t \in [-r, 0]. \end{cases}$$

We consider the following cost function:

$$J(u) = \int_0^T \mathcal{L}(t, x_t^u, x^u(t), u(t))dt,$$

where $\mathcal{L} : J \times C^{1,0}([-r, 0] \times \overline{\Omega}, \mathbb{R}) \times L^2(J \times \Omega, \mathbb{R}) \to \mathbb{R} \cup \{+\infty\}$ for $x \in C^{1,0}([-r, T] \times \overline{\Omega}, \mathbb{R}), u \in L^2(\Omega \times J, \mathbb{R})$. Then

$$\mathcal{L}(t, x_t^u, x^u(t), u(t))(y) = \int_\Omega \int_{-r}^0 |x^u(t+s, y)|^2 ds dy$$

$$+ \int_\Omega |x^u(t, y)|^2 dy + \int_\Omega |u(y, t)|^2 dy.$$

It is easy to verify that all the assumptions in Theorem 4.3 hold. Our result can be used to problem (4.8).

We have the following existence result for optimal controls.

Theorem 4.5. *Under the assumptions in Theorem 4.3, there exists an optimal pair* $u^0 \in L^2(\Omega \times J, \mathbb{R})$ *such that* $J(u^0) \leq J(u)$ *for all* $u \in L^2(\Omega \times J, \mathbb{R})$.

4.2 OPTIMAL FEEDBACK CONTROL

4.2.1 Introduction

Consider the following semilinear fractional feedback control system

$$\begin{cases} {}_0^C D_t^q x(t) = Ax(t) + f(t, x(t), u(t)), & t \in J = [0, T], \\ x(0) = x_0, \end{cases} \tag{4.9}$$

where ${}_0^C D_t^q$ is Caputo fractional derivative of order $q \in (0, 1)$, and $A : D(A) \to X$ is the infinitesimal generator of a compact C_0-semigroup $\{T(t)\}_{t \geq 0}$ in a reflexive Banach space X. The control u takes value from $U[0, T]$, which is a control set, $f : J \times X \times U \to X$ will be specified later.

First, we give the existence of mild solutions of system (4.9). Second, we give the existence result of feasible pairs involving the compactness of operators with the help of the Cesari property and Fillippove theorem. Then, we present the existence of optimal feedback controls for Lagrange problem

(P). We remark that system (4.9) is more complex than the classical first order evolution equation because a fractional derivative has appeared. After overcoming some difficulty from Caputo fractional derivative, we extend the classical results on optimal feedback controls to the case of semilinear fractional evolution equations.

In Subsection 4.2.2, some notations and preparation results are given. In Subsection 4.2.3, the existence of mild solutions and feasible pairs for system (4.9) are presented. At last, the existence of optimal feedback controls for Lagrange problems (P) is proved.

4.2.2 Existence of Feasible Pairs

Denote by X a reflexive Banach space with norm $|\cdot|$, and by U a Polish space which is a separable completely metrizable topological space. Let $C(J, X)$ be Banach space of continuous functions from J to X with the usual sup-norm. Suppose that $A : D(A) \to X$ is the infinitesimal generator of a compact C_0-semigroup $\{T(t)\}_{t \geq 0}$. This means that there exists $M > 0$ such that $\sup_{t \in J} \|T(t)\|_{\mathscr{L}(X)} \leq M$. By

$$O_r(x) = \{y \in X : |y - x| \leq r\},$$

we denote the ball centered at x with the radius $r > 0$.

Definition 4.2. *[156] Let E and F be two metric spaces. A multifunction $F : E \to P(F)$ is said to be pseudo-continuous at $t \in E$ if*

$$\bigcap_{\epsilon > 0} \overline{F(O_\epsilon(t))} = F(t).$$

We say that F is pseudo-continuous on E if it is pseudo-continuous at each point $t \in E$.

We make the following assumptions:

(H_S) X is a reflexive Banach space and U is a Polish space;

(H_A) A is the infinitesimal generator of a compact C_0-semigroup $\{T(t)\}_{t \geq 0}$ on X;

(H_1) $f : J \times X \times U \to X$ is Borel measurable in (t, x, u) and is continuous in (x, u);

(H_2) f satisfies local Lipschitz continuous with respect to x, i.e., for any constant $\rho > 0$, there is a constant $L(\rho) > 0$ such that

$$|f(t, x_1, u) - f(t, x_2, u)| \leq L(\rho)|x_1 - x_2|,$$

for every $x_1, x_2 \in X$, $t \in J$, and uniformly $u \in U$ provided with $|x_1|$, $|x_2| \leq \rho$;

(H_3) for arbitrary $t \in J$, $x \in X$, and $u \in U$, there exists a positive constant $M > 0$ such that

$$|f(t, x, u)| \leq M(1 + |x|);$$

(H_4) for almost all $t \in J$, the set $f(t, x, F(t, x))$ satisfies the following

$$\bigcap_{\delta > 0} \overline{co} f(t, O_\delta(x), F(O_\delta(t, x))) = f(t, x, F(t, x));$$

(H_U) $F : J \times X \to P(U)$ is pseudo-continuous.
   Let

$$U[0, T] = \{u : J \to U, u(\cdot) \text{ is measurable}\}.$$

Then, any element in the set $U[0, T]$ is called a control on J.

In the following, we introduce the following definition of mild solutions for system (4.9).

Definition 4.3. *A mild solution $x \in C(J, X)$ of system (4.9) is defined as a solution of the following integral equation:*

$$x(t) = \mathcal{T}(t)x_0 + \int_0^t (t - \theta)^{q-1} \mathcal{S}(t - \theta) f(\theta, x(\theta), u(\theta)) d\theta, \ t \in J,$$

$$(4.10)$$

where

$$\mathcal{T}(t) = \int_0^\infty \Psi_q(\theta) T(t^q \theta) d\theta, \quad \mathcal{S}(t) = q \int_0^\infty \theta \Psi_q(\theta) T(t^q \theta) d\theta;$$

here, $\Psi_q(\theta)$ is the Wright function (see Definition 1.8).

Any solution $x(\cdot) \in C(J, X)$ of system (4.9) is referred to as a state trajectory of the fractional evolution equation corresponding to the initial state x_0 and the control $u(\cdot)$.

Lemma 4.5. *[302] The operators $\mathscr{T}$ and $\mathscr{S}$ have the following properties:*

(i) *for any fixed $t \geq 0$, $\mathscr{T}(t)$ and $\mathscr{S}(t)$ are linear and bounded operators, i.e., for any $x \in X$,*

$$|\mathscr{T}(t)x| \leq M|x| \text{ and } |\mathscr{S}(t)x| \leq \frac{qM}{\Gamma(1+q)}|x|;$$

(ii) *$\{\mathscr{T}(t)\}_{t\geq 0}$ and $\{\mathscr{S}(t)\}_{t\geq 0}$ are strongly continuous;*
(iii) *for every $t > 0$, $\mathscr{T}(t)$ and $\mathscr{S}(t)$ are also compact operators if $T(t)$ is compact.*

By Lemma 4.5, Lipschitz condition and growth condition of f, and standard method used in our earlier work (see Theorem 3.1 in [250]), we can obtain the following existence and uniqueness of mild solutions for system (4.9). So we omit the proof here.

Theorem 4.6. *Assumptions (H_S), (H_A), (H_1), (H_2), and (H_3) hold. There is a unique mild solution $x \in C(J, X)$ of system (4.9) for any $x_0 \in X$ and $u \in U$, and*

$$\|x\| \leq M,$$

for some constant $M > 0$.

Next, we introduce the following definition of feasible pairs.

Definition 4.4. *A pair (x, u) is said to be feasible if x satisfies (4.10) and*

$$u(t) \in F(t, x(t)), \text{ a.e. } t \in J.$$

Let $[s, v] \subseteq J$,

$$\mathcal{H}[s, v] = \{(x, u) \in C([s, v], X) \times U[s, v] : (x, u) \text{ is feasible}\},$$
$$\mathcal{H}[0, T] = \{(x, u) \in C([0, T], X) \times U[0, T] : (x, u) \text{ is feasible}\}.$$

In order to obtain the existence of feasible pairs we need the following important lemma.

Lemma 4.6. *Assumption* (H_A) *holds. Then operators* $\mathcal{R}_j : L^p(J, X) \to C(J, X)$, $j = 1, 2$ *for some* $\frac{1}{p} < q < 1$, $p > 1$, *given by*

$$(\mathcal{R}_1 h)(t) = \int_0^t (t - s)^{q-1} \mathscr{T}(t - s) h(s) ds,$$

$$(\mathcal{R}_2 h)(t) = \int_0^t (t - s)^{q-1} \mathscr{S}(t - s) h(s) ds,$$

are also compact for $h \in L^p(J, X)$.

Proof. Let $\{h^n\} \subseteq L^p(J, X)$ be bounded. Define $\mathcal{A}_n^j(t) = (\mathcal{R}_j h^n)(t)$, $j = 1, 2$, $t \in J$. One can verify that for any fixed $t \in J$ and $\frac{1}{p} < q < 1$, $|\mathcal{A}_n^j(t)|$ is bounded. By Lemma 4.5, it is not difficult to verify that $\mathcal{A}_n^j(t)$ is compact in X and is also equicontinuous. Due to Lemma 1.2, $\{\mathcal{A}_n^j(t)\}$ is relatively compact in $C(J, X)$. Obviously, $\mathcal{R}_j$ is linear and continuous. Hence, $\mathcal{R}_j$ $(j = 1, 2)$ is a compact operator. $\qquad\square$

To solve the optimal feedback control problem, we need the following result which is an extension of the results corresponding to first order semilinear evolution equations.

Theorem 4.7. *Assumptions* (H_S), (H_A), (H_1)-(H_4), *and* (H_U) *hold. Then for any* $x_0 \in X$ *and* $\frac{1}{p} < q < 1$ *for some* $p > 1$, *the set* $\mathcal{H}[0, T]$ *is nonempty, i.e.,*

$$\mathcal{H}[0, T] \neq \emptyset.$$

Proof. For any $k \geq 0$, let $t_j = \frac{j}{k} T$, $0 \leq j \leq k - 1$. We set

$$u_k(t) = \sum_{j=0}^{k-1} u^j \chi_{[t_j, t_{j+1})}(t), \ t \in J,$$

where $\chi_{[t_j, t_{j+1})}$ is the character function of interval $[t_j, t_{j+1})$. The sequence $\{u^j\}$ is constructed as follows.

First, take $u^0 \in F(0, x^0)$. By Theorem 4.6, there exists a unique $x_k(\cdot)$ which is given by

$$x_k(t) = \mathscr{T}(t) x_0 + \int_0^t (t - \theta)^{q-1} \mathscr{S}(t - \theta) f(\theta, x_k(\theta), u^0(\theta)) d\theta,$$

for $t \in [0, \frac{T}{k}]$. Then, take $u^1 \in F\left(\frac{T}{k}, x_k(\frac{T}{k})\right)$. Repeat this procedure to obtain x_k on $[\frac{T}{k}, \frac{2T}{k}]$, etc. By induction, we end up with the following integral equation

$$x_k(t) = \mathscr{T}(t)x_0 + \int_0^t (t - \theta)^{q-1} \mathscr{S}(t - \theta) f(\theta, x_k(\theta), u_k(\theta))d\theta, \ t \in J,$$

$$(4.11)$$

where $u_k(\cdot) \in F\left(\frac{jT}{k}, x_k(\frac{jT}{k})\right)$, $t \in \left[\frac{jT}{k}, \frac{(j+1)T}{k}\right)$, $0 \le j \le k - 1$.

By Lemma 4.5 and applying the standard singular Gronwall inequality, there exists $M > 0$ such that

$$|x_k(t)| \le M, \ t \in J,$$

Moreover, it comes from (H_3) that

$$|f(t, x_k(t), u_k(t))| \le M, \text{ a.e. } t \in J.$$

From Lemma 4.6, there is a subsequence of $\{x_k\}$, denoted by $\{x_k\}$ again, such that

$$x_k \to \overline{x} \text{ in } C(J, X), \tag{4.12}$$

for some $\overline{x} \in C(J, X)$, and

$$f(\cdot, x_k(\cdot), u_k(\cdot)) \to \overline{f}(\cdot) \text{ in } L^p(J, X), \tag{4.13}$$

for some $\overline{f} \in L^p(J, X)$.

Again, by Lemma 4.6 and (4.11), we have

$$\overline{x}(t) = \mathscr{T}(t)x_0 + \int_0^t (t - \theta)^{q-1} \mathscr{S}(t - \theta) \overline{f}(\theta)d\theta, \ t \in J.$$

By (4.12), for any $\delta > 0$, there exists a $k_0 > 0$ such that

$$x_k(t) \in O_\delta(\overline{x}(t)), \ t \in J, \ k \ge k_0. \tag{4.14}$$

On the other hand, by the definition of $u_k(\cdot)$, for k large enough one can find

$$u_k(t) \in F(t_j, x_k(t_j)) \subset F(O_\delta(t, \overline{x}(t))), \tag{4.15}$$

for all $t \in \left[\frac{jT}{k}, \frac{(j+1)T}{k}\right), 0 \le j \le k - 1$.

Second, by (4.13) and Mazur lemma, let $\alpha_{ij} \ge 0$ and $\Sigma_{j \ge 0} \alpha_{ij} = 1$ such that

$$\psi_l(\cdot) = \sum_{i \ge 1} \alpha_{il} f(\cdot, x_{i+l}(\cdot), u_{i+l}(\cdot)) \to \overline{f}(\cdot) \text{ in } L^p(J, X).$$

Then, there is a subsequence of $\{\psi_l\}$, denoted by $\{\psi_l\}$ again, such that

$$\psi_l(t) \to \overline{f}(t) \text{ in } X, \text{ a.e. } t \in J.$$

Hence, from (4.14) and (4.15), for l large enough,

$$\psi_l(t) \in \text{co} f(t, O_\delta(\overline{x}(t)), F(O_\delta(t, \overline{x}(t)))), \text{ a.e. } t \in J.$$

Thus, for any $\delta > 0$,

$$\overline{f}(t) \in \overline{\text{co}} f(t, O_\delta(\overline{x}(t)), F(O_\delta(t, \overline{x}(t)))), \text{ a.e. } t \in J.$$

By (H_4),

$$\overline{f}(t) \in \overline{\text{co}} f(t, \overline{x}(t), F(t, \overline{x}(t))), \text{ a.e. } t \in J.$$

By (H_U) and Corollary 2.18 of [156], we known that $F(\cdot, \overline{x}(\cdot))$ is Souslin measurable. By the well-known Fillippove theorem (see [18]), there exists a $\overline{u} \in U[0, T]$ such that

$$\overline{u}(t) \in F(t, \overline{x}(t)), \ t \in J,$$

and

$$\overline{f}(t) = \overline{f}(t, \overline{x}(t), \overline{u}(t)), \ t \in J.$$

Therefore, $(\overline{x}, \overline{u})$ is just a feasible pair in $[0, T]$. This completes the proof.

□

4.2.3 Existence of Optimal Feedback Control Pairs

In this subsection, we consider the following Lagrange problem:
(P) find a pair $(x^0, u^0) \in \mathcal{H}[0, T]$ such that

$$\mathcal{J}(x^0, u^0) \leq \mathcal{J}(x, u), \text{ for all } (x, u) \in \mathcal{H}[0, T],$$

where

$$\mathcal{J}(x, u) = \int_0^T \mathcal{L}(t, x(t), u(t)) dt.$$

We impose some assumptions on $\mathcal{L}$:

(L_1) the functional $\mathcal{L} : J \times X \times U \to R \cup \{\infty\}$ is Borel measurable in (t, x, u);

(L_2) $\mathcal{L}(t, \cdot, \cdot)$ is sequentially l.s.c. on $X \times U$ for almost all $t \in J$ and there is a constant $M_1 > 0$ such that

$$\mathcal{L}(t, x, u) \geq -M_1, \ (t, x, u) \in J \times X \times U.$$

For any $(t, x) \in J \times X$, we set

$$\mathcal{W}(t, x) = \{(z^0, z) \in R \times X : z^0 \geq \mathcal{L}(t, x, u),$$
$$z = f(t, x, u), \ u \in F(t, x)\}.$$

In order to prove the existence of optimal control pairs for problem (P), we assume that:

(H_C) for almost all $t \in J$, the map $\mathcal{W}(t, \cdot) : X \to P(R \times X)$ has Cesari property, i.e.,

$$\bigcap_{\delta > 0} \overline{co} \mathcal{W}(t, O_\delta(x)) = \mathcal{W}(t, x),$$

for all $x \in X$.

Theorem 4.8. *Assume that the hypotheses* (H_S), (H_A), (H_1)-(H_4), (H_U), (L_1), (L_2), *and* (H_C) *hold. Then Lagrange problem (P) admits at least one optimal control pair.*

Proof. If $\inf\{\mathcal{J}(x, u) : (x, u) \in \mathcal{H}[0, T]\} = +\infty$, there is nothing to prove. So we assume that $\inf\{\mathcal{J}(x, u) : (x, u) \in \mathcal{H}[0, T]\} = m < +\infty$.

By (L_1) and (L_2) we have $\mathcal{J}(u) \geq m \geq -M_1 > -\infty$. So there exists a sequence $\{x^n, u^n\} \subset \mathcal{H}[0, T]$ such that

$$\mathcal{J}(x^n, u^n) \to m, \quad \text{as } n \to \infty.$$

We denote

$$\mathcal{J}(x^n, u^n) = \int_0^T \mathcal{L}(t, x^n(t), u^n(t))dt$$

and

$$\liminf_{n \to +\infty} \mathcal{J}(x^n, u^n) = m.$$

By the growth condition of f and boundedness of $\{x^n\}$, one can obtain that $\{f(\cdot, x^n(\cdot), u^n(\cdot))\}$ is bounded in $L^p(J, X)$. Without loss of generality, we may assume that

$$f^n(\cdot) = f(\cdot, x^n(\cdot), u^n(\cdot)) \rightharpoonup f(\cdot) \text{ in } L^p(J, X),$$

for some $f(\cdot) \in L^p(J, X)$. By Lemma 4.6, we obtain

$$\begin{aligned}
x^n(t) =& \mathscr{T}(t)x_0 + \int_0^t (t - \theta)^{q-1} \mathscr{S}(t - \theta) f(\theta, x^n(\theta), u^n(\theta))d\theta \\
\to& \mathscr{T}(t)x_0 + \int_0^t (t - \theta)^{q-1} \mathscr{S}(t - \theta) \overline{f}(\theta)d\theta = \overline{x}(t)
\end{aligned}$$

uniformly for $t \in J$, i.e.,

$$x^n(\cdot) \to \overline{x}(\cdot) \text{ in } C(J, X).$$

By Mazur lemma, let $\alpha_{kl} \geq 0$, $\Sigma_{k \geq 1} \alpha_{kl} = 1$, such that

$$\psi_l(\cdot) = \sum_{k \geq 1} \alpha_{kl} f(\cdot, x_{k+l}(\cdot), u_{k+l}(\cdot)) \to \overline{f}(\cdot) \text{ in } L^p(J, X).$$

Set

$$\psi_l^0(\cdot) = \sum_{k \geq 1} \alpha_{kl} \mathcal{L}(\cdot, , x_{k+l}(\cdot), u_{k+l}(\cdot))$$

and

$$\mathcal{L}^0(t) = \liminf_{l \to +\infty} \psi_l^0(t) \geq -M_1, \text{ a.e. } t \in J.$$

For any $\delta > 0$ and l large enough, we have

$$(\psi_l(t), \psi_l^0(t)) \in \mathcal{W}(t, O_\delta(\overline{x}(t))).$$

By (H_C), we have

$$(\mathcal{L}^0(t), \overline{f}(t)) \in \mathcal{W}(t, \overline{x}(t)), \text{ a.e. } t \in J.$$

This means that

$$\begin{cases} \mathcal{L}^0(t) \geq \mathcal{L}(t, \overline{x}(t), u), & t \in J, \\ \overline{f}(t) = f(t, \overline{x}(t), u), & t \in J, \\ u \in F(t, \overline{x}(t)). \end{cases} \quad (4.16)$$

By Filippov theorem (see [18]) again, there is a measurable selection $\overline{u}(\cdot)$ of $F(\cdot, \overline{x}(\cdot))$ such that

$$\begin{cases} \mathcal{L}^0(t) \geq \mathcal{L}(t, \overline{x}(t), \overline{u}(t)), & t \in J, \\ \overline{f}(t) = f(t, \overline{x}(t), \overline{u}(t)), & \text{a.e. } t \in J. \end{cases} \quad (4.17)$$

On the other hand, we have

$$\overline{x}(t) = \mathscr{T}(t)x_0 + \int_0^t (t-\theta)^{q-1} \mathscr{S}(t-\theta) f(\theta, \overline{x}(\theta), \overline{u}(\theta)) d\theta, \ t \in J,$$

and

$$(\overline{x}, \overline{u}) \in \mathcal{H}[0, T].$$

By the well-known Fatou's lemma, we obtain

$$\int_0^T \mathcal{L}^0(t)dt = \int_0^T \liminf_{l \to +\infty} \psi_l^0(t)dt \leq \liminf_{l \to +\infty} \int_0^T \psi_l^0(t)dt,$$

i.e.,

$$\mathcal{J}(\overline{x}, \overline{u}) = \int_0^T \mathcal{L}(t, \overline{x}(t), \overline{u}(t))dt = \inf_{(x,u) \in \mathcal{H}[0,T]} J(x, u) = m.$$

Thus, $(\overline{x}, \overline{u})$ is just an optimal pair. $\qquad \square$

4.3 CONTROLLABILITY

4.3.1 Introduction

Consider the following Sobolev-type fractional evolution system

$$\begin{cases} {}^{C}_{0}D^{q}_{t}(Ex(t)) = Ax(t) + Ef(t, x(t)) + EBu(t), & t \in J = [0, a], \\ Ex(0) = Ex_0, & x_0 \in D(E), \end{cases}$$

$$(4.18)$$

where ${}^{C}_{0}D^{q}_{t}$ is Caputo fractional derivative of order $0 < q < 1$, $A :$ $D(A) \subset X \to X$; here X is a separable Banach space with the norm $|\cdot|$, and $E : D(E) \subset X \to X$ are two closed linear operators and the pair (A, E) generates an exponentially bounded propagation family $\{W(t)\}_{t \geq 0}$ of $D(E)$ to X. The state $x(\cdot)$ takes values in X and the control function $u(\cdot)$ is given in $\mathcal{U}$, the Banach space of admissible control functions, where

$$\mathcal{U} = \begin{cases} L^{p}(J, U), & \text{for } q \in \left(\dfrac{1}{p}, 1\right) \text{ with } 1 < p < \infty, \\ L^{\infty}(J, U), & \text{for } q \in (0, 1), \end{cases}$$

and U is a Banach space with the norm $|\cdot|_{U}$. B is a bounded linear operator from U into $D(E)$, and $f : J \times X \to D(E) \subset X$ will be specified later.

In this section we study the controllability of system (4.18) via the theory of propagation family $\{W(t)\}_{t \geq 0}$ generating by the pair (A, E). Our aim in this section is to present sufficient conditions for the controllability results corresponding to two classes of the possible admissible control sets. To simplify the process, we construct $\{\mathcal{T}_{(A,E)}(t)\}_{t \geq 0}$ and $\{\mathcal{S}_{(A,E)}(t)\}_{t \geq 0}$ associated with the pair (A, E) and give their boundedness and norm continuity in the sense of uniform operator topology.

4.3.2 Characteristic Solution Operators

We recall the concept of exponentially bounded propagation family (see [157]).

Definition 4.5. *A strongly continuous operator family* $\{W(t)\}_{t \geq 0}$ *of* $D(E)$ *to a Banach space* X *satisfying that* $\{W(t)\}_{t \geq 0}$ *is exponentially bounded, which means that there exist* $\omega > 0$ *and* $M > 0$ *such that* $|W(t)x| \leq Me^{\omega t}|x|$ *for any* $x \in D(E)$ *and* $t \geq 0$, *is called an exponentially bounded propagation family for the following abstract degenerate Cauchy* *problem*

$$\begin{cases} (Ex(t))' = Ax(t), & t \in J, \\ Ex(0) = Ex_0, & x_0 \in D(E), \end{cases} \tag{4.19}$$

when $\lambda > \omega$,

$$(\lambda E - A)^{-1} Ex = \int_0^\infty e^{-\lambda t} W(t) x \, dt, \ x \in D(E). \tag{4.20}$$

In this case, we say that problem (4.19) has an exponentially bounded propagation family $\{W(t)\}_{t \geq 0}$.

Moreover, if (4.20) holds, we also say that the pair (A, E) *generates an exponentially bounded propagation family* $\{W(t)\}_{t \geq 0}$.

Remark 4.4. *Since* $D(E) \subset X$ *is dense,* $W(t)$ *can be uniquely extended on* X *as a linear bounded mapping so that* $|W(t)x| \leq M e^{\omega t} |x|$ *for any* $x \in X$ *and* $t \geq 0$. *From now on, we consider such* $W(t)$ *on* X *directly.*

Denote

$$\begin{aligned} \mathscr{T}_{(A,E)}(t) &= \int_0^\infty \Psi_q(\theta) W(t^q \theta) d\theta, \\ \mathscr{S}_{(A,E)}(t) &= q \int_0^\infty \theta \Psi_q(\theta) W(t^q \theta) d\theta, \end{aligned} \tag{4.21}$$

where $\Psi_q(\theta)$ is the Wright function (see Definition 1.8).

Using the similar method in Subsection 2.1.2, we can introduce the following definition of mild solution for system (4.18).

Definition 4.6. *For each* $u \in \mathcal{U}$ *and* $x_0 \in D(E)$, *by a mild solution of system (4.18), we mean a function* $x \in C(J, X)$ *satisfying*

$$\begin{aligned} x(t) = {}& \mathscr{T}_{(A,E)}(t) x_0 + \int_0^t (t-s)^{q-1} \mathscr{S}_{(A,E)}(t-s) f(s, x(s)) \, ds \\ & + \int_0^t (t-s)^{q-1} \mathscr{S}_{(A,E)}(t-s) Bu(s) ds, \ t \in J. \end{aligned}$$

The following results of $\mathscr{T}_{(A,E)}(\cdot)$ and $\mathscr{S}_{(A,E)}(\cdot)$ will be used throughout this section.

Lemma 4.7. *Suppose that the pair* (A, E) *generates an exponentially bounded propagation family* $\{W(t)\}_{t\geq 0}$. *If* $\{W(t)\}_{t\geq 0}$ *is a norm continuous family for* $t > 0$ *and* $\|W(t)\|_{\mathscr{L}(X)} \leq M_1$ *for* $t \geq 0$, *then the following two properties hold:*

(i) *for any fixed* $t \geq 0$, $\mathscr{T}_{(A,E)}(t)$ *and* $\mathscr{S}_{(A,E)}(t)$ *are bounded operators on* X, *i.e., for any* $x \in X$,

$$|\mathscr{T}_{(A,E)}(t)x| \leq M_1|x| \ \ and \ \ |\mathscr{S}_{(A,E)}(t)x| \leq \frac{M_1}{\Gamma(q)}|x|;$$

(ii) $\{\mathscr{T}_{(A,E)}(t)\}_{t\geq 0}$ *and* $\{\mathscr{S}_{(A,E)}(t)\}_{t\geq 0}$ *are norm continuous family for* $t > 0$ *in the sense of uniform operator topology.*

Proof. The first assertion has been proved (see Remark 2.1.3 in [153]). Next, we verify the second assertion. We only need to prove that $\|\mathscr{T}_{(A,E)}(t_1) - \mathscr{T}_{(A,E)}(t_2)\|_{\mathscr{L}(X)}$ and $\|\mathscr{S}_{(A,E)}(t_1) - \mathscr{S}_{(A,E)}(t_2)\|_{\mathscr{L}(X)}$ tend to zero as $t_1 \to t_2$, respectively, in the sense of uniform operator topology.

For $0 < t_1 < t_2 < \infty$, a simple computation implies

$$\|\mathscr{T}_{(A,E)}(t_1) - \mathscr{T}_{(A,E)}(t_2)\|_{\mathscr{L}(X)}$$
$$\leq \int_0^\infty \Psi_q(\theta)\|W(t_1^q\theta) - W(t_2^q\theta)\|_{\mathscr{L}(X)}d\theta, \tag{4.22}$$

$$\|\mathscr{S}_{(A,E)}(t_1) - \mathscr{S}_{(A,E)}(t_2)\|_{\mathscr{L}(X)}$$
$$\leq q\int_0^\infty \theta\Psi_q(\theta)\|W(t_1^q\theta) - W(t_2^q\theta)\|_{\mathscr{L}(X)}d\theta. \tag{4.23}$$

Note that $\|W(t_1^q\theta) - W(t_2^q\theta)\|_{\mathscr{L}(X)} \to 0$ as $t_1 \to t_2$ in the sense of uniform operator topology for any fixed $\theta > 0$. Linking Property 1.11(iii) and (4.22), (4.23), one can obtain the second assertion immediately. The proof is completed. $\qquad\qquad\square$

4.3.3 Controllability Results

In this subsection, we study the controllability of system (4.18) by utilizing the theory of propagation family and techniques of measure of noncompactness.

Definition 4.7. *System (4.18) is said to be controllable on the interval J if for every $x_0 \in D(E)$ and every $x_1 \in D(E)$ there exists a control $u \in \mathcal{U}$ such that the mild solution x of system (4.18) satisfies $x(a) = x_1$.*

We pose the following assumptions:

(H_1) the pair (A, E) generates an exponentially bounded propagation family $\{W(t)\}_{t \geq 0}$ of $D(E)$ to X;

(H_2) $\{W(t)\}_{t \geq 0}$ is norm continuous family for $t > 0$ and $\|W(t)\|_{\mathscr{L}(X)} \leq M_1$ for $t \geq 0$;

(H_3) the control function $u(\cdot)$ takes from $\mathcal{U}$, the Banach space of admissible control functions, either $\mathcal{U} = L^p(J, U)$ for $q \in (\frac{1}{p}, 1)$ with $1 < p < \infty$ or $\mathcal{U} = L^\infty(J, U)$ for $q \in (0, 1)$ where U is Banach space;

(H_4) $B : U \rightarrow D(E)$ is a bounded linear operator and a linear operator $\mathbb{W} : \mathcal{U} \rightarrow X$ defined by $\mathbb{W}u = \int_0^a (a - s)^{q-1} \mathscr{S}_{(A,E)}(a - s) Bu(s) ds$ has a bounded right inverse operator $\mathbb{W}^{-1} : X \rightarrow \mathcal{U}$.

It is easy to see that $\mathbb{W}u \in X$ and $\mathbb{W}$ is well defined due to the following fact:

$$|\mathbb{W}u| = \left| \int_0^a (a - s)^{q-1} \mathscr{S}_{(A,E)}(a - s) Bu(s) ds \right|$$

$$\leq \frac{M_1 \|B\|_{\mathscr{L}(U,X)}}{\Gamma(q)} \int_0^a (a - s)^{q-1} |u(s)|_U ds$$

$$\leq \begin{cases} \dfrac{M_1 \|B\|_{\mathscr{L}(U,X)}}{\Gamma(q)} \left(\dfrac{p-1}{qp-1} a^{\frac{qp-1}{p-1}} \right)^{\frac{p-1}{p}} \|u\|_{L^p J}, \\ \qquad \text{if } q \in \left(\dfrac{1}{p}, 1 \right), \ u \in \mathcal{U} = L^p(J, U), \ 1 < p < \infty, \\ \dfrac{M_1 \|B\|_{\mathscr{L}(U,X)} a^q}{\Gamma(q+1)} \|u\|_{L^\infty J}, \\ \qquad \text{if } q \in (0, 1), \ u \in \mathcal{U} = L^\infty(J, U). \end{cases}$$

Meanwhile,

$$\int_0^t (t - s)^{q-1} |u(s)|_U ds \leq K_q \|u\|_{L^p J}, \qquad (4.24)$$

for any $t \in J$, where

$$
K_q = \begin{cases}
\left(\dfrac{p-1}{qp-1} a^{\frac{qp-1}{p-1}} \right)^{\frac{p-1}{p}} \|u\|_{L^p J}, \\
\qquad \text{if } q \in (\dfrac{1}{p}, 1), \ u \in \mathcal{U} = L^p(J, U), \ 1 < p < \infty, \\
\dfrac{a^q}{q} \|u\|_{L^\infty J}, \ \text{if } q \in (0, 1), \ u \in \mathcal{U} = L^\infty(J, U).
\end{cases}
$$

Next we assume:

(H_5) f satisfies the following two conditions:

  (i) for each $x \in X$ the function $f(\cdot, x) : J \to D(E) \subset X$ is strongly measurable and for each $t \in J$, the function $f(t, \cdot) : X \to D(E) \subset X$ is continuous;

  (ii) for each $k > 0$, there is a measurable function g_k such that

$$
\sup_{|x| \leq k} |f(t, x)| \leq g_k(t), \ \text{with } \|g_k\|_\infty = \sup_{s \in J} g_k(s) < \infty,
$$

  and for some $\gamma > 0$, there exists sufficiently large k_0 such that

$$
\sup_{t \in J} \int_0^t (t-s)^{q-1} g_k(s) ds \leq \gamma k, \ \text{for } k > k_0;
$$

  (iii) there exists a positive constant $L > 0$ such that

$$
\alpha(f(t, D)) \leq L\alpha(D),
$$

  for any bounded set $D \subset X$ and a.e. $t \in J$.

The first step in studying the controllability problem is to determine if an objective can be reached by some suitable control. A standard approach is to transform the controllability problem into a fixed point problem for an appropriate operator in a function space. For the sake of simplicity, we present the standard framework to deal with controllability problems here.

Based on our assumptions, for an arbitrary function $x(\cdot)$, it is suitable to define the following control formula:

$$u(t) = \mathbb{W}^{-1}\Bigg(x_1 - \mathscr{T}_{(A,E)}(a)x_0$$
$$- \int_0^a (a-s)^{q-1}\mathscr{S}_{(A,E)}(a-s)f(s,x(s))ds \Bigg). \tag{4.25}$$

Define the operator $\mathcal{P}$ by

$$(\mathcal{P}x)(t) = \mathscr{T}_{(A,E)}(t)x_0 + \int_0^t (t-s)^{q-1}\mathscr{S}_{(A,E)}(t-s)f(s,x(s))ds$$
$$+ \int_0^t (t-s)^{q-1}\mathscr{S}_{(A,E)}(t-s)Bu(s)ds, \text{ for } t \in J, \tag{4.26}$$

where u is defined by (4.25). It is necessary to show that $\mathcal{P}$ has a fixed point. Clearly, this fixed point is just a mild solution of system (4.18). Further, one can check

$$(\mathcal{P}x)(a) = \mathscr{T}_{(A,E)}(a)x_0 + \int_0^a (a-s)^{q-1}\mathscr{S}_{(A,E)}(a-s)f(s,x(s))ds$$
$$+ \int_0^a (a-s)^{q-1}\mathscr{S}_{(A,E)}(a-s)B\mathbb{W}^{-1}\Bigg(x_1 - \mathscr{T}_{(A,E)}(a)x_0$$
$$- \int_0^a (a-\tau)^{q-1}\mathscr{S}_{(A,E)}(a-\tau)f(\tau,x(\tau))d\tau \Bigg)ds$$
$$= x_1,$$

which means that u steers the fractional system (4.18) from x_0 to x_1 in finite time a. Consequently, we can claim system (4.18) is controllable on J.

For each number $k > 0$, define

$$\mathcal{B}_k = \{ x \in C(J,X) : |x(t)| \leq k, \ t \in J \}.$$

Of course, $\mathcal{B}_k$ is clearly a bounded, closed, convex subset in $C(J,X)$.

Under assumptions (H_1)-(H_5), we will establish some important results as follows.

Lemma 4.8. *Assume that*

$$\rho = \begin{cases} \dfrac{\gamma M_1}{\Gamma(q)}\left(1 + \dfrac{a^{\frac{1}{2}}M_1\|B\|_{\mathscr{L}(U,X)}K_q\|\mathbb{W}^{-1}\|_{\mathscr{L}(X,\mathcal{U})}}{\Gamma(q)}\right) < 1, \\ \qquad\qquad\qquad\qquad\qquad \text{if } \mathcal{U} = L^2(J,U), \\[2mm] \dfrac{\gamma M_1}{\Gamma(q)}\left(1 + \dfrac{M_1\|B\|_{\mathscr{L}(U,X)}K_q\|\mathbb{W}^{-1}\|_{\mathscr{L}(X,\mathcal{U})}}{\Gamma(q)}\right) < 1, \\ \qquad\qquad\qquad\qquad\qquad \text{if } \mathcal{U} = L^\infty(J,U). \end{cases} \qquad (4.27)$$

There exists a constant $K \geq \frac{M^*}{1-\rho}$ *such that* $\mathcal{P}\mathcal{B}_K \subset \mathcal{B}_K$, *where*

$$M^* = \begin{cases} M_1|x_0| + \dfrac{a^{\frac{1}{2}}M_1\|B\|_{\mathscr{L}(U,X)}}{\Gamma(q)}K_q\|\mathbb{W}^{-1}\|_{\mathscr{L}(X,\mathcal{U})}\big(|x_1| + M_1|x_0|\big), \\ \qquad\qquad\qquad\qquad\qquad \text{if } \mathcal{U} = L^2(J,U), \\[2mm] M_1|x_0| + \dfrac{M_1\|B\|_{\mathscr{L}(U,X)}}{\Gamma(q)}K_q\|\mathbb{W}^{-1}\|_{\mathscr{L}(X,\mathcal{U})}\big(|x_1| + M_1|x_0|\big), \\ \qquad\qquad\qquad\qquad\qquad \text{if } \mathcal{U} = L^\infty(J,U). \end{cases}$$

Proof. Let $x \in \mathcal{B}_K$. For $t \in J$, using our assumptions and Lemma 4.7(i), we obtain

$$\begin{aligned} |(\mathcal{P}x)(t)| &\leq M_1|x_0| + \frac{M_1}{\Gamma(q)}\int_0^t (t-s)^{q-1}g_K(s)ds \\ &\quad + \frac{M_1\|B\|_{\mathscr{L}(U,X)}}{\Gamma(q)}\int_0^t (t-s)^{q-1}|u(s)|_U ds \\ &\leq M_1|x_0| + \frac{M_1\gamma K}{\Gamma(q)} + \frac{M_1\|B\|_{\mathscr{L}(U,X)}}{\Gamma(q)}K_q\|u\|_{L^p J} \\ &= \rho K + M^* \\ &\leq K, \end{aligned}$$

where we note that the control u defined in (4.25) satisfies

$$\begin{aligned} |u(t)|_U &\leq \|\mathbb{W}^{-1}\|_{\mathscr{L}(X,\mathcal{U})}\bigg| x_1 - \mathscr{T}_{(A,E)}(a)x_0 \\ &\quad - \int_0^a (a-s)^{q-1}\mathscr{S}_{(A,E)}(a-s)f(s,x(s))ds\bigg| \end{aligned}$$

$$\leq \|\mathbb{W}^{-1}\|_{\mathscr{L}(X,\mathcal{U})} \left(|x_1| + M_1 |x_0| + \frac{M_1}{\Gamma(q)} \gamma K \right),$$

which implies that

$$\|u\|_{L^p J} \leq \begin{cases} a^{\frac{1}{2}} \|\mathbb{W}^{-1}\|_{\mathscr{L}(X,\mathcal{U})} \left(|x_1| + M_1 |x_0| + \frac{M_1}{\Gamma(q)} \gamma K \right), \\ \qquad\qquad\qquad\qquad \text{if } \mathcal{U} = L^2(J,U), \\ \|\mathbb{W}^{-1}\|_{\mathscr{L}(X,\mathcal{U})} \left(|x_1| + M_1 |x_0| + \frac{M_1}{\Gamma(q)} \gamma K \right), \\ \qquad\qquad\qquad\qquad \text{if } \mathcal{U} = L^\infty(J,U). \end{cases} \quad (4.28)$$

Hence, $\mathcal{P}\mathcal{B}_K \subset \mathcal{B}_K$ for any $K \geq \frac{M^*}{1-\rho}$ sufficiently large. The proof is completed. $\qquad\qquad\qquad\qquad\qquad\qquad\qquad\qquad\qquad\qquad\qquad\qquad\qquad \Box$

Lemma 4.9. *The operator $\mathcal{P}$ defined by (4.26) is continuous.*

Proof. Let $\{x_m\}_{m \in N} \subseteq \mathcal{B}_K$ be a sequence such that $x_m \to x$ as $m \to \infty$. Note that $(t-s)^{q-1} f(s, x_m(s)) \to (t-s)^{q-1} f(s, x(s))$ as $m \to \infty$ for very $t \in J$ and almost each $s \in [0, t]$ and

$$(t-s)^{q-1} |f(s, x_m(s)) - f(s, x(s))| \leq 2(t-s)^{q-1} g_K(s).$$

Since $\int_0^t (t-s)^{q-1} g_K(s) \leq \frac{\|g_K\|_\infty}{q}$, by Theorem 1.1, we get

$$|(\mathcal{P}x_m)(t) - (\mathcal{P}x)(t)|$$

$$\leq \frac{M_1}{\Gamma(q)} \int_0^t (t-s)^{q-1} \Bigg(|f(s, x_m(s)) - f(s, x(s))| + \|B\|_{\mathscr{L}(U,X)}$$

$$\times \|\mathbb{W}^{-1}\|_{\mathscr{L}(X,\mathcal{U})} \int_0^a (a-z)^{q-1} |f(z, x_m(z)) - f(z, x(z))| dz \Bigg) ds$$

$$= \frac{M_1}{\Gamma(q)} \int_0^t (t-s)^{q-1} |f(s, x_m(s)) - f(s, x(s))| ds$$

$$+ \frac{M_1 \|B\|_{\mathscr{L}(U,X)} \|\mathbb{W}^{-1}\|_{\mathscr{L}(X,\mathcal{U})} a^q}{\Gamma(q+1)}$$

$$\times \int_0^a (a-s)^{q-1} |f(s, x_m(s)) - f(z, x(s))| ds$$

$$\to 0, \text{ as } m \to \infty,$$

for $t \in J$. This yields that $\mathcal{P}$ is continuous. The proof is completed. $\qquad \Box$

Let α be a Hausdorff *MNC* in X. Consider the measure of noncompactness ν in the space $C(J, X)$ with values in the cone $\mathbb{R}^2$ of the way: for every bounded subset $\Omega \subset C(J, X)$,

$$\nu = (\psi(\Omega), \text{mod}_C(\Omega))$$

where $\psi(\Omega) = \sup_{t \in J} \alpha(\Omega(t))$ and

$$\text{mod}_C(\Omega) = \lim_{\delta \to 0} \sup_{x \in \Omega} \max_{|t_1 - t_2| \le \delta} |x(t_1) - x(t_2)|.$$

Lemma 4.10. *Assume that*

$$\ell L \big(1 + \ell \|B\|_{\mathscr{L}(U,X)} \|\mathbb{W}^{-1}\|_{\mathscr{L}(X,\mathcal{U})} \big) < 1, \tag{4.29}$$

where $\ell = \frac{a^q M_1}{\Gamma(q+1)}$. *If* $\nu(\mathcal{P}(\mathcal{B}_K)) \ge \nu(\mathcal{B}_K)$, *then* $\psi(\mathcal{B}_K) = 0$.

Proof. Clearly, $\mathcal{B}_K \subset C(J, X)$ is nonempty and bounded. For any $t \in J$, we set

$$\Theta(\mathcal{B}_K(t)) = \int_0^t G(s)ds,$$

where a function $s \in [0, t] \multimap G(s)$ is defined as

$$G(s) = \Big\{ (t-s)^{q-1} \mathscr{S}_{(A,E)}(t-s) f(s, x(s))$$
$$+ (t-s)^{q-1} \mathscr{S}_{(A,E)}(t-s) Bu(s) : x \in \mathcal{B}_K \Big\}$$

and $u(t)$ is given by (4.25). It is obvious that G is integrable and integrably bounded. Moreover, a simple computation implies that

$$\alpha(G(s))$$
$$\le \frac{M_1(t-s)^{q-1}}{\Gamma(q)} \alpha \bigg(\Big\{ f(s, x(s)) + B\mathbb{W}^{-1} \Big(x_1 - \mathscr{T}_{(A,E)}(a)x_0$$
$$- \int_0^a (a-s)^{q-1} \mathscr{S}_{(A,E)}(a-s) f(s, x(s))ds \Big) : x \in \mathcal{B}_K \Big\} \bigg)$$
$$\le \frac{M_1(t-s)^{q-1}}{\Gamma(q)} \Big[\alpha(\{f(s, \mathcal{B}_K(s))\}) + \alpha \bigg(\Big\{ B\mathbb{W}^{-1} \Big(x_1 - \mathscr{T}_{(A,E)}(a)x_0$$
$$- \int_0^a (a-s)^{q-1} \mathscr{S}_{(A,E)}(a-s) f(s, \mathcal{B}_K(s))ds \Big) \Big\} \bigg) \Big]$$

$$\leq \frac{M_1(t-s)^{q-1}}{\Gamma(q)} \Big[L\alpha(\mathcal{B}_K(s)) + \frac{M_1}{\Gamma(q)} \|B\|_{\mathscr{L}(U,X)} \|\mathbb{W}^{-1}\|_{\mathscr{L}(X,\mathcal{U})}$$

$$\times \Big(\int_0^a (a-s)^{q-1} L\alpha(\mathcal{B}_K(s)) ds \Big) \Big]$$

$$\leq \frac{M_1 L(t-s)^{q-1}}{\Gamma(q)} \big(1 + \ell \|B\|_{\mathscr{L}(U,X)} \|\mathbb{W}^{-1}\|_{\mathscr{L}(X,\mathcal{U})} \big) \psi(\mathcal{B}_K)$$

$$=: \kappa(s).$$

By Property 1.19, we have

$$\alpha(\Theta(\mathcal{B}_K(t))) \leq \int_0^t \kappa(s) ds$$

$$\leq \ell L \big(1 + \ell \|B\|_{\mathscr{L}(U,X)} \|\mathbb{W}^{-1}\|_{\mathscr{L}(X,\mathcal{U})} \big) \psi(\mathcal{B}_K).$$

Thus,

$$\psi(\mathcal{P}(\mathcal{B}_K(t))) \leq \alpha(\Theta(\mathcal{B}_K(t)))$$

$$\leq \ell L \big(1 + \ell \|B\|_{\mathscr{L}(U,X)} \|\mathbb{W}^{-1}\|_{\mathscr{L}(X,\mathcal{U})} \big) \psi(\mathcal{B}_K),$$

which implies $\psi(\mathcal{B}_K) = 0$ due to condition (4.29) and $\nu(\mathcal{P}(\mathcal{B}_K)) \geq \nu(\mathcal{B}_K)$. The proof is completed. $\qquad\square$

Lemma 4.11. *If* $\nu(\mathcal{P}(\mathcal{B}_K)) \geq \nu(\mathcal{B}_K)$, *then* $\mathrm{mod}_C(\mathcal{B}_K) = 0$.

Proof. To achieve our aim, we need to prove that $\mathcal{P}(\mathcal{B}_K)$ is equicontinuous. Let $x \in \mathcal{B}_K$ and $t', t'' \in J$ be such that $0 < t' < t''$, then

$$|(\mathcal{P}x)(t'') - (\mathcal{P}x)(t')|$$

$$\leq |\mathscr{T}_{(A,E)}(t'')x_0 - \mathscr{T}_{(A,E)}(t')x_0|$$

$$+ \Big| \int_0^{t''} (t''-s)^{q-1} \mathscr{S}_{(A,E)}(t''-s) f(s,x(s)) ds$$

$$- \int_0^{t'} (t'-s)^{q-1} \mathscr{S}_{(A,E)}(t'-s) f(s,x(s)) ds \Big|$$

$$+ \Big| \int_0^{t''} (t''-s)^{q-1} \mathscr{S}_{(A,E)}(t''-s) Bu(s) ds$$

$$- \int_0^{t'} (t'-s)^{q-1} \mathscr{S}_{(A,E)}(t'-s) Bu(s) ds \Big|$$

$$\leq \|\mathscr{T}_{(A,E)}(t'') - \mathscr{T}_{(A,E)}(t')\|_{\mathscr{L}(X)} |x_0|$$

$$+ \int_0^{t''} \left|(t''-s)^{q-1} - (t'-s)^{q-1}\right| |\mathscr{S}_{(A,E)}(t''-s)f(s,x(s))| ds$$

$$+ \int_0^{t'} (t'-s)^{q-1} \left|\left(\mathscr{S}_{(A,E)}(t''-s) - \mathscr{S}_{(A,E)}(t'-s)\right)f(s,x(s))\right| ds$$

$$+ \int_0^{t''} \left|(t''-s)^{q-1} - (t'-s)^{q-1}\right| |\mathscr{S}_{(A,E)}(t''-s)Bu(s)| ds$$

$$+ \int_0^{t'} (t'-s)^{q-1} \left|\left(\mathscr{S}_{(A,E)}(t''-s) - \mathscr{S}_{(A,E)}(t'-s)\right)Bu(s)\right| ds$$

$$+ \int_{t'}^{t''} (t'-s)^{q-1} |\mathscr{S}_{(A,E)}(t''-s)f(s,x(s))| ds$$

$$+ \int_{t'}^{t''} (t'-s)^{q-1} |\mathscr{S}_{(A,E)}(t''-s)Bu(s)| ds$$

$$\leq I_1 + I_2 + I_3 + I_4 + I_5 + I_6 + I_7,$$

where

$$I_1 := \|\mathscr{T}_{(A,E)}(t'') - \mathscr{T}_{(A,E)}(t')\|_{\mathscr{L}(X)} |x_0|,$$

$$I_2 := \frac{M_1}{\Gamma(q)} \int_0^{t''} \left((t'-s)^{q-1} - (t''-s)^{q-1}\right) g_K(s) ds,$$

$$I_3 := \sup_{s \in [0,t']} \|\mathscr{S}_{(A,E)}(t''-s) - \mathscr{S}_{(A,E)}(t'-s)\|_{\mathscr{L}(X)}$$

$$\times \int_0^{t'} (t'-s)^{q-1} g_K(s) ds,$$

$$I_4 := \frac{M_1 \|B\|_{\mathscr{L}(U,X)}}{\Gamma(q)} \int_0^{t''} \left((t'-s)^{q-1} - (t''-s)^{q-1}\right) |u(s)|_U ds,$$

$$I_5 := \sup_{s \in [0,t']} \|\mathscr{S}_{(A,E)}(t''-s) - \mathscr{S}_{(A,E)}(t'-s)\|_{\mathscr{L}(X)}$$

$$\times \|B\|_{\mathscr{L}(U,X)} \int_0^{t'} (t'-s)^{q-1} |u(s)|_U ds,$$

$$I_6 := \frac{M_1}{\Gamma(q)} \int_{t'}^{t''} (t'-s)^{q-1} g_K(s) ds,$$

$$I_7 := \frac{M_1 \|B\|_{\mathscr{L}(U,X)}}{\Gamma(q)} \int_{t'}^{t''} (t'-s)^{q-1} |u(s)|_U ds.$$

Note that Lemma 4.7(ii), $\mathscr{T}_{(A,E)}(t)$, and $\mathscr{S}_{(A,E)}(t)$ are continuous in the uniform operator topology for $t \geq 0$, $\sup_{s \in J} |g_K(s)| < \infty$, and $u(\cdot)$ is bounded by (4.28). We can obtain the terms $I_1, I_3, I_5, I_6, I_7 \to 0$ as $t'' \to t'$. Moreover, applying

$$\int_0^{t''} \left((t'-s)^{q-1} - (t''-s)^{q-1} \right) ds = \frac{t'^q - t''^q + (t''-t')^q}{q},$$

one can check the terms $I_2, I_4 \to 0$ as $t'' \to t'$. Thus, $\mathcal{P}(\mathcal{B}_K)$ is equicontinuous.

Hence, $\mathrm{mod}_C(\mathcal{P}(\mathcal{B}_K)) = 0$. This implies that $\mathrm{mod}_C(\mathcal{B}_K) = 0$ from $\nu(\mathcal{P}(\mathcal{B}_K)) \geq \nu(\mathcal{B}_K)$. The proof is completed. $\qquad\square$

Lemma 4.12. *The operator $\mathcal{P}$ defined by (4.26) is ν-condensing on $\mathcal{B}_K$.*

Proof. It follows from Lemmas 4.10 and 4.11 that $\nu(\mathcal{B}_K) = (0,0)$. The regularity property of v implies the relative compactness of $\mathcal{B}_K$. It follows from Definition 1.24 that $\mathcal{P}$ is ν-condensing on $\mathcal{B}_K$. $\qquad\square$

For $\hat{\lambda} \in (0,1]$, consider a one-parameter family of maps $\mathcal{H} : [0,1] \times C(J,X) \to C(J,X)$ given by

$$(\hat{\lambda}, x) \to \mathcal{H}(\hat{\lambda}, x) = \hat{\lambda}\mathcal{P}(x).$$

Lemma 4.13. *The fixed point set of the family of maps $\mathcal{H}$: $\mathrm{Fix}\mathcal{H} = \{x \in \mathcal{H}(\hat{\lambda}, x) \text{ for some } \hat{\lambda} \in (0,1]\}$ has a priori bounded.*

Proof. The result can be derived by Lemma 4.8 immediately. We omit it here. $\qquad\square$

Now we are ready to state the main results in this subsection.

Theorem 4.9. *Assume that (H_1)-(H_5) are satisfied. Then system (4.18) is controllable on J provided that conditions (4.27) and (4.29) hold.*

Proof. To obtain our conclusion, we need to prove $\mathcal{P}$ has a fixed point in $\mathcal{B}_K$. In fact, it follows from Lemmas 4.8 and 4.12 that $\mathcal{P} : \mathcal{B}_K \to \mathcal{B}_K$ is ν-condensing map. By Theorem 1.8, $\mathcal{P}$ has a fixed point in $\mathcal{B}_K$. This implies that any fixed point of $\mathcal{P}$ is just a mild solution of system (4.18) on J which satisfies $(\mathcal{P}x)(a) = x_1$ with $u(t)$ given by (4.25). Therefore, system (4.18) is controllable on J.   □

Corollary 4.1. *Let the assumptions in Theorem 4.9 be satisfied. The set of mild solutions of system (4.18) is a nonempty and compact subset of $C(J, X)$ with $u(t)$ given by (4.25).*

Proof. It follows from Lemma 4.13 that we can take a closed ball $\mathcal{B}_K$ to contain the set $\mathrm{Fix}\mathcal{H}$ inside itself. Moreover, $\mathcal{P}$ maps $\mathcal{B}_K$ into $C(J, X)$ and is ν-condensing map. By Theorem 1.8, we have the conclusion.   □

4.3.4 Example

Take $X = U = L^2([0, \pi], \mathbb{R})$. We consider the following fractional partial differential equation with control

$$
\begin{cases}
{}^C_0 D_t^{\frac{4}{5}} \left(x(t, y) - x_{yy}(t, y) \right) = x_{yy}(t, y) + \mu t^2 \left(\sin \dfrac{x(t, y)}{t} \right. \\
\qquad \left. - \dfrac{\partial^2}{\partial y^2} \sin \dfrac{x(t, y)}{t} \right) + Bu(t), \quad y \in [0, \pi],\ t \in J_1 = [0, 1], \quad (4.30) \\
x(t, 0) = x(t, \pi) = 0, \qquad\qquad t \geq 0, \\
x(0, y) - x_{yy}(0, y) = x_0(y), \qquad 0 \leq y \leq \pi,
\end{cases}
$$

where $0 < \mu < \infty$.

Define $A : D(A) \subset X \to X$ by $Ax = x_{yy}$ and $E : D(E) \subset X \to X$ by $Ex = x - x_{yy}$, respectively, where each domain $D(A)$, $D(E)$ is given by $\{x \in X : x, x_y$ are absolutely continuous, $x_{yy} \in X$, $x(0) = x(\pi) = 0\}$.

It follows from Theorem 2.2 in [157] that the pair (A, E) can generate a propagation family $\{W(t)\}_{t \geq 0}$ of uniformly bounded and $\{W(t)\}_{t \geq 0}$ is norm continuous for $t > 0$ and $\|W(t)\|_{\mathscr{L}(X)} \leq 1$. Meanwhile, it follows from [158] that A and E can be written as $Ax = -\sum_{n=1}^{\infty} n^2 \langle x, x_n \rangle$, $x \in D(A)$ and $Ex = \sum_{n=1}^{\infty} (1 + n^2) \langle x, x_n \rangle x_n$, $x \in D(E)$, respectively, where $x_n(y) = \sqrt{\frac{2}{\pi}} \sin ny$, $n = 1, 2, \dots$ is the orthonormal set of eigenfunctions of A. Hence, for any $x \in D(E)$, $\lambda > 0$ we obtain

$$
(\lambda E - A)^{-1} Ex = \sum_{n=1}^{\infty} \frac{1 + n^2}{\lambda(1 + n^2) + n^2} \langle x, x_n \rangle x_n
$$

$$= \sum_{n=1}^{\infty} \int_0^{\infty} e^{-\lambda t} e^{-\frac{n^2}{1+n^2}t} \langle x, x_n \rangle x_n dt.$$

Therefore, $\{W(t)\}_{t \geq 0}$ can be generated by $-AE^{-1}$ and written as

$$W(t)x = \sum_{n=1}^{\infty} e^{-\frac{n^2}{1+n^2}t} \langle x, x_n \rangle x_n.$$

Then, $\mathscr{T}_{(A,E)}(\cdot)$ and $\mathscr{S}_{(A,E)}(\cdot)$ can be written as

$$\mathscr{T}_{(A,E)}(t)x = \int_0^{\infty} \Psi_{\frac{4}{5}}(\theta) \sum_{n=1}^{\infty} e^{-\frac{n^2}{1+n^2}t^{\frac{4}{5}}\theta} \langle x, x_n \rangle x_n d\theta,$$

$$\mathscr{S}_{(A,E)}(t)x = \frac{4}{5} \int_0^{\infty} \theta \Psi_{\frac{4}{5}}(\theta) \sum_{n=1}^{\infty} e^{-\frac{n^2}{1+n^2}t^{\frac{4}{5}}\theta} \langle x, x_n \rangle x_n d\theta.$$

Clearly, $\|\mathscr{T}_{(A,E)}(t)\|_{\mathscr{L}(X)} \leq 1$ and $\|\mathscr{S}_{(A,E)}(t)\|_{\mathscr{L}(X)} \leq \frac{1}{\Gamma(\frac{4}{5})}$ for $t \geq 0$.

Next, $B : U \to D(E)$ is defined by $B = bI$, $b > 0$ and $\mathbb{W}$ is defined by

$$\mathbb{W}u = b \int_0^1 (1-s)^{-\frac{1}{5}} \mathscr{S}_{(A,E)}(1-s)u(s,y)ds.$$

Since $q = \frac{4}{5} > \frac{1}{2}$, we can take $p = 2$ and $\mathcal{U} = L^2(J_1, U)$, thus $K_{\frac{4}{5}} = \sqrt{\frac{5}{3}}\|u\|_{L^2[0,1]}$. It is easy to show that $\mathbb{W}$ is surjective. Indeed, if $u(s,y) = x(y) \in \mathcal{U}$, then

$$\mathbb{W}u = b \int_0^1 (1-s)^{-\frac{1}{5}} \frac{4}{5} \int_0^{\infty} \theta \Psi_{\frac{4}{5}}(\theta) \sum_{n=1}^{\infty} e^{-\frac{n^2}{1+n^2}(1-s)^{\frac{4}{5}}\theta} \langle x, x_n \rangle x_n d\theta ds$$

$$= b \int_0^{\infty} \Psi_{\frac{4}{5}}(\theta) \sum_{n=1}^{\infty} \int_0^1 \frac{4}{5}\theta(1-s)^{-\frac{1}{5}} e^{-\frac{n^2}{1+n^2}(1-s)^{\frac{4}{5}}\theta} \langle x, x_n \rangle x_n ds d\theta$$

$$= b \int_0^{\infty} \Psi_{\frac{4}{5}}(\theta) \sum_{n=1}^{\infty} \int_0^1 \frac{1+n^2}{n^2} \frac{d}{ds}\left(e^{-\frac{n^2}{1+n^2}(1-s)^{\frac{4}{5}}\theta}\right) \langle x, x_n \rangle x_n ds d\theta$$

$$= b \int_0^{\infty} \Psi_{\frac{4}{5}}(\theta) \sum_{n=1}^{\infty} \frac{1+n^2}{n^2}\left(1 - e^{-\frac{n^2}{1+n^2}\theta}\right) \langle x, x_n \rangle x_n d\theta$$

$$= b \sum_{n=1}^{\infty} \frac{1+n^2}{n^2}\left[1 - \mathbb{E}_{\frac{4}{5}}\left(-\frac{n^2}{1+n^2}\right)\right] \langle x, x_n \rangle x_n,$$

where $E_{\frac{4}{5}}$ is Mittag-Leffler function (see Definition 1.7). So we can define a right inverse $\mathbb{W}^{-1} : X \to \mathcal{U}$ by

$$(\mathbb{W}^{-1}x)(t, y) = \frac{1}{b} \sum_{n=1}^{\infty} \frac{n^2}{1 + n^2} \frac{\langle x, x_n \rangle x_n}{1 - E_{\frac{4}{5}}\left(-\frac{n^2}{1+n^2}\right)},$$

for $x = \sum_{n=1}^{\infty} \langle x, x_n \rangle x_n$, with

$$\|\mathbb{W}^{-1}\|_{\mathscr{L}(X,\mathcal{U})} = \frac{1}{b\left(1 - E_{\frac{4}{5}}\left(-\frac{n^2}{1+n^2}\right)\right)} \leq \frac{1}{b\left(1 - E_{\frac{4}{5}}\left(-\frac{1}{2}\right)\right)}.$$

Now $f : J_1 \times \mathbb{R} \to \mathbb{R}$ is defined by $f(t, x(t, y)) = \mu t^2 \sin \frac{x(t,y)}{t}$. It is easy to see that f is measurable for the first variable and $f(t, x)$ is continuous for the second variable. Moreover, clearly

$$\limsup_{k \to \infty} \frac{1}{k} \sup_{t \in J_1, |x| \leq k} |f(t, x)| = 0$$

and $\alpha(f(t, D_1)) \leq \mu t \alpha(D_1) \leq \mu \alpha(D_1)$ for any bounded set $D_1 \subset X$ and $t \in J_1$. Hence, $\gamma = 0$ and $L = \mu$.

Define $F : J_1 \times C(J, X) \to D(E)$ by $F(t, z)(y) = f(t, z(y))$. Now, system (4.30) can be abstracted as

$$\begin{cases} {}^{C}_{0}D_t^{\frac{4}{5}}(Ex(t)) = -Ax(t) + EF(t, x(t)) + EBu(t), \ t \in J_1, \\ Ex(0) = Ex_0. \end{cases}$$

From the above discussion, all the assumptions in Theorem 4.9 are satisfied, since $\gamma = 0$, (4.27) holds. Furthermore, (4.29) holds when

$$\frac{\mu}{\Gamma(\frac{9}{5})} \left(1 + \frac{1}{\Gamma(\frac{9}{5})\left(1 - E_{\frac{4}{5}}\left(-\frac{1}{2}\right)\right)}\right) < 1.$$

Then system (4.30) is controllable on J_1.

Finally, one can numerically find that $\mu < 0.229071$. It is key to compute $E_{\frac{4}{5}}\left(-\frac{1}{2}\right)$. We only provide a possible way to compute $E_{\frac{4}{5}}\left(-\frac{1}{2}\right)$. In fact, we can use the definition

$$E_{\frac{4}{5}}\left(-\frac{1}{2}\right) = \sum_{k=0}^{\infty} \frac{(-1)^k}{2^k \Gamma(1 + \frac{5i}{4})} = \sum_{k=0}^{25} \frac{(-1)^k}{2^k \Gamma(1 + \frac{5i}{4})} + \sum_{k=26}^{\infty} \frac{(-1)^k}{2^k \Gamma(1 + \frac{5i}{4})}.$$

Using Mathematica we get

$$\sum_{k=0}^{25} \frac{(-1)^k}{2^k \Gamma(1 + \frac{5i}{4})} \doteq 0.626879.$$

On the other hand, it holds

$$\sum_{k=26}^{\infty} \frac{(-1)^k}{2^k \Gamma(1 + \frac{5i}{4})} \leq \sum_{k=26}^{\infty} \frac{1}{2^k} = \frac{1}{2^{25}} \doteq 2.98023 \times 10^{-8}.$$

Hence, $E_{\frac{4}{5}}\left(-\frac{1}{2}\right) \doteq 0.626879$. By Mathematica, the rest of the computation to estimate μ is given again, since Γ is built in Mathematica.

4.4 APPROXIMATE CONTROLLABILITY

4.4.1 Introduction

Let X be a Hilbert space with a scalar product $\langle \cdot, \cdot \rangle$ and the corresponding norm $| \cdot |$. We consider the following Sobolev-type fractional evolution system:

$$\begin{cases} {}_0^C D_t^q (Ex(t)) + Ax(t) = f(t, x(t)) + Bu(t), \ t \in J = [0, a], \\ x(0) + \sum_{k=1}^{m} a_k x(t_k) = 0, \end{cases} \tag{4.31}$$

where ${}_0^C D_t^q$ is Caputo fractional derivative of order $0 < q < 1$, E and A are two linear operators with domains contained in X and ranges still contained in X, the pre-fixed points t_k satisfies $0 = t_0 < t_1 < t_2 < \cdots < t_m < t_{m+1} = a$ and a_k are real numbers.

In order to guarantee that $-AE^{-1} : X \to X$ generates a semigroup $\{W(t)\}_{t \geq 0}$, we consider that the operators A and E satisfy the following conditions:

(S_1) $A : D(A) \subset X \to X$ and $E : D(E) \subset X \to X$ are linear, A is closed;
(S_2) $D(E) \subset D(A)$ and E is bijective;
(S_3) $E^{-1} : X \to D(E)$ is compact;
$(S_3)'$ $E^{-1} : X \to D(E)$ is continuous.

Now we note

(i) $(S_3)'$ implies that E is closed;
(ii) (S_3) implies $(S_3)'$;
(iii) it follows from (S_1), (S_2), $(S_3)'$ and the closed graph theorem that $-AE^{-1} : X \to X$ is bounded, which generates a uniformly continuous semigroup $\{W(t)\}_{t\geq 0}$ of bounded linear operators from X to itself.

Denote by $\rho(-AE^{-1})$ the resolvent set of $-AE^{-1}$. If we assume that the resolvent $R(\lambda; -AE^{-1})$ is compact, then $\{W(t)\}_{t\geq 0}$ is a compact semigroup (see [199]).

The state $x(t)$ takes values in X and the control function $u(\cdot)$ is given in $\mathcal{U}$, Banach space of admissible control functions, where $\mathcal{U} = L^p(J, U)$, for $q \in (\frac{1}{p}, 1)$ with $1 < p < \infty$ and U is Hilbert space. Moreover, $B \in \mathscr{L}(U, X)$ is a bounded linear operator and $f : J \times X \to X$ will be specified later.

Define the following two operators:

$$
\begin{aligned}
\mathscr{T}_{(A,E)}(t) &= \int_0^\infty \Psi_q(\theta) W(t^q \theta) d\theta, \\
\mathscr{S}_{(A,E)}(t) &= q \int_0^\infty \theta \Psi_q(\theta) W(t^q \theta) d\theta,
\end{aligned}
\tag{4.32}
$$

where $\Psi_q(\theta)$ is the Wright function (see Definition 1.8).

Similar to the proof in Zhou and Jiao [301] and Fečkan et al. [93], the following results can be given.

Lemma 4.14. *Assume that* $\sup_{t\geq 0} \|W(t)\|_{\mathscr{L}(X)} \leq M_1$. *One has the following properties:*

(i) *for any fixed* $t \geq 0$, $\mathscr{T}_{(A,E)}(t)$ *and* $\mathscr{S}_{(A,E)}(t)$ *are linear bounded operators on* X *with*

$$
\|\mathscr{T}_{(A,E)}(t)\|_{\mathscr{L}(X)} \leq M_1 \ and \ \|\mathscr{S}_{(A,E)}(t)\|_{\mathscr{L}(X)} \leq \frac{M_1}{\Gamma(q)};
$$

(ii) *if* $W(t)$ *is compact, then* $\mathscr{T}_{(A,E)}(t)$ *and* $\mathscr{S}_{(A,E)}(t)$ *are compact in* X *for* $t > 0$;
(iii) $\mathscr{T}_{(A,E)} : [0, \infty) \to \mathscr{L}(X)$ *and* $\mathscr{S}_{(A,E)} : [0, \infty) \to \mathscr{L}(X)$ *are continuous.*

Next, we define

$$\mathcal{T}^*_{(A^*,E^*)}(t) = \int_0^\infty \Psi_q(\theta)W^*(t^q\theta)d\theta,$$

$$\mathcal{S}^*_{(A^*,E^*)}(t) = q\int_0^\infty \theta\Psi_q(\theta)W^*(t^q\theta)d\theta,$$

(4.33)

where $\{W^*(t)\}_{t\geq 0}$ is the adjoint semigroup of $\{W(t)\}_{t\geq 0}$.

Using Corollary 10.6 in [199] and the proof of Lemma 2.11 in [251], one has the following results.

Lemma 4.15. *The following properties hold:*

(i) *for any fixed* $t \geq 0$, $\mathcal{T}^*_{(A^*,E^*)}(t)$ *and* $\mathcal{S}^*_{(A^*,E^*)}(t)$ *are linear bounded operators on* X *with*

$$\|\mathcal{T}^*_{(A^*,E^*)}(t)\|_{\mathscr{L}(X)} \leq M_1 \ and \ \|\mathcal{S}^*_{(A^*,E^*)}(t)\|_{\mathscr{L}(X)} \leq \frac{M_1}{\Gamma(q)};$$

(ii) *if* $W(t)$ *is compact, then* $\mathcal{T}^*_{(A^*,E^*)}(t)$ *and* $\mathcal{S}^*_{(A^*,E^*)}(t)$ *are compact in* X *for* $t > 0$.

Using Lemma 10.1 in [199], we have:

Lemma 4.16. E^{*-1} *and* B^* *are bounded operators with* $\|E^{-1}\|_{\mathscr{L}(X)} = \|E^{*-1}\|_{\mathscr{L}(X)}$, $\|B\|_{\mathscr{L}(U,X)} = \|B^*\|_{\mathscr{L}(X,U)}$.

Assume that there exists a continuous linear operator Θ on X given by

$$\Theta = \left(I + \sum_{k=1}^m a_k \mathcal{T}_{(A,E)}(t_k)\right)^{-1},$$

where I is the identity operator.

Remark 4.5. *One can give a sufficient condition to guarantee the existence of* Θ. *For example, assuming* $M_1 \sum_{k=1}^m |a_k| < 1$. *Indeed, applying Neumann lemma, we get*

$$\|\Theta\|_{\mathscr{L}(X)} \leq \frac{1}{1 - M_1 \sum_{k=1}^m |a_k|}.$$

Now we introduce Green function:

$$
\begin{aligned}
& G_{(A,E)}(t,s) \\
& = E^{-1} G^0_{(A,E)}(t,s) \\
& = E^{-1}\left(-\sum_{k=1}^{m} \mathscr{T}_{(A,E)}(t)\chi_k(s)\Theta(t_k - s)^{q-1}\mathscr{S}_{(A,E)}(t_k - s) \right. \qquad (4.34) \\
& \qquad \left. + \chi_t(s)(t - s)^{q-1}\mathscr{S}_{(A,E)}(t - s) \right), \quad \text{for } t, s \in J,
\end{aligned}
$$

where

$$
\chi_k(s) = \begin{cases} a_k, & \text{for } s \in [0, t_k), \\ 0, & \text{for } s \in [t_k, a], \end{cases} \qquad \chi_t(s) = \begin{cases} 1, & \text{for } s \in [0, t), \\ 0, & \text{for } s \in [t, a]. \end{cases}
$$

Hence, we have that $\chi_k(s)(t_k - s)^{q-1} = 0$ for $s \in [t_k, a]$ and $\chi_t(s)(t - s)^{q-1} = 0$ for $s \in [t, a]$.

Now, we introduce the following definition of a suitable mild solution.

Definition 4.8. *For each $u \in \mathcal{U}$, by a mild solution of system (4.31) we mean a function $x \in C(J, X)$ satisfying*

$$
x(t) = \int_0^a G_{(A,E)}(t,s)\big(f(s, x(s)) + Bu(s)\big)ds, \ t \in J.
$$

Remark 4.6. *To explain the above formula, like Lemma 3.1 in [93], one can integrate the first equation of system (4.31) via Laplace transform to derive*

$$
\begin{aligned}
Ex(t) = & \mathscr{T}_{(A,E)}(t)Ex(0) \\
& + \int_0^t (t - s)^{q-1}\mathscr{S}_{(A,E)}(t - s)\big(f(s, x(s)) + Bu(s)\big)ds,
\end{aligned}
$$

which implies that

$$
\begin{aligned}
x(t) = & E^{-1}\mathscr{T}_{(A,E)}(t)Ex(0) \\
& + \int_0^t (t - s)^{q-1}E^{-1}\mathscr{S}_{(A,E)}(t - s)\big(f(s, x(s)) + Bu(s)\big)ds.
\end{aligned}
$$

Now using the nonlocal initial condition in system (4.31) one can solve

$$Ex(0) = -\sum_{k=1}^{m} a_k \Theta \int_0^{t_k} (t_k - s)^{q-1}$$
$$\times \mathscr{S}_{(A,E)}(t_k - s)\big(f(s, x(s)) + Bu(s)\big)ds,$$

which leads to the desired formula of mild solution.

4.4.2 Linear Systems

Consider the following linear system

$$\begin{cases} {}_0^C D_t^q (Ex(t)) = Ax(t) + Bu(t), \quad t \in J, \\ x(0) + \displaystyle\sum_{k=1}^{m} a_k x(t_k) = 0. \end{cases} \quad (4.35)$$

Using the mild solution of (4.35), we get

$$x(a) = \int_0^a G_{(A,E)}(a, s) Bu(s) ds.$$

Define a linear operator $P : \mathcal{U} \to X$ by

$$Pu = x(a) = \int_0^a G_{(A,E)}(a, s) Bu(s) ds.$$

For convenience, we set

$$\ell_1 = \frac{M_1 \|E^{-1}\|_{\mathscr{L}(X)}}{\Gamma(q)}, \quad \ell_2 = M_1 \|\Theta\|_{\mathscr{L}(X)},$$
$$\ell_3 = \|B\|_{\mathscr{L}(U,X)}^2 \ell_1 M_{\Upsilon,k}.$$

By Lemmas 4.14 and 4.15, we obtain

$$|Pu| \leq \int_0^a \|G_{(A,E)}(a, s)\|_{\mathscr{L}(X)} \|B\|_{\mathscr{L}(U,X)} |u(s)|_U ds$$
$$\leq \ell_1 \|B\|_{\mathscr{L}(U,X)} \int_0^a \Big(\ell_2 \sum_{k=1}^{m} \chi_k(s)(t_k - s)^{q-1} \quad (4.36)$$
$$+ \chi_a(s)(a - s)^{q-1}\Big) |u(s)|_U ds$$

$$
\begin{aligned}
=&\ell_1\|B\|_{\mathscr{L}(U,X)}\left(\ell_2\sum_{k=1}^{m}|a_k|\int_0^{t_k}(t_k-s)^{q-1}|u(s)|_U ds\right.\\
&+\left.\int_0^a (a-s)^{q-1}|u(s)|_U ds\right)\\
\leq&\ell_1\|B\|_{\mathscr{L}(U,X)}\left(\ell_2\sum_{k=1}^{m}|a_k|+1\right)\left(\frac{p-1}{qp-1}a^{\frac{qp-1}{p-1}}\right)^{\frac{p-1}{p}}\|u\|_{L^p J}\\
=&M_P\|u\|_{L^p J}.
\end{aligned}
$$

Thus, P is bounded.

Furthermore, (4.35) is approximately controllable if and only if $\overline{P(\mathcal{U})}=X$. This is equivalent to $\ker P^*=\{0\}$. Note that U is a Hilbert space, so then $L^p(J,U)^*=L^{p^*}(J,U^*)=L^{p^*}(J,U)$ for $\frac{1}{p}+\frac{1}{p^*}=1$. Next, we compute P^* as follows. Let $x^*\in X$, then

$$
\begin{aligned}
\langle x^*, x(a)\rangle =&\left\langle x^*, \int_0^a G_{(A,E)}(a,s)Bu(s)ds\right\rangle\\
=&\int_0^a\left\langle B^*G^*_{(A,E)}(a,s)x^*, u(s)\right\rangle ds.
\end{aligned}
$$

Hence, we derive

$$
(P^*x^*)(s)=B^*G^*_{(A,E)}(a,s)x^*,\ s\in J,\ x^*\in X,
$$

where

$$
\begin{aligned}
&G^*_{(A,E)}(a,s)\\
=&\left(-\sum_{k=1}^{m}\chi_k(s)(t_k-s)^{q-1}\mathscr{S}^*_{(A^*,E^*)}(t_k-s)\Theta^*\mathscr{T}^*_{(A^*,E^*)}(a)\right.\\
&\left.+\chi_a(s)(a-s)^{q-1}\mathscr{S}^*_{(A^*,E^*)}(a-s)\right)E^{*-1}.
\end{aligned}
$$

Note $P^*: X\to\mathcal{U}^*$. We need $\mathcal{U}^*=L^{p^*}(J,U)\subset L^p(J,U)=\mathcal{U}$, if we want to compose P and P^*. This is satisfied, when $p\leq p^*=\frac{p}{p-1}$, $1<p\leq 2$. Recall $q\in(\frac{1}{p},1)$ which gives a restriction $\frac{1}{2}<q<1$. Now we can define Gramian controllability operator

$$
\Gamma_0^a=PP^*=\int_0^a G_{(A,E)}(a,s)BB^*G^*_{(A,E)}(a,s)ds.
$$

Noting Lemma 4.16, it is straightforward that Γ_0^a is a linear bounded operator. In fact, it follows from (4.36) that

$$\|\Gamma_0^a\|_{\mathscr{L}(X)} \leq \|B\|_{\mathscr{L}(\mathcal{U},X)}\|P^*\|_{\mathscr{L}(X,\mathcal{U}^*)} \leq M_P^2.$$

Now we recall the following result.

Theorem 4.10. *[170] Assume that $\Gamma : X \to X$ is symmetric. Then the following two conditions are equivalent:*

(i) *$\Gamma : X \to X$ is positive, that is, $\langle x, \Gamma x \rangle > 0$ for all nonzero $x \in X$;*
(ii) *for all $\eta \in X$, $x_\varepsilon(\eta) = \varepsilon(\varepsilon I + \Gamma)^{-1}(\eta)$ strongly converges to zero as $\varepsilon \to 0+$.*

We apply Theorem 4.10 with Γ_0^a. Then for any $x^* \in X$, we have

$$
\begin{aligned}
\langle x^*, \Gamma_0^a x^* \rangle &= \left\langle x^*, \int_0^a G_{(A,E)}(a,s)BB^*G_{(A,E)}^*(a,s)ds\, x^* \right\rangle \\
&= \int_0^a \left| B^*G_{(A,E)}^*(a,s)x^* \right|^2 ds \\
&= \int_0^a |(P^*x^*)(s)|^2 ds.
\end{aligned}
$$

Note $P^* : X \to \mathcal{U}^* = L^{p^*}(J,U) \subset L^p(J,U) = \mathcal{U} \subset L^2(J,U)$, since $1 < p \leq 2$. So the above last integral is well defined. We also get that $\langle x^*, \Gamma_0^a x^* \rangle > 0$ if and only if $P^*x^* \neq 0$, i.e., $x^* \notin \ker P^*$. Consequently, Γ_0^a is positive if and only if $\ker P^* = \{0\}$, i.e., Γ_0^a is positive if and only if the linear system (4.35) is approximately controllable on J. Setting

$$R(\varepsilon; \Gamma_0^a) = (\varepsilon I + \Gamma_0^a)^{-1} : X \to X, \ \varepsilon > 0,$$

by Theorem 4.10, we arrive at the following result (see also [170]).

Theorem 4.11. *Let $\frac{1}{2} < q < 1$. The linear system (4.35) is approximately controllable on J if and only if $\varepsilon R(\varepsilon; \Gamma_0^a) \to 0$ as $\varepsilon \to 0+$ in the strong topology.*

Finally, we note that $R(\varepsilon; \Gamma_0^a)$ is continuous with $\|R(\varepsilon; \Gamma_0^a)\|_{\mathscr{L}(X)} \leq \frac{1}{\varepsilon}$.

4.4.3 Approximate Controllability

In this subsection, we study the approximate controllability of system (4.31) by imposing that the corresponding linear system is approximately controllable and using Schauder fixed point theorem.

Definition 4.9. *Let $x(a; x(0), u)$ be the state value of system (4.31) at terminal time a corresponding to the control $u \in \mathcal{U}$ and nonlocal initial condition $x(0)$. System (4.31) is said to be approximately controllable on the interval J if the closure $\overline{\mathfrak{R}(a, x(0))} = X$. Here, $\mathfrak{R}(a, x(0)) = \{x(a; x(0), u) : u \in \mathcal{U}\}$ is called the reachability set of system (4.31) at terminal time a.*

In the sequel, we introduce the following assumptions:

(H_1) (S_1), (S_2), and (S_3) hold;
(H_2) $f : J \times X \to X$ is continuous such that

$$g_k = \sup_{t \in J, |x| \leq k} |f(t, x)| < \infty \text{ with } \liminf_{k \to \infty} \frac{g_k}{k} = 0;$$

(H_3) system (4.35) is approximately controllable on J.

Recalling condition (H_3) and Theorem 4.11, for any $x \in C(J, X)$ and $h \in X$, we define the following control formula:

$$u_\varepsilon(t, x) = B^* G^*_{(A,E)}(a, t) R(\varepsilon; \Gamma_0^a) \Upsilon(x) \tag{4.37}$$

with

$$\Upsilon(x) = h - \int_0^a G_{(A,E)}(a, s) f(s, x(s)) ds.$$

For each $k > 0$, define

$$\mathcal{B}_k = \{x \in C(J, X) : \|x\| \leq k\}.$$

Of course, $\mathcal{B}_k$ is a bounded, closed, convex subset in $C(J, X)$, which is Banach space with the norm $\| \cdot \|$. Using the above control u in (4.37), we consider an operator $\mathcal{P} : \mathcal{B}_k \to C(J, X)$ given by

$$(\mathcal{P}_\varepsilon x)(t) = \int_0^a G_{(A,E)}(t, s)\big(f(s, x(s)) + B u_\varepsilon(s, x)\big) ds, \text{ for } t \in J. \tag{4.38}$$

Now we present the following important result.

Theorem 4.12. *Let $\frac{1}{2} < q < 1$. Under the assumptions (H_1)-(H_3), for any $\varepsilon > 0$, there exists a $k(\varepsilon) > 0$ such that P_ε has a fixed point in $\mathcal{B}_{k(\varepsilon)}$.*

Proof. We divide the proof into four steps.

Claim 1. For an arbitrary $\varepsilon > 0$, there is a $k = k(\varepsilon) > 0$ such that $\mathcal{P}_\varepsilon(\mathcal{B}_k) \subset \mathcal{B}_k$.

If this was not the case, then for each $k > 0$, there would exist $x \in \mathcal{B}_k$ and $\bar{t}_k \in J$ such that $|(\mathcal{P}_\varepsilon x)(\bar{t}_k)| > k$. Using

$$\|G^*_{(A,E)}(a,t)\|_{\mathscr{L}(X)} = \|G_{(A,E)}(a,t)\|_{\mathscr{L}(X)}$$
$$\leq \ell_1 \left(\ell_2 \sum_{k=1}^{m} \chi_k(t)(t_k - t)^{q-1} + \chi_a(t)(a-t)^{q-1} \right),$$

we derive

$$|\Upsilon(x)| \leq |h| + \frac{\ell_1 a^q}{q} \left(\ell_2 \sum_{k=1}^{m} |a_k| + 1 \right) g_k$$
$$= |h| + M_G g_k$$
$$=: M_{\Upsilon,k},$$

where

$$M_G = \frac{\ell_1 a^q}{q} \left(\ell_2 \sum_{k=1}^{m} |a_k| + 1 \right),$$

which implies

$$|u_\varepsilon(s,x)|_U$$
$$\leq \|B\|_{\mathscr{L}(U,X)} \|G_{(A,E)}(a,s)\|_{\mathscr{L}(X)} \|R(\varepsilon;\Gamma_0^a)\|_{\mathscr{L}(X)} |\Upsilon(x)|$$
$$\leq \frac{\|B\|_{\mathscr{L}(U,X)} \ell_1 M_{\Upsilon,k}}{\varepsilon} \left(\ell_2 \sum_{k=1}^{m} \chi_k(s)(t_k - s)^{q-1} + \chi_a(s)(a-s)^{q-1} \right).$$

Notice that

$$\left(\sum_{i=1}^{n} c_i \right)^2 \leq n \sum_{i=1}^{n} c_i^2 \text{ for } c_i > 0,$$

then we obtain

$$
\begin{aligned}
k <&|(\mathcal{P}_\varepsilon x)(\bar{t}_k)| \\
\le& \int_0^a \|G_{(A,E)}(\bar{t}_k, s)\|_{\mathscr{L}(X)} \left(g_k + \|B\|_{\mathscr{L}(U,X)} |u_\varepsilon(s, x)|\right) ds \\
\le& g_k I_1 + \frac{\ell_1 \ell_3}{\varepsilon} I_2,
\end{aligned}
$$

where

$$
I_1 := \ell_1 \int_0^a \left(\ell_2 \sum_{j=1}^m \chi_j(s)(t_j - s)^{q-1} + \chi_{\bar{t}_k}(s)(\bar{t}_k - s)^{q-1}\right) ds,
$$

$$
\begin{aligned}
I_2 :=& \int_0^a \left(\ell_2 \sum_{j=1}^m \chi_j(s)(t_j - s)^{q-1} + \chi_{\bar{t}_k}(s)(\bar{t}_k - s)^{q-1}\right) \\
&\times \left(\ell_2 \sum_{j=1}^m \chi_j(s)(t_j - s)^{q-1} + \chi_a(s)(a - s)^{q-1}\right) ds.
\end{aligned}
$$

Since $I_1 \le M_G$, and

$$
\begin{aligned}
I_2 \le& \int_0^a \left(\ell_2 \sum_{j=1}^m \chi_j(s)(t_j - s)^{q-1} \right. \\
& \left. + \chi_{\bar{t}_k}(s)(\bar{t}_k - s)^{q-1} + \chi_a(s)(a - s)^{q-1}\right)^2 ds \\
\le& (m + 2) \int_0^a \left(\ell_2^2 \sum_{j=1}^m \chi_j^2(s)(t_j - s)^{2(q-1)} \right. \\
& \left. + \chi_{\bar{t}_k}(s)(\bar{t}_k - s)^{2(q-1)} + \chi_a(s)(a - s)^{2(q-1)}\right) ds \\
\le& \frac{(m + 2)a^{2q-1}}{2q - 1} \left(\ell_2^2 \sum_{j=1}^m a_j^2 + 2\right),
\end{aligned}
$$

then

$$
\begin{aligned}
k \le& M_G g_k + \frac{(m + 2)\ell_1 \ell_3 a^{2q-1}}{\varepsilon(2q - 1)} \left(\ell_2^2 \sum_{j=1}^m a_j^2 + 2\right) \\
\le& M_G g_k \left(1 + \frac{M_G'}{\varepsilon}\right) + \frac{M_G'}{\varepsilon} |h|,
\end{aligned}
$$

where

$$M'_G = \frac{(m+2)\|B\|^2_{\mathscr{L}(U,X)}\ell_1^2 a^{2q-1}}{2q-1}\left(\ell_2^2 \sum_{j=1}^{m} a_j^2 + 2\right).$$

Dividing both sides by k and taking the lower limit as $k \to \infty$, we derive a contradiction $1 \le 0$.

Claim 2. $\mathcal{P}_\varepsilon : C(J,X) \to C(J,X)$ is continuous.

Let $\{x^m\}_{m \in \mathbb{N}} \subseteq C(J,X)$ be a sequence such that $x^m \to x$ as $m \to \infty$. Then $\|f(\cdot,x^m) - f(\cdot,x)\| \to 0$ as $m \to \infty$. Next, following the above estimations, we first obtain

$$|u_\varepsilon^m(s,x^m) - u_\varepsilon(s,x)|_U$$
$$\le \frac{M_G}{\varepsilon}\|B\|_{\mathscr{L}(U,X)}\|G_{(A,E)}(a,s)\|_{\mathscr{L}(X)}\|f(\cdot,x^m) - f(\cdot,x)\|,$$

and then we get

$$|(\mathcal{P}_\varepsilon x^m)(t) - (\mathcal{P}_\varepsilon x)(t)|$$
$$\le \int_0^a \|G_{(A,E)}(t,s)\|_{\mathscr{L}(X)}\left(\|f(\cdot,x^m) - f(\cdot,x)\|\right.$$
$$\left. + \|B\|_{\mathscr{L}(U,X)}|u_\varepsilon^m(s,x^m) - u_\varepsilon(s,x)|_U\right)ds$$
$$\le \int_0^a \|G_{(A,E)}(t,s)\|_{\mathscr{L}(X)}\left(\|f(\cdot,x^m) - f(\cdot,x)\|\right.$$
$$\left. + \frac{M_G}{\varepsilon}\|B\|^2_{\mathscr{L}(U,X)}\|G_{(A,E)}(a,s)\|_{\mathscr{L}(X)}\|f(\cdot,x^m) - f(\cdot,x)\|\right)ds$$
$$\le M_G\left(1 + \frac{M'_G}{\varepsilon}\right)\|f(\cdot,x^m) - f(\cdot,x)\|$$
$$\to 0, \quad \text{as } m \to \infty,$$

for any $t \in J$. This yields that $\mathcal{P}_\varepsilon$ is continuous.

Claim 3. For every fixed $t \in J$, the set $\Pi(t) = \{(\mathcal{P}_\varepsilon x)(t) : x \in \mathcal{B}_k\}$ is relatively compact in X.

Note that

$$(\mathcal{P}_\varepsilon x)(t) = E^{-1}(\mathcal{P}_\varepsilon^0 x)(t), \text{ for } t \in J,$$

where

$$(\mathcal{P}_{\varepsilon}^0 x)(t) = \int_0^a G_{(A,E)}^0(t,s)\big(f(s,x(s)) + Bu_{\varepsilon}(x,s)\big)ds, \text{ for } t \in J.$$

Next, by Claim 1 for $x \in \mathcal{B}_k$, we derive

$$(\mathcal{P}_{\varepsilon}^0 x)(t) \leq M_G \|E^{-1}\|_{\mathscr{L}(X)}^{-1} \left[\left(1 + \frac{M_G'}{\varepsilon}\right)g_k + \frac{M_G'}{\varepsilon}|h|\right].$$

Thus, $\{(\mathcal{P}_{\varepsilon}^0 x)(t) : x \in \mathcal{B}_k\}$ is bounded in X. By (S_3) we know that $\Pi(t) = \{(\mathcal{P}_{\varepsilon}x)(t) : x \in \mathcal{B}_k\}$ is relatively compact in X.

   Claim 4. $\{(\mathcal{P}_{\varepsilon}x) : x \in B_k\}$ is an equicontinuous family of functions on J.

   Let $x \in \mathcal{B}_k$ and $t', t'' \in J$ be such that $t' < t''$. Note that

$$|(\mathcal{P}_{\varepsilon}x)(t'') - (\mathcal{P}_{\varepsilon}x)(t')|$$
$$\leq \int_0^a \|G_{(A,E)}(t'',s) - G_{(A,E)}(t',s)\|_{\mathscr{L}(X)}|f(s,x(s)) + Bu_{\varepsilon}(s,x)|ds,$$

and

$$G_{(A,E)}(t'',s) - G_{(A,E)}(t',s)$$
$$= E^{-1}\bigg(-\sum_{k=1}^m [\mathscr{T}_{(A,E)}(t'') - \mathscr{T}_{(A,E)}(t')]\chi_k(s)$$
$$\times \Theta(t_k - s)^{q-1}\mathscr{S}_{(A,E)}(t_k - s)\bigg)$$
$$+ E^{-1}\Big(\chi_{t'}(s)(t'-s)^{q-1}\big(\mathscr{S}_{(A,E)}(t''-s) - \mathscr{S}_{(A,E)}(t'-s)\big)\Big)$$
$$+ E^{-1}\Big(\big(\chi_{t''}(s)(t''-s)^{q-1} - \chi_{t'}(s)(t'-s)^{q-1}\big)\mathscr{S}_{(A,E)}(t''-s)\Big).$$

Then,

$$|(\mathcal{P}_{\varepsilon}x)(t'') - (\mathcal{P}_{\varepsilon}x)(t')| \leq \sum_{i=1}^3 K_i,$$

where

$$K_1 := \int_0^a \left\|E^{-1}\bigg(\sum_{k=1}^m [\mathscr{T}_{(A,E)}(t'') - \mathscr{T}_{(A,E)}(t')]\chi_k(s)\right.$$

$$\times\,\Theta(t_k - s)^{q-1}\mathscr{S}_{(A,E)}(t_k - s)\Big)\bigg\|_{\mathscr{L}(X)} |f(s,x(s)) + Bu_\varepsilon(s,x)|ds,$$

$$K_2 := \int_0^a \left\| E^{-1}\Big(\chi_{t'}(s)(t'-s)^{q-1}\big(\mathscr{S}_{(A,E)}(t''-s) - \mathscr{S}_{(A,E)}(t'-s)\big)\Big)\right\|$$
$$\times\,|f(s,x(s)) + Bu_\varepsilon(s,x)|ds,$$

$$K_3 := \int_0^a \left\| E^{-1}\Big((\chi_{t''}(s)(t''-s)^{q-1} - \chi_{t'}(s)(t'-s)^{q-1})\right.$$
$$\left.\times\,\mathscr{S}_{(A,E)}(t''-s)\Big)\right\|_{\mathscr{L}(X)} |f(s,x(s)) + Bu_\varepsilon(s,x)|ds.$$

Since

$$K_1 \le \frac{1}{M_1\Gamma(q)}\left(\|E^{-1}\|_{\mathscr{L}(X)}\|\mathscr{T}_{(A,E)}(t'') - \mathscr{T}_{(A,E)}(t')\|_{\mathscr{L}(X)}\right)$$
$$\times \int_0^a \sum_{j=1}^m \ell_2 \chi_j(s)(t_j - s)^{q-1}$$
$$\times \left[g_k + \frac{\ell_3}{\varepsilon}\left(\ell_2 \sum_{j=1}^m \chi_j(s)(t_j - s)^{q-1} + \chi_a(s)(a-s)^{q-1}\right)\right]ds$$
$$\le \frac{1}{M_1}\left(M_G g_k + \frac{M_G'}{\varepsilon}M_{\Upsilon,k}\right)\left(\|\mathscr{T}_{(A,E)}(t'') - \mathscr{T}_{(A,E)}(t')\|_{\mathscr{L}(X)}\right),$$

then

$$K_2 \le \max_{s\in[0,t']}\|\mathscr{S}_{(A,E)}(t''-s) - \mathscr{S}_{(A,E)}(t'-s)\|_{\mathscr{L}(X)}$$
$$\times \|E^{-1}\|_{\mathscr{L}(X)}\int_0^a \chi_{t'}(s)(t'-s)^{q-1}$$
$$\times \left[g_k + \frac{\ell_3}{\varepsilon}\left(\ell_2 \sum_{j=1}^m \chi_j(s)(t_j - s)^{q-1} + \chi_a(s)(a-s)^{q-1}\right)\right]ds$$
$$\le \max_{s\in[0,t']}\|\mathscr{S}_{(A,E)}(t''-s) - \mathscr{S}_{(A,E)}(t'-s)\|_{\mathscr{L}(X)}$$
$$\times \|E^{-1}\|_{\mathscr{L}(X)}\int_0^a \left[\chi_{t'}(s)(t'-s)^{q-1} + g_k\right.$$
$$\left. + \frac{\ell_3}{\varepsilon}\left(\ell_2 \sum_{j=1}^m \chi_j(s)(t_j - s)^{q-1} + \chi_a(s)(a-s)^{q-1}\right)\right]^2 ds$$

$$\leq \max_{s \in [0,t']} \|\mathscr{S}_{(A,E)}(t''-s) - \mathscr{S}_{(A,E)}(t'-s)\|_{\mathscr{L}(X)} \|E^{-1}\|_{\mathscr{L}(X)}$$

$$\times (m+3) \int_0^a \left[\chi_{t'}(s)(t'-s)^{2(q-1)} + g_k^2 \right.$$

$$\left. + \left(\frac{\ell_3}{\varepsilon}\right)^2 \left(\ell_2^2 \sum_{j=1}^m \chi_j^2(s)(t_j-s)^{2(q-1)} + \chi_a(s)(a-s)^{2(q-1)} \right) \right] ds$$

$$= \max_{s \in [0,t']} \|\mathscr{S}_{(A,E)}(t''-s) - \mathscr{S}_{(A,E)}(t'-s)\|_{\mathscr{L}(X)} \|E^{-1}\|_{\mathscr{L}(X)}$$

$$\times \frac{(m+3)a^{2q-1}}{2q-1} \left[1 + g_k^2 + \left(\frac{\ell_3}{\varepsilon}\right)^2 \left(\ell_2^2 \sum_{j=1}^m a_j^2 + 1 \right) \right],$$

and by Hölder inequality, $(c-d)^2 \leq |c-d|(c+d)$, $c,d \geq 0$, and

$$\int_0^{t''} \left| (t''-s)^{2(q-1)} - \chi_{t'}(s)(t'-s)^{2(q-1)} \right| ds$$

$$= \int_0^{t'} \left((t'-s)^{2(q-1)} - (t''-s)^{2(q-1)} \right) ds + \int_{t'}^{t''} (t''-s)^{2(q-1)} ds$$

$$= \frac{t'^{2q-1} - t''^{2q-1} + 2(t''-t')^{2q-1}}{2q-1}$$

$$\leq \frac{2(t''-t')^{2q-1}}{2q-1},$$

we have

$$K_3 \leq \ell_1 \int_0^a \left| \chi_{t''}(s)(t''-s)^{q-1} - \chi_{t'}(s)(t'-s)^{q-1} \right|$$

$$\times \left[g_k + \frac{\ell_3}{\varepsilon} \left(\ell_2 \sum_{j=1}^m \chi_j(s)(t_j-s)^{q-1} + \chi_a(s)(a-s)^{q-1} \right) \right] ds$$

$$\leq \ell_1 \left(\int_0^a \left(\chi_{t''}(s)(t''-s)^{q-1} - \chi_{t'}(s)(t'-s)^{q-1} \right)^2 ds \right)^{\frac{1}{2}}$$

$$\times \left(\int_0^a \left[g_k + \frac{\ell_3}{\varepsilon} \left(\ell_2 \sum_{j=1}^m \chi_j(s)(t_j-s)^{q-1} \right. \right. \right.$$

$$\left. \left. \left. + \chi_a(s)(a-s)^{q-1} \right) \right]^2 ds \right)^{\frac{1}{2}}$$

$$\leq \ell_1 \left(\int_0^{t''} \left| (t''-s)^{2(q-1)} - \chi_{t'}(s)(t'-s)^{2(q-1)} \right| ds \right)^{\frac{1}{2}}$$

$$\times \sqrt{m+2} \left(\int_0^a \left[g_k^2 + \left(\frac{\ell_3}{\varepsilon} \right)^2 \left(\ell_2^2 \sum_{j=1}^m \chi_j^2(s)(t_j - s)^{2(q-1)} \right. \right. \right.$$

$$\left. \left. \left. + \chi_a(s)(a-s)^{2(q-1)} \right) \right] ds \right)^{\frac{1}{2}}$$

$$\leq \ell_1 \left(\frac{2(m+2)(t'' - t')^{2q-1}}{2q-1} \right)^{\frac{1}{2}} \left[g_k^2 a + \frac{\ell_3^2 a^{2q-1}}{\varepsilon^2(2q-1)} \left(\ell_2^2 \sum_{j=1}^m a_j^2 + 1 \right) \right]^{\frac{1}{2}}.$$

By applying Lemma 4.14(iii), we obtain

$$\lim_{t'' \to t'+} \| \mathscr{T}_{(A,E)}(t'') - \mathscr{T}_{(A,E)}(t') \|_{\mathscr{L}(X)} = 0,$$

$$\lim_{t'' \to t'+} \max_{s \in [0,t']} \| \mathscr{S}_{(A,E)}(t'' - s) - \mathscr{S}_{(A,E)}(t' - s) \|_{\mathscr{L}(X)} = 0,$$

so terms K_1 and K_2 tend to zero as $t'' \to t'+$. It is elementary to see that the term K_3 tends to zero as $t'' \to t'+$, as well.

As a result, $\mathcal{P}_\varepsilon$ is compact due to Lemma 1.2. From above, $\mathcal{P}_\varepsilon$ is completely continuous operator for all $\varepsilon > 0$. By using Schauder fixed point theorem, $\mathcal{P}_\varepsilon$ has at least one fixed point which rises at least one mild solution of system (4.31). □

In the sequel, we need the following compactness result.

Lemma 4.17. *Assume that* (S_3) *holds. Then, for any* r *with* $rq > 1$, *the operator* $Q : L^r(J, X) \to X$ *given by*

$$Ql = \int_0^a G_{(A,E)}(a, s)l(s)ds$$

is compact.

Proof. We can write the linear operator Q as

$$Q = E^{-1}Q_0, \tag{4.39}$$

where

$$Q_0 l = \int_0^a G^0_{(A,E)}(a, s)l(s)ds.$$

Then following computations for (4.36), we derive

$$|Q_0 l| \leq \frac{M_1}{\Gamma(q)} \left(\ell_2 \sum_{k=1}^{m} |a_k| + 1 \right) \left(\frac{r-1}{qr-1} a^{\frac{qr-1}{r-1}} \right)^{\frac{r-1}{r}} \|l\|_{L^r J}.$$

So $Q_0 : L^r(J, X) \to X$ is continuous. Then (S_3) and (4.39) imply that Q is compact. The proof is completed.   □

Now we are ready to present the main result.

Theorem 4.13. *Let all the assumptions in Theorem 4.12 be satisfied. Moreover, there exists r with $rq > 1$ and $N \in L^r(J, \mathbb{R}^+)$ such that $|f(t,x)| \leq N(t)$ for all $(t,x) \in J \times X$. Then system (4.31) is approximately controllable on the interval J.*

Proof. By Theorem 4.12, there exists a fixed point x_ε of $\mathcal{P}_\varepsilon$ in $\mathcal{B}_{k(\varepsilon)}$, which is a mild solution of system (4.31) under the control $u_\varepsilon(t, x_\varepsilon)$ in (4.37) and satisfies

$$\begin{aligned}
x_\varepsilon(a) &= \int_0^a G_{(A,E)}(a,s)\big(f(s, x_\varepsilon(s)) + Bu_\varepsilon(s, x_\varepsilon)\big)ds \\
&= \int_0^a G_{(A,E)}(a,s)f(s, x_\varepsilon(s))ds \\
&\quad + \int_0^a G_{(A,E)}(a,s)Bu_\varepsilon(s, x_\varepsilon)ds \\
&= \int_0^a G_{(A,E)}(a,s)f(s, x_\varepsilon(s))ds \qquad (4.40) \\
&\quad + \int_0^a G_{(A,E)}(a,s)BB^*G^*_{(A,E)}(a,s)R(\varepsilon; \Gamma_0^a)\Upsilon(x_\varepsilon)ds \\
&= \int_0^a G_{(A,E)}(a,s)f(s, x_\varepsilon(s))ds + \Gamma_0^a R(\varepsilon; \Gamma_0^a)\Upsilon(x_\varepsilon)ds \\
&= h - \varepsilon R(\varepsilon; \Gamma_0^a)\Upsilon(x_\varepsilon),
\end{aligned}$$

where

$$\Upsilon(x_\varepsilon) = h - \int_0^a G_{(A,E)}(a,z)f(z, x_\varepsilon(z))dz.$$

Furthermore,

$$\int_0^a |f(s, x_\varepsilon(s))|^r ds \leq \int_0^a |N(s)|^r ds \leq \|N\|_{L^r J}^r.$$

From the reflexivity of $L^r(J, X)$, there exists a subsequence $\{f(t, x_{\varepsilon_i}(t))\}_{i=1}^\infty$, where $\varepsilon_i \to 0$ as $i \to \infty$, that converges weakly to $f \in L^r(J, X)$. Let

$$w = h - \int_0^a G_{(A,E)}(a, s) f(s) ds.$$

Since

$$|\Upsilon(x_{\varepsilon_i}) - w| = \left| \int_0^a G_{(A,E)}(a, s)(f(s, x_{\varepsilon_i}(s)) - f(s)) ds \right|, \qquad (4.41)$$

by Lemma 4.17 we find that the right-hand side of (4.41) tends to zero as $i \to \infty$. Thus, it follows from Theorem 4.11, (4.40), and (4.41) that

$$\begin{aligned}
|x_{\varepsilon_i}(a) - h| &\leq |\varepsilon_i R(\varepsilon_i; \Gamma_0^a)(w)| + |\varepsilon_i R(\varepsilon_i; \Gamma_0^a)(w)||\Upsilon(x_{\varepsilon_i}) - w| \\
&\leq |\varepsilon_i R(\varepsilon_i; \Gamma_0^a)(w)| + |w||\Upsilon(x_{\varepsilon_i}) - w| \\
&\to 0, \quad \text{as } i \to \infty.
\end{aligned}$$

This proves the approximate controllability of system (4.31). □

Remark 4.7. *By applying (ii) of Lemmas 4.14 and 4.15, Theorem 4.13 can be extended to the case when (S_3) is replaced by $(S_3)'$, and in addition, compactness of $\{W(t)\}_{t \geq 0}$ is assumed.*

4.4.4 Example

Consider the following fractional partial differential equation with control

$$\begin{cases}
{}_0^C D_t^{\frac{2}{3}} (x(t, y) - x_{yy}(t, y)) = x_{yy}(t, y) \\
\qquad + \mu \cos(2\pi t) \sin x(t, y) + u(t), & y \in [0, \pi],\ t \in J_1 = [0, 1], \\
x(t, 0) = x(t, \pi) = 0, & t \geq 0, \\
x(0, y) + \sum_{k=1}^m a_k x(t_k, y) = 0, & y \in [0, \pi],\ t_k \in J_1,
\end{cases}$$

$$(4.42)$$

where $0 < \mu < \infty$, $q = \frac{2}{3} \in (\frac{1}{2}, 1)$, $a_k \in \mathbb{R}$.

Take $X = U = L^2([0, \pi], \mathbb{R})$ and $p = 2$. Define

$$A : D(A) \subset X \to X, \ Ax = -x_{yy},$$
$$E : D(E) \subset X \to X, \ Ex = x - x_{yy},$$

where $D(A), D(E)$ are given by $\{x \in X : x, x_y \text{ are absolutely continuous,} \ x_{yy} \in X, \ x(0) = x(\pi) = 0\}$. Then A and E can be written as

$$Ax = -\sum_{n=1}^{\infty} n^2 \langle x, x_n \rangle, \ x \in D(A),$$

$$Ex = \sum_{n=1}^{\infty} (1 + n^2) \langle x, x_n \rangle x_n, \ x \in D(E),$$

respectively (see [158]), where $x_n(y) = \sqrt{\frac{2}{\pi}} \sin(ny), \ n = 1, 2, \dots$ is the orthonormal set of eigenfunctions of A. Moreover,

$$E^{-1}x = \sum_{n=1}^{\infty} (1 + n^2)^{-1} \langle x, x_n \rangle x_n, \ x \in D(E)$$

is compact since $\lim_{n \to \infty} \frac{1}{1+n^2} = 0$. Thus, E^{-1} is compact, and bounded. Hence, (H_1) is satisfied. Next, the bounded operator $-AE^{-1}$ generates a C_0- semigroup $\{W(t)\}_{t \geq 0}$ written as $W(t)x = \sum_{n=1}^{\infty} e^{-\frac{n^2}{1+n^2}t} \langle x, x_n \rangle x_n$, with $\|W(t)\|_{\mathscr{L}(X)} \leq e^{-t} \leq 1$.

Furthermore, $\mathscr{T}_{(A,E)}(\cdot)$ and $\mathscr{S}_{(A,E)}(\cdot)$ are now given by

$$\mathscr{T}_{(A,E)}(t)x = \int_0^{\infty} \Psi_{\frac{2}{3}}(\theta) \sum_{n=1}^{\infty} e^{-\frac{n^2}{1+n^2}t^{\frac{2}{3}}\theta} \langle x, x_n \rangle x_n d\theta$$

$$= \sum_{n=1}^{\infty} E_{\frac{2}{3}} \left(-\frac{n^2}{1+n^2} t^{\frac{2}{3}} \right) \langle x, x_n \rangle x_n,$$

$$\mathscr{S}_{(A,E)}(t)x = \frac{2}{3} \int_0^{\infty} \theta \Psi_{\frac{2}{3}}(\theta) \sum_{n=1}^{\infty} e^{-\frac{n^2}{1+n^2}t^{\frac{2}{3}}\theta} \langle x, x_n \rangle x_n d\theta$$

$$= \sum_{n=1}^{\infty} e_{\frac{2}{3}} \left(-\frac{n^2}{1+n^2} t^{\frac{2}{3}} \right) \langle x, x_n \rangle x_n,$$

where $E_{\frac{2}{3}}$ and $e_{\frac{2}{3}}$ are the classical Mittag-Leffler function and generalized Mittag-Leffler function [17, 134], respectively. Thus, $\|\mathscr{T}_{(A,E)}(t)\|_{\mathscr{L}(X)} \leq 1$

and $\|\mathscr{S}_{(A,E)}(t)\|_{\mathscr{L}(X)} \leq \frac{1}{\Gamma(\frac{2}{3})}$ for $t \geq 0$. Supposing $\sum_{k=1}^{m} |a_k| < 1$, by Remark 4.5, a linear operator $\Theta = \left(I + \sum_{k=1}^{m} a_k \mathscr{T}_{(A,E)}(t_k)\right)^{-1}$ exists.

Define an operator $f : J \times X \to X$ by $f(t, x)(y) = \mu \cos(2\pi t) \sin x(y)$. It is easy to verify (H_2). Next, $B : U \to X$ is defined by $B = I$. Now, system (4.42) can be abstracted as

$$\begin{cases} {}_0^C D_t^{\frac{2}{3}} (Ex(t)) = -Ax(t) + f(t, x(t)) + Bu(t), \quad t \in J_1, \\ x(0) + \sum_{k=1}^{m} a_k x(t_k) = 0. \end{cases}$$

Next, $\Gamma_0^1 : X \to X$ has the form

$$\Gamma_0^1 = \int_0^1 G_{(A,E)}(1, s) G_{(A,E)}^*(1, s) ds,$$

where by (4.34) we compute

$$\begin{aligned}
&\langle G_{(A,E)}(1, s)x, x_n \rangle \\
&= E^{-1} \Bigg(-\sum_{k=1}^{m} \mathscr{T}_{(A,E)}(1)\chi_k(s)\Theta(t_k - s)^{-\frac{1}{3}}\mathscr{S}_{(A,E)}(t_k - s) \\
&\quad + \chi_1(s)(1 - s)^{-\frac{1}{3}}\mathscr{S}_{(A,E)}(1 - s) \Bigg) \\
&= \frac{\langle x, x_n \rangle}{1 + n^2} \Bigg[-\sum_{k=1}^{m} \chi_k(s)\Theta(t_k - s)^{-\frac{1}{3}} E_{\frac{2}{3}}\left(-\frac{n^2}{1 + n^2} \right) \\
&\quad \times e_{\frac{2}{3}}\left(-\frac{n^2}{1 + n^2}(t_k - s)^{\frac{2}{3}} \right) \\
&\quad + \chi_1(s)(1 - s)^{-\frac{1}{3}} e_{\frac{2}{3}}\left(-\frac{n^2}{1 + n^2}(1 - s)^{\frac{2}{3}} \right) \Bigg]
\end{aligned} \tag{4.43}$$

for any $x \in X$ and $n \in \mathbb{N}$. To check that Γ_0^1 is positive, we consider the equation

$$\begin{aligned}
\langle \Gamma_0^1 x, x \rangle &= \left\langle \int_0^1 G_{(A,E)}(1, s) G_{(A,E)}^*(1, s) ds, x \right\rangle \\
&= \int_0^1 |G_{(A,E)}^*(1, s)x|^2 ds = 0,
\end{aligned}$$

which implies

$$G^*_{(A,E)}(1,s)x = 0, \ 0 \le s < 1.$$

Then (4.43) gives

$$0 = \left\langle G^*_{(A,E)}(1,s)x, x_n \right\rangle = \frac{\langle x, x_n \rangle}{1 + n^2}(1-s)^{-\frac{1}{3}}e_{\frac{2}{3}}\left(-\frac{n^2}{1+n^2}(1-s)^{\frac{2}{3}}\right)$$

for any $n \in \mathbb{N}$ and $t_m < s < 1$. If $x \neq 0$, then there is an $n_0 \in \mathbb{N}$ such that

$$e_{\frac{2}{3}}\left(-\frac{n_0^2}{1+n_0^2}(1-s)^{\frac{2}{3}}\right) = 0$$

for any $t_m < s < 1$, which is not possible, since $e_{\frac{2}{3}}(0) = \frac{1}{\Gamma(\frac{2}{3})}$. So we obtain $x = 0$ and Γ_0^1 is positive. Finally, we take $r = 2$, so $rq = \frac{4}{3} > 1$, $N(\cdot) = \mu\sqrt{2\pi} \in L^r(J_1, \mathbb{R}^+)$ and obtain

$$|f(t,x)| = \mu\left(\int_0^\pi \cos^2(2\pi t)\sin^2 x(y)dy\right)^{\frac{1}{2}} \le \mu\sqrt{2\pi} = N(t).$$

Summarily, all the assumptions in Theorem 4.13 are satisfied and thus system (4.42) is controllable on J_1.

4.5 TOPOLOGICAL STRUCTURE OF SOLUTION SETS

4.5.1 Introduction

In this section, we assume that X and V are Banach spaces. Consider the following control problem of semilinear fractional delay evolution equation:

$$\begin{cases} {}^C_0D^q_t x(t) = Ax(t) + f(t, x_t) + Bu(t), & t \in [0, b], \\ x(t) = \varphi(t), & t \in [-h, 0], \end{cases} \quad (4.44)$$

where A is a linear closed operator generating a C_0-semigroup $\{T(t)\}_{t \ge 0}$ on X, $h \ge 0$, ${}^C_0D^q_t$ is Caputo fractional derivative of order $q \in (0, 1)$, the state function x takes values in X, the control function u takes values in V, B is a bounded linear operator from V to X, $\varphi \in C([-h, 0], X)$, $x_t \in C([-h, 0], X)$ is defined by $x_t(s) = x(t + s)$ ($s \in [-h, 0]$), and $f : [0, b] \times C([-h, 0], X) \to X$ is, in general, a nonlinear function to be specified later.

In Subsection 4.5.2 we present some preliminaries. Subsection 4.5.3 is devoted to the study of the compactness and R_δ-property of solution set and then the invariance of reachability set under nonlinear perturbations for control system (4.44). Finally, as a sample of application, we present an example to illustrate the feasibility of our results.

4.5.2 Preliminaries

Let $C([a, b], X)$ stand for the Banach space of all continuous functions from $[a, b]$ to X equipped with the sup-norm. Denote by $\| \cdot \|_{-h,0}$ the sup-norm of $C([-h, 0], X)$.

Throughout this section, we assume that A is a linear closed operator generating a uniformly bounded C_0- semigroup $\{T(t)\}_{t \geq 0}$ on X. Write

$$M = \sup_{t \geq 0} \|T(t)\|_{\mathscr{L}(X)}.$$

Let $0 < q < 1$ and let us define two families $\{S_q(t)\}_{t \geq 0}$ and $\{P_q(t)\}_{t \geq 0}$ of linear operators by

$$S_q(t)\omega = \int_0^\infty \Psi_q(s)T(st^q)\omega ds,$$

$$P_q(t)\omega = \int_0^\infty qs\Psi_q(s)T(st^q)\omega ds, \ t \geq 0, \ \omega \in X,$$

where $\Psi_q(s)$ is the Wright function (see Definition 1.8).

Lemma 4.18. *Let $p > 1$ and $pq > 1$. Assume that for $t > 0$, $P_q(t)$ is continuous in the uniform operator topology. Define the operator Φ : $L^p([0, b], X) \to C([0, b], X)$ by*

$$(\Phi g)(t) = \int_0^t (t - s)^{q-1} P_q(t - s)g(s)ds, \quad g \in L^p([0, b], X).$$

Then Φ sends each bounded set to equicontinuous one.

Proof. Since $P_q(t)$ is continuous for $t > 0$ in the uniform operator topology, the lemma can be proved in a standard argument (see [278]). □

We adopt the following definition of mild solutions to control system (4.44).

Definition 4.10. *By a mild solution of control system (4.44), we mean a function $x \in C([-h, b], X)$ satisfying*

$$x(t) = \begin{cases} S_q(t)\varphi(0) + \int_0^t (t-s)^{q-1} \\ \qquad \times P_q(t-s)\big(f(s, x_s) + Bu(s)\big)\mathrm{d}s, & t \in [0, b], \\ \varphi(t), & t \in [-h, 0]. \end{cases}$$

Denote by $x(\cdot, u, f)$ the mild solution of control system (4.44). The set

$$\mathcal{K}_{b,f} = \{x(b, u, f) : u \in L^p([0, b], V)\}$$

is called the reachability set of control system (4.44).

Definition 4.11. *Control system (4.44) is said to be approximately controllable on $[0, b]$ if $\overline{\mathcal{K}_{b,f}} = X$, where $\overline{\mathcal{K}_{b,f}}$ denotes the closure of $\mathcal{K}_{b,f}$.*

In order to derive a priori bounds of mild solutions for the control system (4.44), we also need the following delay Gronwall inequality with singularity.

Lemma 4.19. *Suppose that $x \in C([-h, b], X)$ satisfies the following system:*

$$\begin{cases} |x(t)| \leq a_1 + a_2 \int_0^t (t-s)^{q-1} \|x_s(\cdot)\|_{-h,0}\mathrm{d}s, & t \in [0, b], \\ x(t) = \varphi(t), & t \in [-h, 0], \end{cases}$$

where $\varphi \in C([-h, 0], X)$, $0 < q < 1$ and constants $a_1, a_2 \geq 0$. Then exists a constant $N > 0$ (independent of a_1 and φ) such that

$$|x(t)| \leq N(a_1 + \|\varphi\|_{-h,0}), \qquad t \in [-h, b].$$

Let $p > 1$ be given. The following is our standing assumptions on f:

(H_1) $f : [0, b] \times C([-h, 0], X) \to X$ is continuous;
(H_2) there exists $\eta \in C([0, b], \mathbb{R}^+)$ such that

$$|f(t, v)| \leq \eta(t)(1 + \|v\|_{-h,0})$$

for all $t \in [0, b]$ and $v \in C([-h, 0], X)$;

(H_3) there exists $k \in L^p([0, b], \mathbb{R}^+)$ such that

$$\alpha(f(t, \Omega)) \leq k(t) \sup_{s \in [-h, 0]} \alpha(\Omega(s))$$

for a.e. $t \in [0, b]$ and all bounded subset $\Omega \subset C([-h, 0], X)$, where α is Hausdorff MNC in X.

Remark 4.8. *Assumption (H_3) can be deduced from assumption (H_2) if X is a finite dimensional space.*

We write, for each $u \in L^p([0, b], V)$,

$$\Theta(u) = \{x(\cdot, u, f) \in C([-h, b], X) : x(\cdot, u, f) \text{ is}$$

$$\text{the mild solution of (4.44)}\}.$$

The compactness and R_δ-property of solution set for the control system (4.44) is characterized in the following.

Theorem 4.14. *Let $pq > 1$ and assumptions (H_1)-(H_3) be satisfied. Suppose that $P_q(t)$ is continuous in the uniform operator topology for all $t > 0$. Given $u \in L^p([0, b], V)$. Then $\Theta(u)$ is nonempty and compact. If, in addition, $P_q(t)$ is compact for $t > 0$, then $\Theta(u)$ is an R_δ-set.*

Proof. Given $u \in L^p([0, b], V)$. The proof will be divided into two steps.

Step 1. We construct the solution map as follows: for each $x \in C([-h, b], X)$,

$$\mathcal{P}^u(x)(t) = \begin{cases} S_q(t)\varphi(0) + \displaystyle\int_0^t (t - s)^{q-1} \\ \qquad \times P_q(t - s)\big(f(s, x_s) + Bu(s)\big)ds, & t \in [0, b], \\ \varphi(t), & t \in [-h, 0]. \end{cases}$$

It is clear that $x \in \Theta(u)$ if and only if x is a fixed point of $\mathcal{P}^u$. Our purpose is to show that $\mathcal{P}^u$ admits at least one fixed point by making use of Theorem 1.8. To this end, let us first observe, by (H_1), (H_2), and Theorem 1.1, that $\mathcal{P}^u$, mapping $C([-h, b], X)$ into itself, is continuous.

In the sequel, let us introduce the following MNC in $C([-h, b], X)$: for every bounded set $\Omega \subset C([-h, b], X)$,

$$\nu(\Omega) = \max_{D \in \triangle(\Omega)} (\beta(D), \mathrm{mod}_C(D)),$$

where

$$\beta(D) = \sup_{t \in [0,b]} e^{-Lt} \alpha(D(t)), \quad L \geq 0,$$

$$\mathrm{mod}(D) = \lim_{\delta \to 0} \sup_{x \in D} \max_{|t_1 - t_2| \leq \delta} |x(t_1) - x(t_2)|,$$

and $\triangle(\Omega)$ stands for the collection of all countable subsets of Ω and the maximum is taken in the sense of the partial order in the cone $\overline{\mathbb{R}^2_+}$.

We claim that $\mathcal{P}^u$ is ν-condensing. To illustrate this, we argue by contradiction. Let Ω be a bounded subset of $C([-h, b], X)$ that is not relatively compact such that

$$\nu(\mathcal{P}^u(\Omega)) \geq \nu(\Omega). \tag{4.45}$$

From the definition of ν there exists a sequence $\{z^n\} \subset \mathcal{P}^u(\Omega)$ such that

$$\nu(\mathcal{P}^u(\Omega)) = (\beta(\{z^n\}), \mathrm{mod}_C(\{z^n\})).$$

Then we can take a sequence $\{x^n\} \subset \Omega$ such that

$$z^n(t) = \begin{cases} S_q(t)\varphi(0) + \displaystyle\int_0^t (t-s)^{q-1} \\ \qquad \times P_q(t-s)(f(s, x_s^n) + Bu(s))ds, & t \in [0, b], \\ \varphi(t), & t \in [-h, 0], \end{cases}$$

for $n \geq 1$. Let us write

$$\sigma(t) = \frac{2M}{\Gamma(q)} \int_0^t (t-s)^{q-1} k(s) e^{Ls} ds, \quad t \in [0, b],$$

and choose $L > 0$ large enough such that

$$\sup_{t \in [0,b]} e^{-Lt} \sigma(t) < 1. \tag{4.46}$$

For every $t \in [-h, 0]$, it is easy to see that

$$\alpha(\{z^n(t)\}) = 0.$$

Also, by (H_2), we have that for every $t \in [0, b]$, $s < t$,

$$\left|(t - s)^{q-1} P_q(t - s) f(s, x_s^n)\right|$$
$$\leq \frac{M}{\Gamma(q)} (t - s)^{q-1} \eta(s) (1 + \|x_s^n(\cdot)\|_{-h,0}), \quad n \geq 1,$$

which together with the boundedness of $\{x^n\}$ and the fact $(t - \cdot)^{q-1} \eta(\cdot) \in L^1([0, b], \mathbb{R})$ implies that the set

$$G_f(s) = (t - s)^{q-1} P_q(t - s) f(s, \{x_s^n\})$$

is integrably bounded. Moreover, from (H_3) it follows that for every $t \in [0, b]$, $s < t$,

$$\begin{aligned}
\alpha(G_f(s)) &\leq \frac{M}{\Gamma(q)} (t - s)^{q-1} k(s) \sup_{s' \in [-h, 0]} \alpha(\{x_s^n(s')\}) \\
&\leq \frac{M}{\Gamma(q)} (t - s)^{q-1} k(s) e^{Ls} \\
&\quad \times \sup_{s+s' \in [-h, b]} e^{-L(s+s')} \alpha(\{x^n(s + s')\}) \\
&= \frac{M}{\Gamma(q)} (t - s)^{q-1} k(s) e^{Ls} \beta(\{x^n\}).
\end{aligned}$$

Accordingly, we see

$$\alpha(G_f(s)) \leq \frac{M}{\Gamma(q)} (t - s)^{q-1} k(s) e^{Ls} \beta(\Omega), \quad \text{for every } t \in [0, b], s < t,$$

where $(t - \cdot)^{q-1} k(\cdot) \in L^1([0, b], \mathbb{R})$. Hence, an application of Property 1.18 yields

$$\alpha \left(\int_0^t G_f(s) ds + (\Phi(Bu))(t) \right) \leq \sigma(t) \beta(\{x^n\})$$

for each $t \in [0, b]$. This enables us to obtain

$$\beta(\{z^n\}) \leq \sup_{t \in [0, b]} e^{-Lt} \sigma(t) \beta(\{x^n\}),$$

which together with (4.45) implies that

$$\beta(\{x^n\}) \leq \beta(\{z^n\}) \leq \sup_{t \in [0, b]} e^{-Lt} \sigma(t) \beta(\{x^n\}).$$

Therefore, from (4.46) it follows that

$$\beta(\{z^n\}) = 0.$$

On the other hand, from (H_2) it follows that the set $f(\cdot, \{x^n\})$ is bounded in $L^p([0, b], X)$. This together with Lemma 4.18 yields that $\mathrm{mod}_C(\{z^n\}) = 0$.

Summarizing the above, we obtain $\beta(\Omega) = 0$, which is a contradiction. This in turn proves that $\mathcal{P}^u$ is ν-condensing.

Next, we take $x \in C([-h, b], X)$ with $x = \lambda \mathcal{P}^u(x)$ for some $0 < \lambda \le 1$. Then it follows from (H_2) that for each $t \in [0, b]$,

$$
\begin{aligned}
|x(t)| \le & M|\varphi(0)| + \frac{qM}{\Gamma(1+q)} \int_0^t (t-s)^{q-1} \eta(s)(1 + \|x_s(\cdot)\|_{-h,0}) ds \\
& + \frac{qM}{\Gamma(1+q)} \int_0^t (t-s)^{q-1} |Bu(s)| ds \\
\le & a_1 + a_2 \int_0^t (t-s)^{q-1} \|x_s(\cdot)\|_{-h,0} ds,
\end{aligned}
$$

where

$$
\begin{aligned}
a_1 =& M|\varphi(0)| + \frac{Mb^q}{\Gamma(1+q)} \sup_{s \in [0,b]} \eta(s) \\
& + \frac{qMb^{q-\frac{1}{p}}}{\Gamma(1+q)} \left(\frac{p-1}{pq-1}\right)^{\frac{p-1}{p}} \|Bu\|_{L^p[0,b]}, \\
a_2 =& \frac{qM}{\Gamma(1+q)} \sup_{s \in [0,b]} \eta(s).
\end{aligned}
$$

Moreover, note that for each $t \in [-h, 0]$, $x(t) = \lambda \varphi(t)$. Hence, an application of Lemma 4.19 enables us to obtain that there exists a constant $N > 0$ (independent a_1 and φ) such that

$$|x(t)| \le N(a_1 + \|\varphi\|_{-h,0}), \quad t \in [-h, b]. \tag{4.47}$$

Assume that Ω is a closed ball centered at origin with radius $r_0 > N(a_1 + \|\varphi\|_{-h,0})$. Now, applying Theorem 1.8 to $\mathcal{P}^u$ and Ω, we conclude that $\Theta(u)$ is nonempty and compact.

Step 2. Given $\varepsilon_n \in (0,1)$ with $\varepsilon_n \to 0$ as $n \to \infty$. By (H_1) there exists a sequence $\{f_n\}$ of locally Lipschitz functions from $[0, b] \times C([-h, b], X)$ to X for which the estimate

$$|f_n(t, v) - f(t, v)| < \varepsilon_n \tag{4.48}$$

remains true for all $t \in [0, b]$ and $v \in C([-h, b], X)$. We here used the Lasota-Yorke's approximation (see [105]).

On $C([-h, b], X)$ we define the approximation operator $\mathcal{P}_n^u$ by

$$\mathcal{P}_n^u(x)(t) = \begin{cases} S_q(t)\varphi(0) + \displaystyle\int_0^t (t-s)^{q-1} \\ \qquad \times P_q(t-s)\big(f_n(s, x_s) + Bu(s)\big)ds, & t \in [0, b], \\ \varphi(t), & t \in [-h, 0], \end{cases}$$

for $n \geq 1$ and every $x \in C([-h, b], X)$. It is easy to see that $\mathcal{P}_n^u$ are well defined. Also, note that for each $x \in C([-h, b], X)$,

$$|(I - \mathcal{P}_n^u)(x)(t) - (I - \mathcal{P}^u)(x)(t)| = 0, \quad \text{for } t \in [-h, 0],$$

and

$$|(I - \mathcal{P}_n^u)(x)(t) - (I - \mathcal{P}^u)(x)(t)|$$
$$\leq \frac{qM}{\Gamma(1+q)} \int_0^t (t-s)^{q-1} |f_n(s, x_s) - f(s, x_s)| ds$$
$$\leq \frac{Mb^q}{\Gamma(1+q)} \varepsilon_n, \quad \text{for } t \in [0, b]$$

due to (4.48). Accordingly, we obtain

$$I - \mathcal{P}_n^u \to I - \mathcal{P}^u, \quad \text{as } n \to \infty,$$

uniformly on $C([-h, b], X)$. Moreover, making use of (4.48) and (H_2) we obtain

$$|f_n(t, v)| \leq 1 + \eta(t)(1 + \|v\|_{-h,0}), \quad n \geq 1, \tag{4.49}$$

for all $(t, v) \in [0, b] \times C([-h, b], X)$, which implies that for any bounded sequence $\{x^m\} \subset C([-h, b], X)$, the set

$$G_{f_n}(s) = (t-s)^{q-1} P_q(t-s) f_n(s, \{x_s^m\})$$

is integrably bounded. And then, from the compactness of $P_q(t)$ for $t > 0$ it follows that $\alpha(G_{f_n}(s)) = 0$ for every $t \in [0, b]$, $s < t$. Hence, by Property 1.18 one finds that

$$\alpha\left(\int_0^t G_{f_n}(s)ds + (\Phi(Bu))(t)\right) = 0,$$

for each $t \in [0, b]$. Thus, using an argument similar to that as in Step 1 we see that for each $y \in C([-h, b], X)$, the equation

$$(I - \mathcal{P}_n^u)(x) = y \tag{4.50}$$

admits at least one mild solution. Since f_n is locally Lipschitz, the solution to (4.50) is unique.

Given $n \geq 1$. We process to prove that $I - \mathcal{P}^u$ and $I - \mathcal{P}_n^u$ are proper. We first observe that $I - \mathcal{P}_n^u$ is continuous. Let $K \subset C([-h, b], X)$ be a compact set and $(I - \mathcal{P}_n^u)(\Omega) = K$. In the sequel, it suffices to show that Ω is a compact set in $C([-h, b], X)$.

From the continuity of $I - \mathcal{P}_n^u$ and closedness of K it is easy to see that Ω is closed. Let $\{\widetilde{x}^m\} \subset \Omega$, one can take a sequence $\{y^m\} \subset K$ such that

$$\widetilde{x}^m - \mathcal{P}_n^u(\widetilde{x}^m) = y^m,$$

that is,

$$\widetilde{x}^m(t) = \begin{cases} S_q(t)\varphi(0) + y^m(t) + \displaystyle\int_0^t (t-s)^{q-1} \\ \quad \times P_q(t-s)\big(f_n(s, \widetilde{x}_s^m) + Bu(s)\big)ds, & t \in [0, b], \\ \varphi(t) + y^m(t), & t \in [-h, 0], \end{cases}$$

for $n \geq 1$. We use Lemma 4.19 to deduce, in view of the boundedness of $\{y^m\}$ and (4.49), that $\{\widetilde{x}^m\}$ is bounded in $C([-h, b], X)$. This enables us to obtain that $f_n(\cdot, \{\widetilde{x}^m\})$ is bounded in $L^p([0, b], X)$ due to (4.49). Therefore, an application of Lemma 4.18 gives the equicontinuity of $\{\widetilde{x}^m\}$. Also, a similar argument as above shows that

$$\alpha(\Phi(f_n(t, \{\widetilde{x}_t^m\}) + Bu(t))) = 0$$

for each $t \in [0, b]$. Since for each $t \in [-h, b]$, $\{y^m(t)\}$ is relatively compact in X, we conclude that $\alpha(\{\widetilde{x}^m(t)\}) = 0$ for each $t \in [-h, b]$. This proves that $\{\widetilde{x}^m(t)\}$ is relatively compact for each $t \in [-h, b]$.

Now, an application of Lemma 1.2 yields that $\{\widetilde{x}^m\}$ is relatively compact in $C([-h, b], X)$. Accordingly, Ω is compact. Hence, we conclude that $I - \mathcal{P}_n^u$ is proper. A similar argument as above enables us to obtain that $I - \mathcal{P}^u$ is also proper.

Finally, by applying Theorem 1.11, we conclude that

$$\Theta(u) = (I - \mathcal{P}^u)^{-1}(0)$$

is an R_δ-set. The proof is completed. $\square$

Remark 4.9. *Let us note that the compactness of $P_q(t)$ for $t > 0$ implies that $P_q(t)$ is continuous in the uniform operator topology for $t > 0$ (see Lemma 2.1 in [278]).*

Remark 4.10. *As can be seen, no assumption (H_3) is involved in proving the existence and R_δ-property if $P_q(t)$ is compact for $t > 0$.*

We are in a position to prove that the reachability set is invariant under nonlinear perturbations.

Theorem 4.15. *Let $pq > 1$. Suppose that assumptions (H_1) and (H_2) are satisfied and $P_q(t)$ is compact for $t > 0$. Suppose in addition that*

(H_4) there exists $\psi \in L^p([0, b], V)$ such that $(\Phi(B\psi))(b) = (\Phi\phi)(b)$ for each $\phi \in L^p([0, b], X)$.

Then there exists $r' > 0$ such that the reachability set of the control system (4.44) is invariant under nonlinear perturbations, i.e., $\mathcal{K}_{b,f} = \mathcal{K}_{b,0}$, provided $\|B\|_{\mathscr{L}(V,X)} < r'$.

Proof. We proceed in three steps.

Step 1. As shown in Theorem 4.14, for each $u \in L^p([0, b], V)$, $\Theta(u)$ is a nonempty, compact, and R_δ-set. In this step, our objective is to show that the multivalued map Θ is an R_δ-map. To illustrate this, it suffices to show, by Definition 1.28, that Θ is u.s.c. Let us first show that Θ is quasicompact. In fact, if $\Omega \subset L^p([0, b], V)$ is a bounded set and sequence $\{x^n\} \subset \Theta(\Omega)$, then an argument similar to that in Step 2 of Theorem 4.14 enables us to deduce that $\{x^n\}$ is relatively compact. This in particular implies that Θ is quasicompact.

In the sequel, in view of Lemma 1.7 it only needs to show that Θ is closed. Let $u^n \to u$ in $L^p([0,b], V)$ and $x^n \in \Theta(u^n)$, $x^n \to x$ in $C([-h, b], X)$. It is easy to see that x^n verify the following integral equation:

$$x^n(t) = \begin{cases} S_q(t)\varphi(0) + \displaystyle\int_0^t (t-s)^{q-1} \\ \qquad \times P_q(t-s)\big(f(s, x_s^n) + Bu^n(s)\big)ds, & t \in [0, b], \\ \varphi(t), & t \in [-h, 0], \end{cases} \tag{4.51}$$

for $n \geq 1$. In view of (H_1), we have $f(s, x_s^n) \to f(s, x_s)$ for all $s \in [0, b]$. Thus, from the facts that $f(\cdot, x_\cdot^n)$ is L^p-integrably bounded and B is a bounded linear operator from V to X, we infer, by Theorem 1.1, that

$$f(\cdot, x_\cdot^n) + Bu^n(\cdot) \to f(\cdot, x_\cdot) + Bu(\cdot) \quad \text{in } L^p([0,b], X).$$

Passing in (4.51) to the limit as $n \to \infty$, we see that x verifies the integral equation

$$x(t) = \begin{cases} S_q(t)\varphi(0) + \displaystyle\int_0^t (t-s)^{q-1} \\ \qquad \times P_q(t-s)\big(f(s, x_s) + Bu(s)\big)ds, & t \in [0, b], \\ \varphi(t), & t \in [-h, 0], \end{cases}$$

which implies that $x \in \Theta(u)$, as desired.

Step 2. According to (H_4), there exists a continuous map $S : L^p([0, b], X) \to L^p([0, b], V)$ such that for any $\phi \in L^p([0, b], X)$,

$$(\Phi(BS\phi))(b) + (\Phi\phi)(b) = 0 \tag{4.52}$$

and

$$\|S\phi\|_{L^p[0,b]} \leq d\|\phi\|_{L^p[0,b]}, \tag{4.53}$$

where d is a positive number (see [216]).

Now, consider the multivalued map $F : L^p([0, b], V) \to P(L^p([0, b], V))$ denoted by

$$F(u) = S \circ H_f \circ \Theta(u_0 + u),$$

where $u_0 \in L^p([0, b], V)$ is specified later and the map H_f : $C([-h, b], X) \to L^p([0, b], X)$ is defined by

$$(H_f x)(t) = f(t, x_t), \quad \text{for each } t \in [0, b], x \in C([-h, b], X).$$

Note that H_f is a single-valued continuous map due to (H_1). Hence, it is an R_δ-map. Clearly, S is also an R_δ-map. Moreover, as proved in Step 1, Θ is an R_δ-map.

In the sequel, we shall use Theorem 1.12 to show that F admits at least one fixed point. Let $u \in L^p([0, b], V)$ and $\tilde{u} \in F(u)$. For $x' \in \Theta(u_0 + u)$, it follows from (4.53) and (H_2) that

$$\begin{aligned}
\|\tilde{u}\|_{L^p[0,b]} &\leq d\|H_f(x')\|_{L^p[0,b]} \\
&\leq d\left(\int_0^b |\eta(s)|^p (1 + \|x_s'(\cdot)\|_{-h,0})^p ds \right)^{\frac{1}{p}} \\
&\leq db^{\frac{1}{p}} \sup_{s \in [0,b]} \eta(s) \left(1 + \|x'\|_{-h,b} \right),
\end{aligned}$$

which together with (4.47) imply that

$$\|\tilde{u}\|_{L^p[0,b]} \leq M_1 + M_2 \|B\|_{\mathscr{L}(V,X)} \left(\|u_0\|_{L^p[0,b]} + \|u\|_{L^p[0,b]} \right),$$

where

$$\begin{aligned}
M_1 =& db^{\frac{1}{p}} \sup_{s \in [0,b]} \eta(s) \left[1 + N \left(M|\varphi(0)| \right. \right. \\
& \left. \left. + \frac{Mb^q}{\Gamma(1+q)} \sup_{s \in [0,b]} \eta(s) + \|\varphi\|_{-h,0} \right) \right], \\
M_2 =& \frac{dMNb^q}{\Gamma(q)} \sup_{s \in [0,b]} \eta(s) \left(\frac{p-1}{pq-1} \right)^{\frac{p-1}{p}}.
\end{aligned}$$

Taking $r' = \frac{1}{M_2}$ and $\|B\|_{\mathscr{L}(V,X)} < r'$, one finds that there exists $r > 0$ such that for all $u \in L^p([0, b], V)$ satisfying $\|u\|_{L^p[0,b]} \leq r$,

$$\|\tilde{u}\|_{L^p[0,b]} \leq r.$$

This implies $F(D_r) \subset D_r$, where D_r denotes the closed ball in $L^p([0, b], V)$ centered at origin with radius r.

Set $K = F(D_r) \subset D_r$. In view of the compactness of $\Theta(u_0 + D_r)$ and continuity of S, H_f, we obtain that K is a compact set. At the end of this step, applying Theorem 1.12 we conclude that F admits a fixed point.

Step 3. In this step, we shall verify $\mathcal{K}_{b,f} = \mathcal{K}_{b,0}$. To the end, let us take $x(b, u_0, 0) \in \mathcal{K}_{b,0}$ and $u^* \in Fix(F)$ with $F(\cdot) = S \circ H_f \circ \Theta(u_0 + \cdot)$. It is clear

$$x(b, u_0, 0) = Q(b)\varphi(0) + (\Phi(Bu_0))(b). \tag{4.54}$$

Let

$$u^* = SH_f(x), \tag{4.55}$$

where $x \in \Theta(u_0 + u^*)$. Then from (4.52), (4.54), and (4.55) it follows

$$\begin{aligned} x(b, u_0 + u^*, f) &= Q(b)\varphi(0) + \big(\Phi(H_f(x) + B(u_0 + u^*))\big)(b) \\ &= Q(b)\varphi(0) + (\Phi(Bu_0))(b) + \big(\Phi(H_f(x) + Bu^*)\big)(b) \\ &= x(b, u_0, 0) + \big(\Phi(H_f(x) + BSH_f(x))\big)(b) \\ &= x(b, u_0, 0). \end{aligned}$$

From which we see that $\mathcal{K}_{b,0} \subset \mathcal{K}_{b,f}$.

Next, let us take $x(b, u, f) \in \mathcal{K}_{b,f}$ and notice

$$x(b, u, f) = Q(b)\varphi(0) + \big(\Phi(H_f(x) + Bu)\big)(b). \tag{4.56}$$

Set $\hat{u} = u - SH_f(x)$. Then from (4.52) and (4.56) we obtain

$$\begin{aligned} & x(b, \hat{u}, 0) \\ &= Q(b)\varphi(0) + (\Phi(B\hat{u}))(b) \\ &= Q(b)\varphi(0) + \big(\Phi(Bu - BSH_f(x))\big)(b) \\ &= Q(b)\varphi(0) + \big(\Phi(H_f(x) + Bu)\big)(b) - \big(\Phi(H_f(x) + BSH_f(x))\big)(b) \\ &= x(b, u, f), \end{aligned}$$

which enables us to conclude that $\mathcal{K}_{b,f} \subset \mathcal{K}_{b,0}$. The proof is then complete. $\qquad\square$

Remark 4.11. *It is noted that assumption (H_4) is fulfilled if B is surjective.*

In order to characterize the approximate controllability for control system (4.44), we need to introduce the relevant operator as follows:

$$\mathcal{W} = \int_0^b (b - s)^{q-1} P(b - s) B B^* P^* (b - s) ds,$$

where B^* and $P^*(t)$ stand for the adjoints of B and $P_q(t)$, respectively. It is straightforward that the operator $\mathcal{W}$ is a linear bounded operator.

Similar to the proof of Theorem 2 in [172], we give the following lemma.

Lemma 4.20. *Suppose that*

(H_5) $\xi(\xi I + \mathcal{W})^{-1} \to 0$ *as* $\xi \to 0+$ *in the strong operator topology.*

Then the linear control problem

$$\begin{cases} {}_0^C D_t^q (x(t)) = Ax(t) + Bu(t), & t \in [0, b], \\ x(t) = \varphi(t), & t \in [-h, 0]. \end{cases} \tag{4.57}$$

is approximately controllable.

Finally, we have the following assertion.

Theorem 4.16. *Let the hypotheses in Theorem 4.15 hold. Suppose in addition that X, V are Hilbert spaces and (H_5) holds. Then control system (4.44) is approximately controllable.*

Proof. By Lemma 4.20, we get that the linear control problem (4.57) is approximately controllable, which implies that $\overline{\mathcal{K}_{b,0}} = X$. On the other hand, by Theorem 4.15, we know $\mathcal{K}_{b,f} = \mathcal{K}_{b,0}$. Hence, we obtain $\overline{\mathcal{K}_{b,f}} = X$. This proves that control system (4.44) is approximately controllable, which then completes the proof. $\qquad\qquad\square$

4.5.3 Example

As a sample of application, we consider a control problem of fractional differential equation with time delay in this subsection. Such example does not aim at generality but indicates how our theorem can be applied to a concrete problem.

Let us consider the control problem of fractional differential equation with delay in the form

$$
\begin{cases}
{}^{C}_{0}D^{q}_{t}x(t,\xi) = \dfrac{\partial^2 x(t,\xi)}{\partial \xi^2} + \kappa u(t,\xi) \\
\qquad + t^{\frac{1}{k}}\sin(|x_t(\theta,\xi)|), & t \in [0,1],\ \xi \in [0,\pi], \qquad (4.58)\\
x(t,0) = x(t,\pi) = 0, & t \in [0,1], \\
x(t,\xi) = \varphi(t,\xi), & t \in [-h,0],
\end{cases}
$$

where $\frac{1}{2} < q < 1$, $k > 1$ is a constant, $x_t(\theta,\xi) = x(t+\theta,\xi)$, $\theta \in [-h,0]$, φ is continuous, and κ is a real parameter.

Take $X = V = L^2([0,\pi],\mathbb{R})$. Let $A : D(A) \subset X \to X$ be operator defined by $A\omega = \dfrac{\partial^2 \omega}{\partial \xi^2}$ with domain

$$
D(A) = \{\omega \in X : \omega, \omega' \text{ are absolutely continuous,}
$$
$$
\omega'' \in X, \text{ and } \omega(0) = \omega(\pi) = 0\}.
$$

It is known that A has a discrete spectrum and the eigenvalues are $-n^2$, $n \in \mathbb{N}^+$, with the corresponding normalized eigenvectors $\vartheta_n(\xi) = \sqrt{\frac{2}{\pi}}\sin(n\xi)$, $0 \le \xi \le \pi$. Moreover, A generates a compact, analytic semigroup $\{T(t)\}_{t\ge 0}$ on X :

$$
T(t)\omega = \sum_{n=1}^{\infty} e^{-n^2 t}(\omega,\vartheta_n)\vartheta_n, \quad \|T(t)\|_{\mathscr{L}(X)} \le e^{-t}, \text{ for all } t \ge 0
$$

(see [115]). Therefore, we have that for $\omega \in X$,

$$
S_q(t)\omega = \sum_{n=1}^{\infty} E_q(-n^2 t^q)(\omega,\vartheta_n)\vartheta_n, \quad \|S_q(t)\|_{\mathscr{L}(X)} \le 1, \text{ for all } t \ge 0,
$$
$$
P_q(t)\omega = \sum_{n=1}^{\infty} e_q(-n^2 t^q)(\omega,\vartheta_n)\vartheta_n, \quad \|P_q(t)\|_{\mathscr{L}(X)} \le \frac{1}{\Gamma(q)}, \text{ for all } t \ge 0,
$$

where $E_q(t)$ and $e_q(t)$ are Mittag-Leffler functions (see Definition 1.7).

According to the compactness of $T(t)$ for $t > 0$, we know that $S_q(t)$ and $P_q(t)$ are compact operators for $t > 0$. Moreover, from [278], one can see that for $t > 0$, $S_q(t)$ and $P_q(t)$ are continuous in the uniform operator topology.

Define

$$u(t)(\xi) = u(t, \xi),$$
$$x(t)(\xi) = x(t, \xi),$$
$$f(t, v)(\xi) = t^{\frac{1}{k}} \sin(|v(t)(\xi)|), \quad v \in C([-h, b], X).$$

It is easy to see that f is continuous from $[0, 1] \times C([-h, b], X)$ to X. Furthermore, for each $t \in [0, 1]$ and $v \in C([-h, b], X)$, one has

$$|f(t, v)| \le \eta(t)(1 + \|v\|_{-h, 0}),$$

where $\eta(t) = t^{\frac{1}{k}} \in C([0, 1])$.

Finally, let us define a bounded linear operator $B : L^2([0, 1], V) \to L^2([0, 1], X)$ by $(Bu)(t) = \kappa u(t)$ for $u(\cdot) \in L^2([0, 1], V)$. It is clear that B is surjective.

In summary, hypotheses (H_1), (H_2), and (H_4) are satisfied. Hence, we obtain, by Theorems 4.14 and 4.15, that

(i)   the set of all mild solutions for the control problem (4.58) is nonempty and compact for each control function $u \in L^2([0, 1], V)$;

(ii)   the set of all mild solutions for the control problem (4.58) is an R_δ-set for each control function $u \in L^2([0, 1], V)$;

(iii)   when $|\kappa|$ is sufficiently small, the reachability set for the control system (4.58) is invariant under nonlinear perturbation.

4.6 NOTES AND REMARKS

The delay evolution systems is an important class of distributed parameter systems, and optimal control of infinite dimensional systems is a remarkable subject in control theory.

In the last years, fractional evolution systems in infinite dimensional spaces attracted many authors (see, e.g., [13, 24, 35, 37, 38, 60, 62, 89, 113, 122, 125, 186–188, 193, 302] and the references therein). When the fractional differential equations describe the performance index and system dynamics, an optimal control problem reduces to a fractional optimal control problem. The fractional optimal control of a distributed system is a fractional optimal control for which system dynamics are defined with partial fractional differential equations. There has been very little work in the area of fractional optimal control problem in infinite dimensional spaces

[197, 248, 266], especially optimal controls of fractional finite time delay evolution system.

In addition, control systems are most often based on the principle of feedback, whereby the signal to be controlled is compared to a desired reference signal and the discrepancy used to compute corrective control action [98, 181]. Optimal feedback control of semilinear evolution equations in Banach spaces has been studied [130, 281]. However, optimal feedback control problems for fractional evolution equations in Banach spaces has not been studied extensively.

Sobolev-type evolution equations often arise in various applications such as in the flow of fluid through fissured rocks, thermodynamics, and shear in second order fluids. In the past decade, many researchers have studied the existence and controllability of the mild solutions for Cauchy problem of all kinds of Sobolev-type evolution equations under the various conditions on the pair (A, E). After reviewing these interesting results, the reader can find that $D(A) \subset D(E)$, boundedness or compactness of E^{-1} are posed (see [2, 22, 23]). In particular, Li et al. [153] obtained new existence results for Sobolev-type fractional evolution equations by virtue of the theory of propagation family which generated by the pair (A, E) via the techniques of the measure of noncompactness and the condensing maps. The restrict conditions on the $D(A)$, $D(E)$, and E^{-1} are removed.

There are some interesting and important controllability results concerning the semilinear differential systems involving Caputo fractional derivative. For example, Fečkan et al. [93] initiated to study complete controllability of a class of Sobolev-type fractional functional evolution equations by constructing two new characteristic solution operators via the well-known Schauder fixed point theorem.

As is known to all, the problems of exact and approximate controllability are to be distinguished. In general infinite dimensional spaces, the concept of exact controllability, is usually too strong (see [227]). Therefore, the class of evolution equations consisting of fractional diffusion equations must be treated by the weaker concept of controllability, namely approximate controllability. Recently, many works report approximate controllability results; we refer readers to [71, 170, 207] and the references therein. Meanwhile, Sakthivel and Ren [209], Debbouche and Torres [75], Mahmudov [171], and Mahmudov and Zorlu [173, 174] pay attention to study approximate controllability of different types of fractional evolution systems.

It is known that the characterizations of solution sets consisting of compactness, acyclicity, and R_δ-property are useful in the study of the corresponding equations or inclusions. Bader and Kryszewski [19] proved that the set consisting of all mild solutions for a constrained semilinear differential inclusion is a nonempty, compact, and R_δ-set and gave its applications to the periodic problem and to the existence of equilibria. Recently, Andres and Pavlačková [12] have studied the R_δ-structure of solution sets for a fully linearized system of second order ODEs and then obtained an existence result for the corresponding semilinear system by using information about the structure and a fixed-point index technique in Fréchet spaces.

Let us note that by proving that the reachability set is invariant under non-linear perturbations, some controllability problems have been investigated; we refer readers to [191, 192] and the references therein. However, there exists a limitation among these results, that is, in these works the authors assumed the uniqueness of the solution, which implies that the solution set is single-pointed and the Lipschitz continuity on the nonlinearity is involved. More precisely, under a more general class of nonlinearities f which does not guarantee the uniqueness of a mild solution for the control system (4.44), we shall show the invariance of its reachability set under nonlinear perturbations by making use of the information of the topological structure, which in particular implies that the control system (4.44) is approximately controllable if the corresponding linear problem is approximately controllable. We emphasize that the lack of uniqueness prevents us from using the well-known tools such as Banach and Schauder fixed point theorems to show the invariance of reachability set. This difficulty leads us to study the topological structure of solution set, before applying a fixed point theory for multivalued maps with non-convex values.

The results in Section 4.1 are adopted from Wang, Wei, and Zhou [249]. Section 4.2 is adopted from Wang, Zhou, and Wei [270]. The results in Section 4.3 are taken from Wang, Fečkan, and Zhou [242]. The material in Section 4.4 is due to Wang, Fečkan, and Zhou [245]. The results in Section 4.5 are adopted from Wang, Xiang, and Zhou [277].

CHAPTER 5

Fractional Stochastic Evolution Inclusions

Abstract In this chapter, we investigate the existence and the topological structure of solution sets for fractional stochastic evolution inclusions in Hilbert spaces.

Keywords Fractional stochastic inclusions, Brownian motions, Multivalued maps, Weak topology, Existence, Compact R_δ-set.

5.1 EXISTENCE OF MILD SOLUTIONS

5.1.1 Introduction

Consider the following nonlinear stochastic evolution inclusions:

$$\begin{cases} {}^{C}_{0}D^{q}_{t}x(t) \in Ax(t) + f(t, x(t)) + \Sigma(t, x(t))\dfrac{dW(t)}{dt}, \ t \in [0, b], \\ x(0) = x_0, \end{cases} \quad (5.1)$$

where ${}^{C}_{0}D^{q}_{t}$ is Caputo fractional derivative of order $q \in (\frac{1}{2}, 1)$, A is the infinitesimal generator of a C_0-semigroup $\{T(t)\}_{t\geq 0}$ in a Hilbert space $\mathcal{H}$ with inner product $(\cdot, \cdot)$ and norm $|\cdot|$, the state $x(\cdot)$ takes value in $\mathcal{H}$, $f : J \times \mathcal{H} \to \mathcal{H}$, $\Sigma : J \times \mathcal{H} \multimap \mathcal{H}$ is a nonempty, bounded, closed, and convex multivalued map; $\{W(t)\}_{t\geq 0}$ is a given $\mathcal{K}$-valued Brownian motion or Wiener process with a finite trace nuclear covariance operator $Q \geq 0$; here $\mathcal{K}$ is a Hilbert space with inner product $(\cdot, \cdot)_{\mathcal{K}}$ and norm $|\cdot|_{\mathcal{K}}$.

In this subsection, we introduce notations, definitions, and preliminary facts which are used throughout this chapter.

Let $(\Omega, \mathscr{F}, \mathbb{P})$ be a complete probability space equipped with a normal filtration $\mathscr{F}_t$, $t \in [0, b]$ satisfying the usual conditions (i.e., right continuous and $\mathscr{F}_0$ containing all $\mathbb{P}$-null sets). We consider Q-Wiener process on $(\Omega, \mathscr{F}, \mathbb{P})$ with the linear bounded covariance operator Q such that $trQ < \infty$. We assume that there exists a complete orthonormal system $\{e_n\}_{n\geq 1}$ on $\mathcal{K}$, a bounded sequence of nonnegative real numbers $\{\lambda_n\}$ such that $Qe_n = \lambda_n e_n$, $n = 1, 2, ...$, and a sequence $\{W_n\}_{n\geq 1}$ of independent

Fractional Evolution Equations and Inclusions: Analysis and Control. http://dx.doi.org/10.1016/B978-0-12-804277-9.50005-5

Brownian motions such that

$$(W(t), e)_{\mathcal{K}} = \sum_{n=1}^{\infty} \sqrt{\lambda_n}(e_n, e)_{\mathcal{K}} W_n(t), \quad e \in \mathcal{K}, \ t \in [0, b],$$

and $\mathscr{F}_t = \mathscr{F}_t^\omega$, where $\mathscr{F}_t^\omega$ is the sigma algebra generated by $\{W(s) : 0 \le s \le t\}$. Let $\mathscr{L}_2^0 = \mathscr{L}_2(Q^{\frac{1}{2}}\mathcal{K}, \mathcal{H})$ be the space of all Hilbert-Schmidt operators from $Q^{\frac{1}{2}}\mathcal{K}$ to $\mathcal{H}$ with the inner product $(\Psi, \Upsilon)_{\mathscr{L}_2^0} = tr[\Psi Q \Upsilon^*]$.

Let us fix a real number p, $p \ge 2$ and denote by $L^p(\Omega, \mathcal{H})$ Banach space of all $\mathscr{F}_b$-measurable p-integrable random variables with the norm $\|x(\cdot)\| = (\mathbb{E}|x(\cdot, \omega)|^p)^{\frac{1}{p}}$, where the expectation $\mathbb{E}(\cdot)$ is denoted by $\mathbb{E}(h) = \int_\Omega h(\omega)d\mathbb{P}$.

Let $C([0, b], L^p(\Omega, \mathcal{H}))$ be Banach space of continuous maps from $[0, b]$ into $L^p(\Omega, \mathcal{H})$ satisfying $\sup_{t \in [0, b]} \|x(t)\|^p < \infty$. Let $\mathcal{C}$ be a closed subspace of $C([0, b], L^p(\Omega, \mathcal{H}))$ consisting of measurable and $\mathscr{F}_t$-adapted $\mathcal{H}$-valued process $x \in C([0, b], L^p(\Omega, \mathcal{H}))$ endowed with the norm

$$\|x\|_{\mathcal{C}} = \left(\sup_{t \in [0, b]} \|x(t)\|^p \right)^{\frac{1}{p}}.$$

Let $\mathcal{H}_w$ denote the space $\mathcal{H}$ endowed with the weak topology. For a set $D \subset \mathcal{H}$, the symbol $\overline{D}^w$ denotes the weak closure of D. We recall (see [43]) that a sequence $\{x_n\} \subset C([0, b], L^p(\Omega, \mathcal{H}))$ weakly converges to an element $x \in C([0, b], L^p(\Omega, \mathcal{H}))$ if and only if

(i) there exists $N > 0$ such that, for every $n \in \mathbb{N}$ and $t \in [0, b]$, $\|x_n(t)\| \le N$;

(ii) for every $t \in [0, b]$, $x_n(t) \rightharpoonup x(t)$ in $L^p(\Omega, \mathcal{H})$.

Lemma 5.1. *[205] For any $p \ge 2$ and for arbitrary $\mathscr{L}_2^0$-valued predictable process $\varphi(\cdot)$ such that*

$$\sup_{t \in [0, b]} \mathbb{E} \left| \int_0^t \varphi(s)dW(s) \right|^p \le C_p \left(\int_0^t \left(\mathbb{E}\|\varphi(s)\|_{\mathscr{L}_2^0}^p \right)^{\frac{2}{p}} ds \right)^{\frac{p}{2}},$$

where $C_p = \left(\dfrac{p(p-1)}{2} \right)^{\frac{p}{2}}$.

The following approximation estimate, which is similar to Theorem 4.2.1 in [131], will be used in the sequel.

Lemma 5.2. *Let the sequence* $\{\sigma_n\} \subset L^p([0, b], \mathscr{L}_2^0)$ *be integrably bounded:*

$$\|\sigma_n(t)\|_{\mathscr{L}_2^0} \leq \nu(t), \quad for\ a.e.\ t \in [0, b],$$

where $\nu \in L^p([0, b], \mathbb{R}^+)$. *Assume that*

$$\beta(\{\sigma_n(t)\}_{n=1}^{\infty}) \leq k(t)$$

for a.e. $t \in [0, b]$, *where* $k \in L^p([0, b], \mathbb{R}^+)$. *Then for every* $\delta > 0$, *there exists a compact set* $K_\delta \subset \mathscr{L}_2^0$, *a set* $m_\delta \subset [0, b]$, $mes(m_\delta) < \delta$, *and a set of functions* $G_\delta \subset L^p([0, b], \mathscr{L}_2^0)$ *with values in* K_δ *such that for every* $n \geq 1$ *there exists* $g_n \in G_\delta$ *for which*

$$\|\sigma_n(t) - g_n(t)\|_{\mathscr{L}_2^0} \leq 2k(t) + \delta, \quad t \in [0, b] \setminus m_\delta.$$

5.1.2 Statement of Problem

We study the fractional stochastic evolution inclusion (5.1) under the following assumptions:

(H_A) the operator A generates a C_0-semigroup $\{T(t)\}_{t \geq 0}$ in $\mathcal{H}$, and there exists a constant $M_1 \geq 1$ such that

$$\sup_{t \in J} \|T(t)\|_{\mathscr{L}(\mathcal{H})} \leq M_1;$$

(H_1) the function $f(t, \cdot) : \mathcal{H} \to \mathcal{H}$ is weakly sequentially continuous for each $t \in [0, b]$ and maps bounded sets into bounded sets;

$(H_1)'$ the function $f : [0, b] \times \mathcal{H} \to \mathcal{H}$ satisfies locally Lipschitz condition and there exist two positive constants c_0, c_1 such that

$$|f(t, x)|^p \leq c_0 |x|^p + c_1.$$

The multivalued map $\Sigma : [0, b] \times \mathcal{H} \multimap \mathcal{H}$ has closed bounded and convex values and satisfies the following conditions:

(H_2) $\Sigma(\cdot, x) : [0, b] \multimap \mathcal{H}$ has a measurable selection for every $x \in \mathcal{H}$, i.e., there exists a measurable function $\sigma : [0, b] \to \mathcal{H}$ such that $\sigma(t) \in F(t, x)$ for a.e. $t \in [0, b]$;

(H_3) $\Sigma(t, \cdot) : \mathcal{H} \multimap \mathcal{H}$ is weakly sequentially closed for a.e. $t \in [0, b]$, i.e., it has a weakly sequentially closed graph;

$(H_3)'$ $\Sigma(t, \cdot) : \mathcal{H} \multimap \mathcal{H}$ is weakly u.s.c.;

(H_4) for every $r > 0$, there exists a function $\mu_r \in L^1([0, b], \mathbb{R}^+)$ such that for each $x \in \mathcal{H}$, $|x|^p \leq r$

$$\|\Sigma(t, x)\|^p_{\mathscr{L}^0_2} \leq \mu_r(t) \text{ for a.e. } t \in [0, b],$$

where $p \geq \max \left\{2, \frac{1}{q}, \frac{2q}{2q-1}\right\}$ and

$$\|\Sigma(t, x)\|_{\mathscr{L}^0_2} = \sup\{\|\sigma(t)\|_{\mathscr{L}^0_2} : \sigma \in \Sigma(t, x)\};$$

$(H_4)'$ there exists a function $\alpha(t) \in L^1([0, b], \mathbb{R}^+)$ such that

$$\|\Sigma(t, x)\|^p_{\mathscr{L}^0_2} \leq \alpha(t)(1 + |x|^p) \text{ for a.e. } t \in [0, b], \ x \in \mathcal{H}.$$

Given $x \in \mathcal{C}$, let us denote

$$Sel_\Sigma(x) = \{\sigma \in L^p([0, b], \mathscr{L}^0_2) : \sigma(t) \in \Sigma(t, x(t)), \text{ for a.e. } t \in [0, b]\}.$$

The set $Sel_\Sigma(x)$ is always nonempty as Lemmas 5.3 and 5.4 below show.

Lemma 5.3. *Assume that the multivalued map Σ satisfies conditions (H_2)-(H_4). Then the set $Sel_\Sigma(x)$ is nonempty for any $x \in \mathcal{C}$.*

Proof. Let $x \in \mathcal{C}$, by the uniform continuity of x, there exists a sequence $\{x_n\}$ of step functions, $x_n : [0, b] \to L^p(\Omega, \mathcal{H})$, such that

$$\sup_{t \in [0, b]} \|x_n(t) - x(t)\|^p \to 0, \quad \text{as } n \to \infty. \tag{5.2}$$

Hence, by (H_2), there exists a sequence of functions $\{\sigma_n\}$ such that $\sigma_n(t) \in \Sigma(t, x_n(t))$ for a.e. $t \in [0, b]$, and $\sigma_n : [0, b] \to \mathscr{L}^0_2$ is measurable for any $n \in \mathbb{N}$. From (5.2), there exists a bounded set $E \subset L^p(\Omega, \mathcal{H})$ such that $x_n(t), x(t) \in E$, for any $t \in [0, b]$ and $n \in \mathbb{N}$, and by (H_4) there exists $\mu_r \in L^p([0, b], \mathbb{R}^+)$ such that

$$\|\sigma_n(t)\|^p_{\mathscr{L}^0_2} \leq \|\Sigma(t, x)\|^p_{\mathscr{L}^0_2} \leq \mu_r(t), \quad \forall n \in \mathbb{N}, \text{ for a.e. } t \in [0, b].$$

Hence, $\{\sigma_n\} \subset L^p([0, b], \mathscr{L}^0_2)$ is bounded and uniformly integrable and $\{\sigma_n(t)\}$ is bounded in $\mathscr{L}^0_2$ for a.e. $t \in [0, b]$. According to the reflexivity of the space $\mathscr{L}^0_2$ and Lemma 1.4, we have the existence of a subsequence, denoted as the sequence, such that $\sigma_n \rightharpoonup \sigma \in L^p([0, b], \mathscr{L}^0_2)$.

By Mazur lemma, we obtain a sequence

$$\tilde{\sigma}_n = \sum_{i=0}^{k_n} \lambda_{n,i}\sigma_{n+i}, \quad \lambda_{n,i} \geq 0, \quad \sum_{i=0}^{k_n} \lambda_{n,i} = 1$$

such that $\tilde{\sigma}_n \to \sigma$ in $L^p([0,b], \mathscr{L}_2^0)$ and, up to subsequence, $\tilde{\sigma}_n(t) \to \sigma(t)$ for all $t \in [0,b]$. By (H_3), the multivalued map $\Sigma(t,\cdot)$ is locally weakly compact for a.e. $t \in [0,b]$, i.e., for a.e. $t \in [0,b]$ and every $x \in \mathcal{H}$ there is a neighborhood V of x such that the restriction of $\Sigma(t,\cdot)$ to V is weakly compact. Hence, by (H_3) and the locally weak compactness, we easily get that $\Sigma(t,\cdot) : \mathcal{H}_w \multimap \mathcal{H}_w$ is u.s.c. for a.e. $t \in [0,b]$. Thus, $\Sigma(t,\cdot) : \mathcal{H} \multimap \mathcal{H}_w$ is u.s.c. for a.e. $t \in [0,b]$.

To conclude we only need to prove that $\sigma(t) \in \Sigma(t, x(t))$ for a.e. $t \in [0,b]$. Indeed, let the Lebesgue measure of N_0 be zero such that $\Sigma(t,\cdot) : \mathcal{H} \multimap \mathcal{H}_w$ is u.s.c., $\sigma_n(t) \in \Sigma(t, x_n(t))$ and $\tilde{\sigma}_n(t) \to \sigma(t)$ for all $t \in [0,b] \setminus N_0$ and $n \in \mathbb{N}$.

Fix $t_0 \notin N_0$ and assume, by contradiction, that $\sigma(t_0) \notin \Sigma(t_0, x(t_0))$. Since $\Sigma(t_0, x(t_0))$ is closed and convex, from Hahn-Banach theorem there is a weakly open convex set $V \supset \Sigma(t_0, x(t_0))$ satisfying $\sigma(t_0) \notin \overline{V}$. Since $\Sigma(t_0, \cdot) : \mathcal{H} \multimap \mathcal{H}_w$ is u.s.c., we can find a neighborhood U of $x(t_0)$ such that $\Sigma(t_0, x) \subset V$ for all $x \in U$. The convergence $x_n(t_0) \rightharpoonup x(t_0)$ as $n \to \infty$ implies the existence of $n_0 \in \mathbb{N}$ such that $x_n(t_0) \in U$ for all $n > n_0$. Therefore, $\sigma_n(t_0) \in \Sigma(t_0, x_n(t_0)) \subset V$ for all $n > n_0$. Since V is convex, we also have that $\tilde{\sigma}_n(t_0) \in V$ for all $n > n_0$ and, by the convergence, we arrive to the contradictory conclusion that $\sigma(t_0) \in \overline{V}$. We obtain that $\sigma(t) \in \Sigma(t, x(t))$ for a.e. $t \in [0,b]$. $\square$

Lemma 5.4. *Let conditions (H_2), $(H_3)'$, and $(H_4)'$ be satisfied. Then $Sel_\Sigma(x)$ is weakly u.s.c. with nonempty, convex, and weakly compact values.*

Proof. Let $x \in \mathcal{C}$, by the uniform continuity of x, there exists a sequence $\{x_n\}$ of step functions, $x_n : [0,b] \to L^p(\Omega, \mathcal{H})$, such that

$$\sup_{t \in [0,b]} \|x_n(t) - x(t)\|^p \to 0, \quad \text{as } n \to \infty.$$

Hence, by (H_2), there exists a sequence of functions $\{\sigma_n\}$ such that $\sigma_n(t) \in \Sigma(t, x_n(t))$ for a.e. $t \in [0,b]$, and $\sigma_n : [0,b] \to \mathscr{L}_2^0$ is measurable for any $n \in \mathbb{N}$.

Moreover, in view of $(H_4)'$, we have that $\{\sigma_n\} \subset L^p([0, b], \mathscr{L}_2^0)$ is bounded and uniformly integrable and $\{\sigma_n(t)\}$ is bounded in $\mathscr{L}_2^0$ for a.e. $t \in [0, b]$. With the same reason as Lemma 5.3, we obtain a sequence $\tilde{\sigma}_n \in \text{co}\{\sigma_k, k \geq n\}$ for $n \geq 1$ such that $\tilde{\sigma}_n \to \sigma$ in $L^p([0, b], \mathscr{L}_2^0)$ and, up to subsequence, $\tilde{\sigma}_n(t) \to \sigma(t)$ for a.e. $t \in [0, b]$ and $\sigma_n(t) \in \Sigma(t, x_n(t))$ for all $n \geq 1$.

Denote by $\mathcal{N}$ the set of all $t \in [0, b]$ such that $\tilde{\sigma}_n(t) \to \sigma(t)$ in $\mathscr{L}_2^0$ and $\sigma_n(t) \in \Sigma(t, x_n(t))$ for all $n \geq 1$. Let $x^* \in \mathscr{L}_2^0$, $\varepsilon > 0$. From $(H_3)'$, it follows immediately that $\langle x^*, \Sigma(t, \cdot) \rangle : \mathcal{H} \to P(\mathbb{R})$ is u.s.c. with compact convex values, so ε-δ u.s.c. with compact convex values. Accordingly, we have

$$\langle x^*, \tilde{\sigma}_n(t) \rangle \in \text{co}\{\langle x^*, \sigma_k(t) \rangle, k \geq n\} \subset \langle x^*, \Sigma(t, x_n(t)) \rangle$$
$$\subset \langle x^*, \Sigma(t, x(t)) \rangle + (-\varepsilon, \varepsilon).$$

Therefore, we obtain that $\langle x^*, \sigma(t) \rangle \in \langle x^*, \Sigma(t, x) \rangle$ for each $x^* \in \mathscr{L}_2^0$ and $t \in \mathcal{N}$. Since Σ has convex and closed values, we conclude that $\sigma(t) \in \Sigma(t, x(t))$ for each $t \in \mathcal{N}$, which implies $\sigma \in Sel_\Sigma(x)$. This proves the desired result.

Finally, the similar argument (with $\{x_n\} \subset \mathcal{C}$ instead of the step functions) together with Lemma 1.5 shows that $Sel_\Sigma(x)$ is weakly u.s.c. with convex and weakly compact values, completing the proof. $\square$

Let us first introduce two families of operators on $\mathcal{H}$:

$$S_q(t) = \int_0^\infty \Psi_q(\theta) T(t^q \theta) d\theta, \quad \text{for } t \geq 0,$$

$$K_q(t) = \int_0^\infty q\theta \Psi_q(\theta) T(t^q \theta) d\theta, \quad \text{for } t \geq 0,$$

where $\Psi_q(\theta)$ is the Wright function (see Definition 1.8).

Definition 5.1. *A stochastic process $x \in \mathcal{C}$ is said to be a mild solution of inclusion (5.1) if $x(0) = x_0$ and there exists $\sigma(t) \in Sel_\Sigma(x)(t)$ satisfying the following integral equation:*

$$x(t) = S_q(t)x_0 + \int_0^t (t - s)^{q-1} K_q(t - s) f(s, x(s)) ds$$

$$+ \int_0^t (t-s)^{q-1} K_q(t-s)\sigma(s)dW(s).$$

Lemma 5.5. *The operators $S_q(t)$ and $K_q(t)$ have the following properties:*

(i) *for each fixed $t \geq 0$, $S_q(t)$ and $K_q(t)$ are linear and bounded operators, i.e., for any $x \in \mathcal{H}$,*

$$|S_q(t)x| \leq M_1|x| \ \text{ and } \ |K_q(t)x| \leq \frac{M_1|x|}{\Gamma(q)};$$

(ii) *$\{S_q(t)\}_{t\geq 0}$ and $\{K_q(t)\}_{t\geq 0}$ are strongly continuous;*
(iii) *$\{S_q(t)\}_{t\geq 0}$ is compact, if $\{T(t)\}_{t\geq 0}$ is compact.*

Remark 5.1. *For any $x \in \mathcal{C}$, define solution multioperator $\mathcal{F} : \mathcal{C} \to P(\mathcal{C})$ as follows:*

$$\mathcal{F} = S \circ Sel_\Sigma,$$

where

$$S(\sigma) = S_q(t)x_0 + \int_0^t (t-s)^{q-1} K_q(t-s)f(s, x(s))ds$$
$$+ \int_0^t (t-s)^{q-1} K_q(t-s)\sigma(s)dW(s).$$

It is easy to verify that the fixed points of the multioperator $\mathcal{F}$ are mild solutions of inclusion (5.1).

5.1.3 Existence

We study the existence for fractional stochastic evolution inclusion (5.1).

Fix $n \in \mathbb{N}$, consider $Q_n = \{x \in \mathcal{C} : \|x\|_{\mathcal{C}}^p \leq n\}$, and denote by $\mathcal{F}_n = \mathcal{F}|_{Q_n} : Q_n \to P(\mathcal{C})$ the restriction of the multioperator $\mathcal{F}$ on the set Q_n. We describe some properties of $\mathcal{F}_n$.

Lemma 5.6. *The multioperator $\mathcal{F}_n$ has a weakly sequentially closed graph.*

Proof. Let $\{x_m\} \subset Q_n$ and $\{y_m\} \subset C$ satisfy $y_m \in \mathcal{F}_n(x_m)$ for all m and $x_m \rightharpoonup x$, $y_m \rightharpoonup y$ in C, we will prove that $y \in \mathcal{F}_n(x)$.

Since $x_m \in Q_n$ for all m and $x_m(t) \rightharpoonup x(t)$ for every $t \in [0, b]$, it follows that

$$\|x(t)\| \leq \liminf_{m \to \infty} \|x_m(t)\| \leq n^{\frac{1}{p}},$$

for all t (see [47]). The fact that $y_m \in \mathcal{F}(x_m)$ means that there exists a sequence $\{\sigma_m\}$, $\sigma_m \in Sel_\Sigma(x_m)$, such that for every $t \in [0, b]$,

$$y_m(t) = S_q(t)x_0 + \int_0^t (t-s)^{q-1} K_q(t-s)f(s, x_m(s))ds$$
$$+ \int_0^t (t-s)^{q-1} K_q(t-s)\sigma_m(s)dW(s).$$

We observe that, according to (H_4), $\|\sigma_m(t)\|_{\mathscr{L}_2^0}^p \leq \mu_n(t)$ for a.e. t and every m, i.e., $\{\sigma_m\}$ is bounded and uniformly integrable and $\{\sigma_m(t)\}$ is bounded in $\mathscr{L}_2^0$ for a.e. $t \in [0, b]$. Hence, by the reflexivity of the space $\mathscr{L}_2^0$ and Lemma 1.4, we have the existence of a subsequence, denoted as the sequence, and a function σ such that $\sigma_m \rightharpoonup \sigma$ in $L^p([0, b], \mathscr{L}_2^0)$.

Moreover, we have

$$\int_0^t (t-s)^{q-1} K_q(t-s)\sigma_m(s)dW(s) \rightharpoonup \int_0^t (t-s)^{q-1} K_q(t-s)\sigma(s)dW(s).$$

Indeed, let $x' : L^p(\Omega, \mathcal{H}) \to \mathbb{R}$ be a linear continuous operator. We first prove that the operator

$$h \to \int_0^t (t-s)^{q-1} K_q(t-s)h(s)dW(s)$$

is linear and continuous operator from $L^p([0, b], \mathscr{L}_2^0)$ to $L^p(\Omega, \mathcal{H})$.

For any $h_m, h \in L^p([0, b], \mathscr{L}_2^0)$ and $h_m \to h$ $(m \to \infty)$, using (H_4) and Lemma 5.1, we get for each $t \in [0, b]$,

$$\mathbb{E}\left| \int_0^t (t-s)^{q-1} K_q(t-s)[h_m(s) - h(s)]dW(s) \right|^p$$
$$\leq C_p N(q) \left(\int_0^t (t-s)^{2(q-1)} \left(\mathbb{E}\|h_m(s) - h(s)\|_{\mathscr{L}_2^0}^p \right)^{\frac{2}{p}} ds \right)^{\frac{p}{2}}$$

$$\leq C_p N(q) \left(\frac{p-2}{a_1}\right)^{\frac{p-2}{2}} b^{\frac{a_1}{2}} \int_0^t \mathbb{E}\|h_m(s) - h(s)\|_{\mathscr{L}_2^0}^p ds$$
$$\to 0, \text{ as m} \to \infty,$$

where $a_1 = 2qp - p - 2$ and $N(q) = \left(\frac{M_1}{\Gamma(q)}\right)^p$. Hence, the operator

$$h \to \int_0^t (t-s)^{q-1} K_q(t-s) h(s) dW(s)$$

is continuous. Thus we have that the operator

$$h \to x' \circ \int_0^t (t-s)^{q-1} K_q(t-s) h(s) dW(s)$$

is a linear and continuous operator from $L^p([0, b], \mathscr{L}_2^0)$ to $\mathbb{R}$ for all $t \in [0, b]$. Then, from the definition of the weak convergence, we have for every $t \in [0, b]$

$$x' \circ \int_0^t (t-s)^{q-1} K_q(t-s) \sigma_m(s) dW(s)$$
$$\to x' \circ \int_0^t (t-s)^{q-1} K_q(t-s) \sigma(s) dW(s).$$

On the other hand, $f(s, x_m(s)) \rightharpoonup f(s, x(s))$ due to hypothesis (H_1). By the similar method, we also prove that the operator

$$g \to \int_0^t (t-s)^{q-1} K_q(t-s) g(s) ds$$

is linear and continuous operator from $L^p([0, b], L^p(\Omega, \mathcal{H}))$ to $L^p(\Omega, \mathcal{H})$. Thus,

$$\int_0^t (t-s)^{q-1} K_q(t-s) f(s, x_m(s)) ds \rightharpoonup \int_0^t (t-s)^{q-1} K_q(t-s) f(s, x(s)) ds.$$

From the above mentioned, we have

$$y_m(t) \rightharpoonup S_q(t) x_0 + \int_0^t (t-s)^{q-1} K_q(t-s) f(s, x(s)) ds$$
$$+ \int_0^t (t-s)^{q-1} K_q(t-s) \sigma(s) dW(s) = y^*(t), \quad \forall t \in [0, b],$$

which implies, for the uniqueness of the weak limit in $L^p(\Omega, \mathcal{H})$, that $y^*(t) = y(t)$ for all $t \in [0, b]$.

By using the similar method in Lemma 5.3, we can prove that $\sigma(t) \in \Sigma(t, x(t))$ for a.e. $t \in [0, b]$. The proof is completed. $\qquad\square$

Lemma 5.7. *The multioperator $\mathcal{F}_n$ is weakly compact.*

Proof. We first prove that $\mathcal{F}_n(Q_n)$ is relatively weakly sequentially compact.

Let $\{x_m\} \subset Q_n$ and $\{y_m\} \subset \mathcal{C}$ satisfy $y_m \in \mathcal{F}_n(x_m)$ for all m. By the definition of the multioperator $\mathcal{F}_n$, there exists a sequence $\{\sigma_m\}$, $\sigma_m \in Sel_\Sigma(x_m)$, such that for all $t \in [0, b]$,

$$y_m(t) = S_q(t)x_0 + \int_0^t (t - s)^{q-1} K_q(t - s) f(s, x_m(s)) ds$$
$$+ \int_0^t (t - s)^{q-1} K_q(t - s) \sigma_m(s) dW(s).$$

Further, as the reason for Lemma 5.6, we have that there exists a subsequence, denoted as the sequence, and a function σ such that $\sigma_m \rightharpoonup \sigma$ in $L^p([0, b], \mathscr{L}_2^0)$. Since the operator f maps bounded sets into bounded sets and Q_n is bounded, we obtain that $f(s, x_m(s)) \rightharpoonup \overline{f}(s) \in \mathcal{H}$ up to subsequence. Therefore,

$$y_m(t) \rightharpoonup l(t) = S_q(t)x_0 + \int_0^t (t - s)^{q-1} K_q(t - s) \overline{f}(s) ds$$
$$+ \int_0^t (t - s)^{q-1} K_q(t - s) \sigma(s) dW(s), \quad \forall t \in [0, b].$$

Furthermore, by (H_A), (H_1), and (H_4), we have

$$\mathbb{E}|y_m(t)|^p \leq 3^{p-1}\mathbb{E}|S_q(t)x_0|^p$$
$$+ 3^{p-1}\mathbb{E}\left| \int_0^t (t - s)^{q-1} K_q(t - s) f(s, x_m(s)) ds \right|^p$$
$$+ 3^{p-1}\mathbb{E}\left| \int_0^t (t - s)^{q-1} K_q(t - s) \sigma_m(s) dW(s) \right|^p$$
$$\leq 3^{p-1} M_1^p \mathbb{E}|x_0|^p + 3^{p-1} N(q)\mathbb{E}\left| \int_0^t (t - s)^{q-1} f(s, x_m(s)) ds \right|^p$$

$$+ C_p 3^{p-1} N(q) \left(\int_0^t (t-s)^{2(q-1)} \left(\mathbb{E} \|\sigma_m(s)\|_{\mathscr{L}_2^0}^p \right)^{\frac{2}{p}} ds \right)^{\frac{p}{2}}$$

$$\leq 3^{p-1} M_1^p \mathbb{E} |x_0|^p$$

$$+ 3^{p-1} N(q) \left(\frac{p-1}{pq-1} \right)^{p-1} t^{pq-1} \int_0^t \mathbb{E} |f(s, x_m(s))|^p ds$$

$$+ C_p 3^{p-1} N(q) \left(\int_0^t (t-s)^{2(q-1)} \mu_n^{\frac{2}{p}}(s) ds \right)^{\frac{p}{2}}$$

$$\leq 3^{p-1} M_1^p \|x_0\|^p + 3^{p-1} N(q) \left(\frac{p-1}{pq-1} \right)^{p-1} b^{pq} l$$

$$+ 3^{p-1} C_p N(q) \left(\frac{p-2}{a_1} \right)^{\frac{p-2}{2}} b^{\frac{a_1}{2}} \|\mu_n\|_{L[0,b]},$$

for all $m \in \mathbb{N}$ and $t \in [0,b]$, where $l = \max |f(t,x)|^p$. Thus

$$\|y_m(t)\|^p \leq 3^{p-1} M_1^p \|x_0\|^p + 3^{p-1} N(q) \left(\frac{p-1}{pq-1} \right)^{p-1} b^{pq} l$$

$$+ 3^{p-1} C_p N(q) \left(\frac{p-2}{a_1} \right)^{\frac{p-2}{2}} b^{\frac{a_1}{2}} \|\mu_n\|_{L[0,b]}$$

$$\leq N,$$

for all $m \in \mathbb{N}$ and for a.e. $t \in [0,b]$ and some N.

Recalling the weak convergence of $C([0,b], L^p(\Omega, \mathcal{H}))$, it is easy to prove that $y_m \rightharpoonup l$ in $\mathcal{C}$. Thus $\mathcal{F}_n(Q_n)$ is relatively weakly sequentially compact, hence relatively weakly compact by Theorem 1.3. $\square$

Lemma 5.8. *The multioperator $\mathcal{F}_n$ has convex and weakly compact values.*

Proof. Fix $x \in Q_n$. Since Σ is convex valued, from the linearity of the integral, and the operators $S_q(t)$ and $K_q(t)$, it follows that the set $\mathcal{F}_n(x)$ is convex. The weak compactness of $\mathcal{F}_n(x)$ follows by Lemmas 5.6 and 5.7. $\square$

Now we state the main results of this section.

Theorem 5.1. *Assume that (H_A) and (H_1)-(H_4) hold. Moreover,*

$$\liminf_{n\to\infty} \frac{1}{n} \int_0^t \mu_n(s)ds = 0. \tag{5.3}$$

Then inclusion (5.1) has at least one mild solution.

Proof. We show that there exists $n \in \mathbb{N}$ such that the operator $\mathcal{F}_n$ maps the ball Q_n into itself.

Assume, on the contrary, that there exist sequences $\{z_n\}$, $\{y_n\}$ such that $z_n \in Q_n, y_n \in \mathcal{F}_n(z_n)$, and $y_n \notin Q_n, \forall n \in \mathbb{N}$. Then there exists a sequence $\{\sigma_n\} \subset L^p([0,b], \mathscr{L}_2^0), \sigma_n(s) \in \Sigma(s, z_n(s)), \forall n \in \mathbb{N}$ and a.e. $s \in [0,b]$ such that

$$y_n(t) = S_q(t)x_0 + \int_0^t (t-s)^{q-1} K_q(t-s)f(s, x_n(s))ds$$
$$+ \int_0^t (t-s)^{q-1} K_q(t-s)\sigma_n(s)dW(s), \quad \forall t \in [0,b].$$

As the reason for Lemma 5.7, we have

$$1 < \frac{\|y\|_{\mathcal{C}}^p}{n} \le \frac{1}{n}\left[3^{p-1}M_1^p\|x_0\|^p + 3^{p-1}N(q)\left(\frac{p-1}{pq-1}\right)^{p-1}b^{pq}l\right]$$
$$+ \frac{1}{n}3^{p-1}C_pN(q)\left(\frac{p-2}{a_1}\right)^{\frac{p-2}{2}}b^{\frac{a_1}{2}}\int_0^t \mu_n(s)ds, \quad n \in \mathbb{N},$$

which contradicts (5.3).

Now, fix $n \in \mathbb{N}$ such that $\mathcal{F}_n(Q_n) \subsetneqq Q_n$. By Lemma 5.7, the set $V_n = \overline{\mathcal{F}_n(Q_n)}^w$ is weakly compact. Let now $\tilde{V}_n = \overline{co}(V_n)$, where $\overline{co}(V_n)$ denotes the closed convex hull of V_n. By Theorem 1.4, $\tilde{V}_n$ is a weakly compact set. Moreover, from the fact that $\mathcal{F}_n(Q_n) \subset Q_n$ and that Q_n is a convex closed set we have that $\tilde{V}_n \subset Q_n$, and hence

$$\mathcal{F}_n(\tilde{V}_n) = \mathcal{F}_n(\overline{co}(\mathcal{F}_n(Q_n))) \subseteq \mathcal{F}_n(Q_n) \subseteq \overline{\mathcal{F}_n(Q_n)}^w = V_n \subset \tilde{V}_n.$$

In view of Lemma 5.6, $\mathcal{F}_n$ has a weakly sequentially closed graph. Thus from Theorem 1.6, inclusion (5.1) has a solution. The proof is completed.

$\square$

Remark 5.2. *Suppose, for example, that there exist $\alpha(t) \in L^1([0, b], \mathbb{R}^+)$ and a nondecreasing function $\varrho : [0, +\infty) \rightarrow [0, +\infty)$ such that $\mathbb{E}\|\Sigma(t, x)\|_{\mathscr{L}_2^0}^p \leq \alpha(t)\varrho(\|x\|_C)$ for a.e. $t \in [0, b]$ and every $x \in C$. Then condition (5.3) is equivalent to*

$$\liminf_{n \to \infty} \frac{\varrho(n)}{n} = 0.$$

Theorem 5.2. *Assume that (H_A), (H_1)-(H_3), and $(H_4)'$ hold. If*

$$3^{p-1} C_p N(q) \left(\frac{p-2}{a_1} \right)^{\frac{p-2}{2}} b^{\frac{a_1}{2}} \|\alpha\|_{L[0,b]} < 1, \tag{5.4}$$

then inclusion (5.1) has at least one mild solution.

Proof. As the reason for Theorem 5.1, and assume that there exist $\{z_n\}$, $\{y_n\}$ such that $z_n \in Q_n$, $y_n \in \mathcal{F}_n(z_n)$, and $y_n \notin Q_n$, $\forall\, n \in \mathbb{N}$, we would get

$$
\begin{aligned}
\mathbb{E}|y_n(t)|^p \leq{} & 3^{p-1} M_1^p \|x_0\|^p + 3^{p-1} N(q) \left(\frac{p-1}{pq-1} \right)^{p-1} b^{pq} l \\
& + C_p 3^{p-1} N(q) \left(\int_0^t (t-s)^{2(q-1)} \alpha^{\frac{2}{p}}(t) \mathbb{E}(1 + |z_n(s)|^p)^{\frac{2}{p}} ds \right)^{\frac{p}{2}} \\
\leq{} & 3^{p-1} M_1^p \|x_0\|^p + 3^{p-1} N(q) \left(\frac{p-1}{pq-1} \right)^{p-1} b^{pq} l \\
& + 3^{p-1} C_p N(q) \left(\frac{p-2}{a_1} \right)^{\frac{p-2}{2}} b^{\frac{a_1}{2}} \int_0^t \alpha(t)(1 + \mathbb{E}|z_n(s)|^p) ds \\
\leq{} & 3^{p-1} M_1^p \|x_0\|^p + 3^{p-1} N(q) \left(\frac{p-1}{pq-1} \right)^{p-1} b^{pq} l \\
& + 3^{p-1} C_p N(q) \left(\frac{p-2}{a_1} \right)^{\frac{p-2}{2}} b^{\frac{a_1}{2}} (1 + n) \|\alpha\|_{L[0,b]}, \quad n \in \mathbb{N},
\end{aligned}
$$

so

$$
\begin{aligned}
n < \|y_n\|_C^p \leq{} & 3^{p-1} M_1^p \|x_0\|^p + 3^{p-1} N(q) \left(\frac{p-1}{pq-1} \right)^{p-1} b^{pq} l \\
& + 3^{p-1} C_p N(q) \left(\frac{p-2}{a_1} \right)^{\frac{p-2}{2}} b^{\frac{a_1}{2}} (1 + n) \|\alpha\|_{L[0,b]},
\end{aligned}
$$

which contradicts (5.4).

The conclusion then follows by Theorem 1.6, like Theorem 5.1.   □

Furthermore, we also consider superlinear growth condition, as next theorem shows.

Theorem 5.3. *Assume that* (H_A), (H_1), *and* (H_2) *hold. In addition, suppose that*

$(H_4)''$ *there exists* $\alpha \in L^1([0, b], \mathbb{R}^+)$ *and a nondecreasing function* $\rho : [0, +\infty) \to [0, +\infty)$ *such that*

$$\mathbb{E}\|\Sigma(t, x)\|^p_{\mathscr{L}^0_2} \leq \alpha(t)\rho(\|x\|_\mathcal{C}), \text{ for a.e. } t \in [0, b], \ \forall \, x \in \mathcal{C}.$$

Furthermore, there exists a constant $R > 0$ *such that*

$$3^{p-1}M_1^p\|x_0\|^p + 3^{p-1}N(q)$$
$$\times \left[\left(\frac{p-1}{pq-1} \right)^{p-1} b^{pq}l + C_p \left(\frac{p-2}{a_1} \right)^{\frac{p-2}{2}} b^{\frac{a_1}{2}} \|\alpha\|_{L[0,b]}\rho(R) \right] < R.$$

Then inclusion (5.1) has at least one mild solution.

Proof. It is sufficient to prove that the operator $\mathcal{F}$ maps the ball Q_R into itself. In fact, given any $z \in Q_R$ and $y \in \mathcal{F}(z)$, it holds

$$\mathbb{E}|y_n(t)|^p \leq 3^{p-1}M_1^p\|x_0\|^p + 3^{p-1}N(q)\left(\frac{p-1}{pq-1} \right)^{p-1} b^{pq}l$$
$$+ 3^{p-1}C_pN(q)\left(\frac{p-2}{a_1} \right)^{\frac{p-2}{2}} b^{\frac{a_1}{2}} \int_0^b \alpha(s)\rho(\|z\|_\mathcal{C})ds$$
$$\leq 3^{p-1}M_1^p\|x_0\|^p + 3^{p-1}N(q)\left(\frac{p-1}{pq-1} \right)^{p-1} b^{pq}l$$
$$+ 3^{p-1}C_pN(q)\left(\frac{p-2}{a_1} \right)^{\frac{p-2}{2}} b^{\frac{a_1}{2}} \|\alpha\|_{L[0,b]}\rho(R),$$

which implies

$$\|y_n\|^p_\mathcal{C} \leq 3^{p-1}M_1^p\|x_0\|^p + 3^{p-1}N(q)\left[\left(\frac{p-1}{pq-1} \right)^{p-1} b^{pq}l \right.$$
$$\left. + C_p \left(\frac{p-2}{a_1} \right)^{\frac{p-2}{2}} b^{\frac{a_1}{2}} \|\alpha\|_{L[0,b]}\rho(R) \right] \leq R.$$

The conclusion then follows by Theorem 1.6, like Theorem 5.1.   □

5.2 TOPOLOGICAL STRUCTURE OF SOLUTION SETS

In this section, we study the topological structure of solution sets in cases that $T(t)$ is compact and noncompact, respectively.

We present an approximation lemma in the following, which is a slightly modified version of Lemma 3.3 in [65].

Lemma 5.9. *Let hypotheses* (H_2), $(H_3)'$, *and* $(H_4)'$ *be satisfied. Then there exists a sequence* $\Sigma_n : [0, b] \times \mathcal{H} \to P_{cl,cv}(\mathcal{H})$ *such that*

(i) $\Sigma(t, x) \subset \Sigma_{n+1}(t, x) \subset \Sigma_n(t, x) \subset \overline{co}(\Sigma(t, B_{3^{1-n}}(x)))$, $n \geq 1$, *for each* $t \in [0, b]$ *and* $x \in \mathcal{H}$;

(ii) $\|\Sigma_n(t, x)\|_{\mathscr{L}_2^0}^p \leq r_1(t)(3 + |x|^p)$, $n \geq 1$, *for a.e.* $t \in [0, b]$ *and each* $x \in \mathcal{H}$;

(iii) *there exists* $E \subset [0, b]$ *with* $mes(E) = 0$ *such that for each* $x^* \in \mathcal{H}$, $\varepsilon > 0$ *and* $x \in \mathcal{H}$, *there exists* $N > 0$ *such that for all* $n \geq N$,

$$\langle x^*, \Sigma_n(t, x) \rangle \subset \langle x^*, \Sigma(t, x) \rangle + (-\varepsilon, \varepsilon);$$

(iv) $\Sigma_n(t, \cdot) : \mathcal{H} \to P_{cl,cv}(\mathcal{H})$ *is continuous for a.e.* $t \in [0, b]$ *with respect to Hausdorff metric for each* $n \geq 1$;

(v) *for each* $n \geq 1$, *there exists a selection* $\tilde{\sigma}_n : [0, b] \times X \to \mathcal{H}$ *of* Σ_n *such that* $\tilde{\sigma}_n(\cdot, x)$ *is measurable for each* $x \in \mathcal{H}$ *and for any compact subset* $\mathscr{D} \subset \mathcal{H}$ *there exist constants* $C_V > 0$ *and* $\delta > 0$ *for which the estimate*

$$\|\tilde{\sigma}_n(t, x_1) - \tilde{\sigma}_n(t, x_2)\|_{\mathscr{L}_2^0}^p \leq C_V \alpha(t) |x_1 - x_2|^p$$

holds for a.e. $t \in [0, b]$ *and each* $x_1, x_2 \in V$ *with* $V := \mathscr{D} + B_\delta(0)$;

(vi) Σ_n *verifies condition* $(H_3)'$ *with* Σ_n *instead of* Σ *for each* $n \geq 1$.

5.2.1 Compact Semigroup Case

In this subsection, we study the topological structure of solution sets in cases that $T(t)$ is compact. The following compactness characterizations of solution sets to inclusion (5.1) will be useful.

Lemma 5.10. *Suppose that* $T(t)$ *is compact for* $t > 0$. *Let* $D \subset L^p(\Omega, \mathcal{H})$ *be relatively compact and* $K \subset L^p([0, b], \mathscr{L}_2^0)$-*integrable bound-*

ed, that is,

$$\|\sigma(t, x)\|^p_{\mathscr{L}^0_2} \leq \gamma(t)$$

for all $\sigma \in K$ and a.e. $t \in [0, b]$, where $\gamma \in L^1([0, b], \mathbb{R}^+)$. Then the set of mild solutions

$$\{x(\cdot, x_0, \sigma) : x_0 \in D, \sigma \in K\}$$

is relatively compact in $\mathcal{C}$.

Proof. Write

$$\Delta(D \times K) = \{x(\cdot, x_0, \sigma) : x_0 \in D, \sigma \in K\}.$$

Let $t \in [0, b]$ be arbitrary and $\varepsilon > 0$ small enough. Define the operator $\Psi_\varepsilon : \Delta(D \times K)(t) \to L^p(\Omega, \mathcal{H})$ by

$$
\begin{aligned}
\Psi_\varepsilon x(t) =& S_q(t)x_0 + T(\varepsilon^q \delta) \int_0^{t-\varepsilon} \int_\delta^\infty q\theta(t-s)^{q-1} \xi_q(\theta) \\
& \times T((t-s)^q \theta - \varepsilon^q \delta) f(s, x(s)) d\theta ds \\
& + T(\varepsilon^q \delta) \int_0^{t-\varepsilon} \int_\delta^\infty q\theta(t-s)^{q-1} \xi_q(\theta) \\
& \times T((t-s)^q \theta - \varepsilon^q \delta) \sigma(s) d\theta dW(s).
\end{aligned}
$$

Using the compactness of $T(t)$ for $t > 0$, we deduce that the set $\{\Psi_\varepsilon x(t) : x \in \Delta(D \times K)(t)\}$ is relatively compact in $L^p(\Omega, \mathcal{H})$ for every $\varepsilon, 0 < \varepsilon < t$. Moreover, for every $x \in \Delta(D \times K)$, we have

$$
\begin{aligned}
& \mathbb{E}|\Psi_\varepsilon x(t) - x(t)|^p \\
\leq & 4^{p-1} \mathbb{E} \left| \int_0^t \int_0^\delta q\theta(t-s)^{q-1} \xi_q(\theta) T((t-s)^q \theta) f(s, x(s)) d\theta ds \right|^p \\
& + 4^{p-1} \mathbb{E} \left| \int_{t-\varepsilon}^t \int_\delta^\infty q\theta(t-s)^{q-1} \xi_q(\theta) T((t-s)^q \theta) f(s, x(s)) d\theta ds \right|^p \\
& + 4^{p-1} \mathbb{E} \left| \int_0^t \int_0^\delta q\theta(t-s)^{q-1} \xi_q(\theta) T((t-s)^q \theta) \sigma(s) d\theta dW(s) \right|^p \\
& + 4^{p-1} \mathbb{E} \left| \int_{t-\varepsilon}^t \int_\delta^\infty q\theta(t-s)^{q-1} \xi_q(\theta) T((t-s)^q \theta) \sigma(s) d\theta dW(s) \right|^p \\
\leq & 4^{p-1} \left(M_1 \int_0^\delta q\theta \xi_q(\theta) d\theta \right)^p \mathbb{E} \left| \int_0^t (t-s)^{q-1} f(s, x(s)) ds \right|^p
\end{aligned}
$$

$$+ 4^{p-1} \left(M_1 \int_\delta^\infty q\theta\xi_q(\theta)d\theta \right)^p \mathbb{E} \left| \int_{t-\varepsilon}^t (t-s)^{q-1} f(s, x(s))ds \right|^p$$

$$+ 4^{p-1} C_p \left(M_1 \int_0^\delta q\theta\xi_q(\theta)d\theta \right)^p$$

$$\times \left(\int_0^t (t-s)^{2(q-1)} \left(\mathbb{E}\|\sigma(s)\|_{\mathscr{L}_2^0}^p \right)^{\frac{2}{p}} ds \right)^{\frac{p}{2}}$$

$$+ 4^{p-1} C_p \left(M_1 \int_\delta^\infty q\theta\xi_q(\theta)d\theta \right)^p$$

$$\times \left(\int_{t-\varepsilon}^t (t-s)^{2(q-1)} \left(\mathbb{E}\|\sigma(s)\|_{\mathscr{L}_2^0}^p \right)^{\frac{2}{p}} ds \right)^{\frac{p}{2}}$$

$$\leq 4^{p-1} \left(M_1 \int_0^\delta q\theta\xi_q(\theta)d\theta \right)^p \left(\frac{p-1}{pq-1} \right)^{p-1} t^{pq-1} \int_0^t \mathbb{E}|f(s, x(s))|^p ds$$

$$+ C_p 4^{p-1} \left(M_1 \int_0^\delta q\theta\xi_q(\theta)d\theta \right)^p \left(\frac{p-2}{a_1} \right)^{\frac{p-2}{2}} b^{\frac{a_1}{2}} \|\gamma\|_{L[0,b]}$$

$$+ 4^{p-1} N(q) \left(\frac{p-1}{pq-1} \right)^{p-1} \varepsilon^{pq-1} \int_{t-\varepsilon}^t \mathbb{E}|f(s, x(s))|^p ds$$

$$+ C_p 4^{p-1} N(q) \left(\frac{p-2}{a_1} \right)^{\frac{p-2}{2}} \varepsilon^{\frac{a_1}{2}} \int_{t-\varepsilon}^t \gamma(t)ds$$

$$\to 0, \quad \text{as } \varepsilon \to 0.$$

Then, we obtain $\|\Psi_\varepsilon x(t) - x(t)\|_C \to 0$, which proves that the identity operator $I : \Delta(D \times K)(t) \to \Delta(D \times K)(t)$ is a compact operator, which yields that the set $\Delta(D \times K)(t)$ is relatively compact in $L^p(\Omega, \mathcal{H})$ for each $t \in (0, b]$.

We proceed to verify that the set $\Delta(D \times K)$ is equicontinuous on $(0, b]$. Taking $0 < t_1 < t_2 \leq b$. For each $x \in \Delta(D \times K)$, we obtain

$$\mathbb{E}|x(t_2) - x(t_1)|^p$$

$$\leq 3^{p-1} \left(\mathbb{E}|[S_q(t_2) - S_q(t_1)]x_0|^p + \sum_{i=1}^2 \mathbb{E}|I_i(t_2) - I_i(t_1)|^p \right),$$

where

$$I_1(t) = \int_0^t (t-s)^{q-1} K_q(t-s) f(s, x(s))ds,$$

$$I_2(t) = \int_0^t (t - s)^{q-1} K_q(t - s)\sigma(s)dW(s).$$

From the strong continuity of $S_q(t)$, it is clear that the first term goes to zero as $t_2 - t_1 \to 0$.

Next, it follows from assumptions on the theorem that

$\mathbb{E}|I_1(t_2) - I_1(t_1)|^p$

$\leq 4^{p-1}\mathbb{E}\left|\int_{t_1}^{t_2} (t_2 - s)^{q-1} K_q(t_2 - s)f(s, x(s))ds\right|^p$

$+ 4^{p-1}\mathbb{E}\left|\int_0^{t_1-\delta} (t_1 - s)^{q-1}\left(K_q(t_2 - s) - K_q(t_1 - s)\right)f(s, x(s))ds\right|^p$

$+ 4^{p-1}\mathbb{E}\left|\int_{t_1-\delta}^{t_1} (t_1 - s)^{q-1}\left(K_q(t_2 - s) - K_q(t_1 - s)\right)f(s, x(s))ds\right|^p$

$+ 4^{p-1}\mathbb{E}\left|\int_0^{t_1} \left((t_2 - s)^{q-1} - (t_1 - s)^{q-1}\right)K_q(t_2 - s)f(s, x(s))ds\right|^p$

$\leq 4^{p-1}N(q)\left(\int_{t_1}^{t_2} (t_2 - s)^{\frac{p(q-1)}{p-1}}ds\right)^{p-1}\int_{t_1}^{t_2} \mathbb{E}|f(s, x(s))|^p ds$

$+ 4^{p-1}\sup_{s\in[0,t_1-\delta]}\|K_q(t_2 - s) - K_q(t_1 - s)\|_{\mathscr{L}(\mathcal{H})}^p$

$\times \left(\int_0^{t_1-\delta} (t_1 - s)^{\frac{p(q-1)}{p-1}}ds\right)^{p-1}\int_0^{t_1-\delta} \mathbb{E}|f(s, x(s))|^p ds$

$+ 8^{p-1}N(q)\left(\int_{t_1-\delta}^{t_1} (t_1 - s)^{\frac{p(q-1)}{p-1}}ds\right)^{p-1}\int_{t_1-\delta}^{t_1} \mathbb{E}|f(s, x(s))|^p ds$

$+ 4^{p-1}N(q)\left(\int_0^{t_1} \left((t_2 - s)^{q-1} - (t_1 - s)^{q-1}\right)^{\frac{p}{p-1}}ds\right)^{p-1}$

$\times \int_0^{t_1} \mathbb{E}|f(s, x(s))|^p ds$

$\leq 4^{p-1}N(q)\left(\frac{p-1}{pq-1}\right)^{p-1}(t_2 - t_1)^{pq-1}\int_{t_1}^{t_2} \mathbb{E}|f(s, x(s))|^p ds$

$+ 4^{p-1}\sup_{s\in[0,t_1-\delta]}\|K_q(t_2 - s) - K_q(t_1 - s)\|_{\mathscr{L}(\mathcal{H})}^p\left(\frac{p-1}{pq-1}\right)^{p-1}$

$\times \left(t_1^{pq-1} - \delta^{pq-1}\right)\int_0^{t_1} \mathbb{E}|f(s, x(s))|^p ds$

$$+ 8^{p-1} N(q) \left(\frac{p-1}{pq-1} \right)^{p-1} \delta^{pq-1} \int_{t_1-\delta}^{t_1} \mathbb{E}|f(s, x(s))|^p ds$$

$$+ 4^{p-1} N(q) \left(\frac{p-1}{pq-1} \right)^{p-1} \left(t_2^{\frac{pq-1}{p-1}} - t_1^{\frac{pq-1}{p-1}} - (t_2 - t_1)^{\frac{pq-1}{p-1}} \right)^{p-1}$$

$$\times \int_0^{t_1} \mathbb{E}|f(s, x(s))|^p ds.$$

Further, we obtain

$$\mathbb{E}|I_2(t_2) - I_2(t_1)|^p$$

$$\leq 4^{p-1} \mathbb{E} \left| \int_{t_1}^{t_2} (t_2 - s)^{q-1} K_q(t_2 - s) \sigma(s) dW(s) \right|^p$$

$$+ 4^{p-1} \mathbb{E} \left| \int_0^{t_1-\delta} (t_1 - s)^{q-1} \big(K_q(t_2 - s) - K_q(t_1 - s) \big) \sigma(s) dW(s) \right|^p$$

$$+ 4^{p-1} \mathbb{E} \left| \int_{t_1-\delta}^{t} (t_1 - s)^{q-1} \big(K_q(t_2 - s) - K_q(t_1 - s) \big) \sigma(s) dW(s) \right|^p$$

$$+ 4^{p-1} \mathbb{E} \left| \int_0^{t_1} \big((t_2 - s)^{q-1} - (t_1 - s)^{q-1} \big) K_q(t_2 - s) \sigma(s) dW(s) \right|^p$$

$$\leq 4^{p-1} C_p N(q) \left(\int_{t_1}^{t_2} (t_2 - s)^{2(q-1)} \big(\mathbb{E} \|\sigma(s)\|_{\mathscr{L}_2^0}^p \big)^{\frac{2}{p}} ds \right)^{\frac{p}{2}}$$

$$+ 4^{p-1} C_p \sup_{s \in [0, t_1-\delta]} \| K_q(t_2 - s) - K_q(t_1 - s) \|_{\mathscr{L}(\mathcal{H})}^p$$

$$\times \left(\int_0^{t_1-\delta} (t_1 - s)^{2(q-1)} \big(\mathbb{E} \|\sigma(s)\|_{\mathscr{L}_2^0}^p \big)^{\frac{2}{p}} ds \right)^{\frac{p}{2}}$$

$$+ 8^{p-1} C_p N(q) \left(\int_{t_1-\delta}^{t} (t_1 - s)^{2(q-1)} \big(\mathbb{E} \|\sigma(s)\|_{\mathscr{L}_2^0}^p \big)^{\frac{2}{p}} ds \right)^{\frac{p}{2}} + 4^{p-1} C_p$$

$$\times N(q) \left(\int_0^{t_1} \big((t_2 - s)^{q-1} - (t_1 - s)^{q-1} \big)^2 \big(\mathbb{E} \|\sigma(s)\|_{\mathscr{L}_2^0}^p \big)^{\frac{2}{p}} ds \right)^{\frac{p}{2}}$$

$$\leq 4^{p-1} C_p N(q) \left(\int_{t_1}^{t_2} (t_2 - s)^{2(q-1)} \gamma^{\frac{2}{p}}(s) ds \right)^{\frac{p}{2}}$$

$$+ 4^{p-1} C_p \sup_{s \in [0, t_1-\delta]} \| K_q(t_2 - s) - K_q(t_1 - s) \|_{\mathscr{L}(\mathcal{H})}^p$$

$$\times \left(\int_0^{t_1-\delta} (t_1 - s)^{2(q-1)} \gamma^{\frac{2}{p}}(s) ds \right)^{\frac{p}{2}}$$

$$+ 8^{p-1} C_p N(q) \left(\int_{t_1-\delta}^{t} (t_1 - s)^{2(q-1)} \gamma^{\frac{2}{p}}(s) ds \right)^{\frac{p}{2}}$$

$$+ 4^{p-1} C_p N(q) \left(\int_{0}^{t_1} \left((t_2 - s)^{q-1} - (t_1 - s)^{q-1} \right)^2 \gamma^{\frac{2}{p}}(s) ds \right)^{\frac{p}{2}}$$

$$\leq 4^{p-1} C_p N(q) \left(\frac{p-2}{a_1} \right)^{\frac{p-2}{2}} (t_2 - t_1)^{\frac{a_1}{2}} \int_{t_1}^{t_2} \gamma(s) ds$$

$$+ 4^{p-1} C_p \sup_{s \in [0, t_1-\delta]} \| K_q(t_2 - s) - K_q(t_1 - s) \|_{\mathscr{L}(\mathcal{H})}^{p} \left(\frac{p-2}{a_1} \right)^{\frac{p-2}{2}}$$

$$\times \left(t_1^{\frac{a_1}{2}} - \delta^{\frac{a_1}{2}} \right) \int_{0}^{t_1-\delta} \gamma(s) ds$$

$$+ 8^{p-1} C_p N(q) \left(\frac{p-2}{a_1} \right)^{\frac{p-2}{2}} \delta^{\frac{a_1}{2}} \int_{t_1-\delta}^{t_1} \gamma(s) ds + 4^{p-1} C_p N(q)$$

$$\times \left(\frac{p-2}{a_1} \right)^{\frac{p-2}{2}} \left(t_2^{\frac{a_1}{p-2}} - t_1^{\frac{a_1}{p-2}} - (t_2 - t_1)^{\frac{a_1}{p-2}} \right)^{\frac{p-2}{2}} \int_{0}^{t_1} \gamma(s) ds.$$

Therefore, for $t_2 - t_1$ sufficiently small, the right-hand side of above two inequalities tends to zero, since $T(t)$ is strongly continuous, and the compactness of $T(t)$ $(t > 0)$ implies the continuity in the uniform operator topology.

Moreover, we see by the relative compactness of D that these limits remain true uniformly for $x \in \Delta(D \times K)$. That is to say

$$\| x(t_1) - x(t_2) \|^p \to 0, \quad \text{as } t_2 - t_1 \to 0$$

uniformly for $x \in \Delta(D \times K)$, and hence we get the desired result.

Thus, an application of Lemma 1.2 justifies that $\Delta(D \times K)$ is relatively compact in $\mathcal{C}$. The proof is completed.   □

Let $c \in [0, b)$. Under the condition of $(H_1)'$, we consider the singular integral equation of the form

$$x(t) = \phi(t) + \int_{c}^{t} (t - s)^{q-1} K_q(t - s) f(s, x(s)) ds$$

$$+ \int_{c}^{t} (t - s)^{q-1} K_q(t - s) \tilde{\sigma}(s, x(s)) dW(s), \quad \text{for } t \in [c, b].$$
(5.5)

Similar to the proof of Lemma 3.2 in [280], we can get the following lemma.

Lemma 5.11. *Let* $p > 2$, $pq > 1$, $\tilde{\sigma}(\cdot, x)$ *be* L^p-*integrable for every* $x \in \mathcal{H}$. *Assume that* $\{T(t)\}_{t \geq 0}$ *is compact. Suppose in addition that*

(i) *for any compact subset* $K \subset \mathcal{H}$, *there exist* $\delta > 0$ *and* $L_K \in L^1([c, b], \mathbb{R}^+)$ *such that*

$$\|\tilde{\sigma}(t, x_1) - \tilde{\sigma}(t, x_2)\|_{\mathscr{L}_2^0}^p \leq L_K(t)|x_1 - x_2|^p,$$

for a.e. $t \in [c, b]$ *and each* x_1, $x_2 \in B_\delta(K) = K + B_\delta(0)$;
(ii) *there exists* $\gamma_1(t) \in L^1([c, b], \mathbb{R}^+)$ *such that* $\|\tilde{\sigma}(t, x)\|_{\mathscr{L}_2^0}^p \leq \gamma_1(t)(c' + |x|^p)$ *for a.e.* $t \in [c, b]$ *and every* $x \in \mathcal{H}$, *where* c' *is arbitrary, but fixed.*

Then the integral equation (5.5) admits a unique solution for every $\phi(t) \in C([c, b], L^p(\Omega, \mathcal{H}))$. *Moreover, the solution of (5.5) depends continuously on* ϕ.

Let us also present the following approximation result, the proof of which is similar to the proof of Lemma 2.4 in [279].

Lemma 5.12. *Suppose* $\{T(t)\}_{t \geq 0}$ *is compact and* $f(t, x)$ *is continuous. If the two sequences* $\{\sigma_m\} \subset L^p([0, b], \mathscr{L}_2^0)$ *and* $\{x_m\} \subset C$, *where* x_m *is a mild solution of the stochastic problem:*

$$\begin{cases} {}_0^C D_t^q x_m(t) = A x_m(t) + f(t, x_m(t)) + \sigma_m(t)\dfrac{dW(t)}{dt}, & t \in (0, b], \\ x_m(0) = x_0, \end{cases}$$

$\sigma_m \rightharpoonup \sigma$ *in* $L^p([0, b], \mathscr{L}_2^0)$ *and* $x_m \to x$ *in* C, *then* x *is a mild solution of the limit problem:*

$$\begin{cases} {}_0^C D_t^q x(t) = A x(t) + f(t, x(t)) + \sigma(t)\dfrac{dW(t)}{dt}, & t \in (0, b], \\ x(0) = x_0. \end{cases}$$

Theorem 5.4. *Let conditions* (H_A), $(H_1)'$, (H_2), $(H_3)'$, *and* $(H_4)'$ *be satisfied. Suppose in addition that* $T(t)$ *is compact for* $t > 0$. *Then the solution set of inclusion (5.1) for fixed* $x_0 \in \mathcal{H}$ *is a nonempty compact subset of* C. *Moreover, it is a compact* R_δ-*set.*

Proof. We will proceed in two steps.

Claim 1. The solution set of inclusion (5.1) is nonempty.

Set

$$\mathcal{M}_0 = \{x \in \mathcal{C} : \|x(t)\|^p \le \psi(t), \ t \in [0, b]\},$$

where $\psi(t)$ is the solution of the integral equation

$$\psi(t) = \rho_3 + \rho_1 \int_0^t c_0 \psi(s) ds + \rho_2 \int_0^t \alpha(s) \psi(s) ds,$$

in which ρ_1, ρ_2, and ρ_3 are defined as

$$\rho_1 = 3^{p-1} N(q) \left(\frac{p-1}{pq-1} \right)^{p-1} b^{pq-1},$$

$$\rho_2 = C_p 3^{p-1} N(q) \left(\frac{p-2}{a_1} \right)^{\frac{p-2}{2}} b^{\frac{a_1}{2}},$$

$$\rho_3 = 3^{p-1} M_1^p \|x_0\|^p + \rho_1 c_1 + \rho_2 \|\alpha\|_{L[0,b]}.$$

It is clear that $\mathcal{M}_0$ is closed and convex of $\mathcal{C}$. We first show that $\mathcal{F}(\mathcal{M}_0) \subset \mathcal{M}_0$. Indeed, taking $x \in \mathcal{M}_0$ and $y(t) \in \mathcal{F}(x)$, we have

$$\mathbb{E}|y(t)|^p$$

$$\le 3^{p-1} \mathbb{E}|S_q(t) x_0|^p + 3^{p-1} \mathbb{E} \left| \int_0^t (t-s)^{q-1} K_q(t-s) f(s, x(s)) ds \right|^p$$

$$+ 3^{p-1} \mathbb{E} \left| \int_0^t (t-s)^{q-1} K_q(t-s) \sigma(s) dW(s) \right|^p$$

$$\le 3^{p-1} M_1^p \mathbb{E}|x_0|^p + 3^{p-1} N(q) \left(\frac{p-1}{pq-1} \right)^{p-1} t^{pq-1} \int_0^t \mathbb{E}|f(s, x(s))|^p ds$$

$$+ C_p 3^{p-1} N(q) \left(\int_0^t (t-s)^{2(q-1)} \alpha^{\frac{2}{p}}(t) 1 + \psi(s))^{\frac{2}{p}} ds \right)^{\frac{p}{2}}$$

$$\le 3^{p-1} M_1^p \mathbb{E}|x_0|^p + 3^{p-1} N(q) \left(\frac{p-1}{pq-1} \right)^{p-1} b^{pq-1} \int_0^t (c_0 \psi(s) + c_1) ds$$

$$+ C_p 3^{p-1} N(q) \left(\frac{p-2}{a_1} \right)^{\frac{p-2}{2}} b^{\frac{a_1}{2}} \int_0^t \alpha(s)(1 + \psi(s)) ds$$

$$\le \rho_3 + \rho_1 \int_0^t c_0 \psi(s) ds + \rho_2 \int_0^t \alpha(s) \psi(s) ds$$

$$\leq \psi(t),$$

which implies that $\|y(t)\|^p \leq \psi(t)$. Thus $y \in \mathcal{M}_0$. Set $\widetilde{\mathcal{M}} = \overline{co}\mathcal{F}(\mathcal{M}_0)$. It is clear that $\widetilde{\mathcal{M}}$ is a closed, bounded, and convex set. Also, $\widetilde{\mathcal{M}}$ is invariant under $\mathcal{F}$, i.e., $\mathcal{F}(\widetilde{\mathcal{M}}) \subset \widetilde{\mathcal{M}}$. Moreover, using Lemma 5.10 enables us to find that $\widetilde{\mathcal{M}}$ is relative compact in $\mathcal{C}$.

We consider $\mathcal{F} : \widetilde{\mathcal{M}} \to P(\widetilde{\mathcal{M}})$. In order to apply the fixed point principle given by Theorem 1.13, it remains to show that $\mathcal{F}$ is u.s.c. with contractible values.

By Lemma 1.7, it suffices to show that $\mathcal{F}$ has closed graph. Let $x_n \subset \widetilde{\mathcal{M}}$ with $x_n \to x$ and $y_n \in \mathcal{F}(x_n)$ with $y_n \to y$. We shall prove that $y \in \mathcal{F}(x)$. By the definition of $\mathcal{F}$, there exist $\sigma_n \in Sel_\Sigma(x_n)$ such that

$$y_n(t) = S_q(t)x_0 + \int_0^t (t-s)^{q-1} K_q(t-s) f(s, x_n(s)) ds$$
$$+ \int_0^t (t-s)^{q-1} K_q(t-s) \sigma_n(s) dW(s), \quad \forall\, t \in [0, b].$$

We need to prove that there exists $\sigma \in Sel_\Sigma(x)$ such that for a.e. $t \in [0, b]$,

$$y(t) = S_q(t)x_0 + \int_0^t (t-s)^{q-1} K_q(t-s) f(s, x(s)) ds$$
$$+ \int_0^t (t-s)^{q-1} K_q(t-s) \sigma(s) dW(s).$$

First, we should prove

$$\int_0^t (t-s)^{q-1} K_q(t-s) f(s, x_n(s)) ds$$
$$\to \int_0^t (t-s)^{q-1} K_q(t-s) f(s, x(s)) ds.$$

Since $x_n \subset \widetilde{\mathcal{M}}$ and $x_n \to x$, using $(H_1)'$, we have $f(t, x_n(t)) \to f(s, x(s))$, as $n \to \infty$. On the one hand, we get for each $t \in [0, b]$,

$$\int_0^t \mathbb{E}|f(s, x_n(s)) - f(s, x(s))|^p ds \leq 2^p \left(c_0 \int_0^t \psi(s) ds + c_1 t \right) < \infty.$$

By Theorem 1.1, we get

$$
\mathbb{E}\left| \int_0^t (t-s)^{q-1} K_q(t-s)[f(s, x_n(s)) - f(s, x(s))]ds \right|^p
$$

$$
\leq N(q)\mathbb{E}\left(\int_0^t (t-s)^{q-1} |f(s, x_n(s)) - f(s, x(s))|ds \right)^p
$$

$$
\leq N(q)\left(\int_0^t (t-s)^{\frac{p(q-1)}{p-1}} ds \right)^{p-1} \int_0^t \mathbb{E}|f(s, x_n(s)) - f(s, x(s))|^p ds
$$

$$
\leq N(q)\left(\frac{p-1}{pq-1} \right)^{p-1} b^{pq-1} \int_0^t \mathbb{E}|f(s, x_n(s)) - f(s, x(s))|^p ds
$$

$$
\to 0, \text{ as } \to \infty.
$$

Therefore,

$$
\int_0^t (t-s)^{q-1} K_q(t-s) f(s, x_n(s))ds
$$

$$
\to \int_0^t (t-s)^{q-1} K_q(t-s) f(s, x(s))ds.
$$

Noticing that Sel_Σ is weakly u.s.c. with weakly compact and convex values due to Lemma 5.4, an application of Lemma 1.5 yields that there exist $\sigma \in Sel_\Sigma(x)$ and a subsequence of σ_n, still denoted by σ_n, such that $\sigma_n \rightharpoonup \sigma$ in $L^p([0, b], \mathscr{L}_2^0)$. From this and Lemma 5.12, we see that $y(t) = S(\sigma)$ and $y \in \mathcal{F}(x)$. It follows that $\mathcal{F}$ is closed.

After that, we can show that $\mathcal{F}$ has contractible values. Given $x \in \widetilde{\mathcal{M}}$. Fix $\sigma^* \in Sel_\Sigma(x)$ and put $y^* = S(\sigma^*)$. Define a function $h : [0, 1] \times \mathcal{F}(x) \to \mathcal{F}(x)$ as

$$
h(\lambda, y)(t) = \begin{cases} y(t), & t \in [0, \lambda b], \\ x(t, \lambda b, y(\lambda b)), & t \in [\lambda b, b], \end{cases}
$$

for each $(\lambda, y) \in [0, 1] \times \mathcal{F}(x)$, where

$$
x(t, \lambda b, y(\lambda b)) = S_q(t)x_0 + \int_0^{\lambda b} (t-s)^{q-1} K_q(t-s) f(s, y(s))ds
$$

$$
+ \int_0^{\lambda b} (t-s)^{q-1} K_q(t-s)\sigma(s)dW(s)
$$

$$+ \int_{\lambda b}^{t} (t - s)^{q-1} K_q(t - s) f(s, x(s)) ds$$

$$+ \int_{\lambda b}^{t} (t - s)^{q-1} K_q(t - s) \sigma^*(s)(s) dW(s);$$

here, $\sigma \in Sel_\Sigma(y)$ is such that $y = S(\sigma)$.

It is easy to see that h is well defined. Also, it is clear that

$$h(0, y) = y^*, \quad h(1, y) = y, \text{ on } \mathcal{F}(x).$$

Moreover, it follows readily that h is continuous. Thus, we have proved that $\mathcal{F}(x)$ is contractible.

Claim 2. Let $\Theta(x_0)$ denote the set of all mild solutions of inclusion (5.1). Then $\Theta(x_0)$ is also an R_δ-set.

To this aim, let us consider the fractional stochastic evolution inclusion

$$\begin{cases} {}_0^C D_t^q x(t) \in Ax(t) + f(t, x(t)) + \Sigma_n(t) \dfrac{dW(t)}{dt}, & t \in [0, b], \\ x(0) = x_0. \end{cases} \tag{5.6}$$

Let $\Theta^n(x_0)$ denote the set of all mild solutions of inclusion (5.6).

From (ii) and (vi), it follows that $\{\Sigma_n\}$ verifies conditions (H_2), $(H_3)'$, and $(H_4)'$ for each $n \geq 1$. Then from Lemma 5.4 one finds that Sel_{Σ_n} is nonempty weakly u.s.c. with convex and weakly compact values. Moreover, one can see from the above arguments that the solution set of inclusion (5.6) is nonempty and compact in $\mathcal{C}$ for each $n \geq 1$.

We show that $\Theta^n(x_0)$ is contractible for all $n \geq 1$. To do this, for any $\lambda \in [0, 1]$, let $x \in \Theta^n(x_0)$ and $\tilde{\sigma}_n$ be the selection of $\Sigma_n, n \geq 1$. We deal with the existence and uniqueness of solutions to the integral equation

$$y(t) = S_q(t) x_0 + \int_0^{\lambda b} (t - s)^{q-1} K_q(t - s) f(s, x(s)) ds$$

$$+ \int_0^{\lambda b} (t - s)^{q-1} K_q(t - s) \sigma^x(s) dW(s) \tag{5.7}$$

$$+ \int_{\lambda b}^{t} (t-s)^{q-1} K_q(t-s) f(s, y(s)) ds$$

$$+ \int_{\lambda b}^{t} (t-s)^{q-1} K_q(t-s) \tilde{\sigma}_n(s, y(s)) dW(s),$$

where $\sigma^x \in Sel_{\Sigma_n}(x)$. Moreover, it follows from Lemma 5.9(ii) that $\|\tilde{\sigma}_n(t,x)\|_{\mathscr{L}_2^0}^p \le r_1(t)(3 + |x|^p)$ for a.e. $t \in [0, b]$ and $x \in \mathcal{H}$. Therefore, noticing Lemma 5.9(ii) one finds, together with Lemma 5.11, that equation (5.7) has a unique solution on $[\lambda b, b]$, denoted by $y(t, \lambda b, x(\lambda b))$.

We define $h : [0, 1] \times \Theta^n(x_0) \to \Theta^n(x_0)$ by

$$h(\lambda, x)(t) = \begin{cases} x(t), & t \in [0, \lambda b], \\ y(t, \lambda b, x(\lambda b)), & t \in [\lambda b, b], \end{cases}$$

for each $(\lambda, x) \in [0, 1] \times \Theta^n(x_0)$.

It is easy to see that h is well defined. Also, it is clear that

$$h(0, x) = y(t, 0, x_0), \quad h(1, x) = x, \text{ on } \Theta^n(x_0).$$

Moreover, it follows readily that h is continuous. Thus, we have proved that $\Theta^n(x_0)$ is contractible for each $n \ge 1$.

Finally, in view of Lemma 5.9(i), it is easy to verify that $\Theta(x_0) \subset \cdots \subset \Theta^n(x_0) \cdots \subset \Theta^2(x_0) \subset \Theta^1(x_0)$. Similar arguments as the proof in Theorem 3.6, we can prove that $\Theta(x_0) = \bigcap_{n \ge 1} \Theta^n(x_0)$.

Consequently, we conclude that $\Theta(x_0)$ is an R_δ-set. The proof is completed. $\square$

Example 5.1. In the following, an example is presented for the existence results to the following stochastic differential inclusion:

$$\begin{cases} {}_0^C D_t^q z(t, \xi) \in \dfrac{\partial^2 z(t, \xi)}{\partial \xi^2} + \hat{f}(t, z(t, \xi)) \\ \qquad\qquad + G(t, z(t, \xi)) \dfrac{dW(t)}{dt}, & t \in [0, 1], \ \xi \in [0, \pi], \quad (5.8) \\ z(t, 0) = z(t, \pi) = 0, & t \in [0, 1], \\ z(0, \xi) = x_0(\xi), & \xi \in [0, \pi], \end{cases}$$

where $W(t)$ is real standard Brownian motion, $\hat{f} : [0,1] \times \mathcal{H} \to \mathcal{H}$ satisfies $(H_1)'$.

Setting $\mathcal{H} = L^2([0,1], \mathbb{R}^+)$, $W(t)$ stands for a standard Brownian motion in $\mathcal{H}$ defined on a stochastic basis $(\Omega, \mathscr{F}, \mathbb{P})$. Consider the operator $A : D(A) \subset \mathcal{H} \to \mathcal{H}$ defined by

$$D(A) = \{z \in \mathcal{H} : z, z' \text{ are absolutely continuous,}$$
$$z'' \in \mathcal{H}, \ z(0) = z(\pi) = 0\},$$
$$Az = z''.$$

Then

$$Az = \sum_{n=1}^{\infty} n^2 (z, z_n) z_n,$$

where $z_n(t) = \sqrt{\frac{2}{\pi}} \sin(nt)$, $n = 1, 2, \dots$ is the orthogonal set of eigenvectors of A. It is well known that A generates a compact, analytic semigroup $\{T(t)\}_{t \geq 0}$ in $\mathcal{H}$.

Now, we assume that

$$g_i : [0, b] \times \mathcal{H} \to \mathcal{H}, \ i = 1, 2$$

be such that

(F_1) g_1 is l.s.c. and g_2 is u.s.c.;
(F_2) $g_1(t, z) \leq g_2(t, z)$ for each $(t, z) \in [0, b] \times \mathcal{H}$;
(F_3) there exist $\alpha_1, \alpha_2 \in L^\infty([0, b], \mathbb{R})$ such that

$$\|g_i(t, z)\|_{L_2^q}^p \leq \alpha_1(t)|z|^p + \alpha_2(t), \quad i = 1, 2,$$

for each $(t, z) \in [0, b] \times \mathcal{H}$.

Let $\Sigma(t, z) = [g_1(t, z), g_2(t, z)]$. From our assumptions on (F_1)-(F_3), it follows readily that the multivalued function $\Sigma(\cdot, \cdot) : [0, b] \times \overline{\Omega} \to 2^{\mathcal{H}}$ satisfies (H_1), $(H_3)'$, and $(H_4)'$.

Then inclusion (5.8) can be reformulated as

$$\begin{cases} {}^C_0 D_t^q x(t) \in Ax(t) + f(t, x(t)) + \Sigma(t, z(t, x(t))) \dfrac{dW(t)}{dt}, & t \in [0, 1], \\ x(0) = x_0, \end{cases}$$

where $x(t)(\xi) = z(t, \xi), \Sigma(t, x(t))(\xi) = G(t, z(t, \xi))$.

Thus, all the assumptions in Theorem 5.4 are satisfied, our result can be used to inclusion (5.8).

5.2.2 Noncompact Semigroup Case

We study the fractional evolution inclusion (5.1) under the following assumptions:

(H_5) there exist functions $k_1(t)$, $k_2(t) \in L^p([0, b], \mathbb{R}^+)$ such that

$$\beta(\Sigma(t, D)) \leq k_1(t)\beta(D) \text{ and } \beta(f(t, D)) \leq k_2(t)\beta(D),$$

for every bounded set $D \subset L^p(\Omega, \mathcal{H})$.

Define the generalized Cauchy operators $\Pi_1 : L^p([0, b], L^p(\Omega, \mathcal{H})) \to \mathcal{C}$ and $\Pi_2 : L^p([0, b], \mathscr{L}_2^0) \to \mathcal{C}$ as follows:

$$\Pi_1(f)(t) = \int_0^t (t - s)^{q-1} K_q(t - s) f(s) ds,$$

$$\Pi_2(\sigma)(t) = \int_0^t (t - s)^{q-1} K_q(t - s) \sigma(s) dW(s).$$

In the following, we present some properties of the operator Π_2.

Lemma 5.13. *The operator* Π_2 *satisfies properties (i) and (ii):*

(i) there exists a constant $c_2 > 0$ such that

$$\mathbb{E}|\Pi_2(\sigma_1)(t) - \Pi_2(\sigma_2)(t)|^p \leq c_2 \int_0^t \mathbb{E}\|\sigma_1(s) - \sigma_2(s)\|_{\mathscr{L}_2^0}^p ds, \ t \in [0, b]$$

for every $\sigma_1, \sigma_2 \in L^p([0, b], \mathscr{L}_2^0)$;

(ii) for each compact set $K \subset \mathscr{L}_2^0$ and sequence $\{\sigma_n\} \subset L^p([0, b], \mathscr{L}_2^0)$ such that $\{\sigma_n(t)\} \subset K$ for a.e. $t \in [0, b]$, the weak convergence $\sigma_n \rightharpoonup \sigma$ implies the convergence $\Pi_2(\sigma_n) \to \Pi_2(\sigma)$.

Proof. (i) By calculation, we have

$$\mathbb{E}|\Pi_2(\sigma_1)(t) - \Pi_2(\sigma_2)(t)|^p$$

$$=\mathbb{E}\left|\int_0^t (t-s)^{q-1} K_q(t-s)(\sigma_1(s) - \sigma_2(s)) dW(s)\right|^p$$

$$\leq C_p N(q)\left(\int_0^t (t-s)^{2(q-1)}\left(\mathbb{E}\|\sigma_1(s) - \sigma_2(s)\|^p_{\mathscr{L}_2^0}\right)^{\frac{2}{p}} ds\right)^{\frac{p}{2}}$$

$$\leq C_p N(q)\left(\int_0^t (t-s)^{2(q-1)\times\frac{p}{p-2}} ds\right)^{\frac{p-2}{p}}\int_0^t \mathbb{E}\|\sigma_1(s) - \sigma_2(s)\|^p_{\mathscr{L}_2^0} ds$$

$$\leq C_p N(q)\left(\frac{p-2}{a_1}\right)^{\frac{p-2}{2}} b^{\frac{a_1}{2}} \int_0^t \mathbb{E}\|\sigma_1(s) - \sigma_2(s)\|^p_{\mathscr{L}_2^0} ds$$

$$= c_2 \int_0^t \mathbb{E}\|\sigma_1(s) - \sigma_2(s)\|^p_{\mathscr{L}_2^0} ds,$$

where

$$c_2 = C_p N(q)\left(\frac{p-2}{a_1}\right)^{\frac{p-2}{2}} b^{\frac{a_1}{2}}.$$

(ii) First, we prove that for every $t \in [0, b]$ the sequence $\{\Pi_2(\sigma_n)(t)\} \subset L^p(\Omega, \mathcal{H})$ is relatively compact. Indeed, noting that $K \subset \mathscr{L}_2^0$ is compact and $\{\sigma_n(t)\} \subset K$, $\{\sigma_n(t)\}$ has a convergent subsequence $\{\sigma_{n_k}(t)\}$, denoted by $\sigma_{n_k}(t) \to \sigma_0(t) \in K$.

Since

$$\int_0^t \mathbb{E}\|\sigma_{n_k}(s) - \sigma_0(s)\|^p_{\mathscr{L}_2^0} ds$$

$$\leq 2^{p-1}\left(\int_0^t \mathbb{E}\|\sigma_{n_k}(s)\|^p_{\mathscr{L}_2^0} ds + \int_0^t \mathbb{E}\|\sigma_0(s)\|^p_{\mathscr{L}_2^0} ds\right)$$

$$< \infty,$$

and by Theorem 1.1, we have

$$\mathbb{E}|\Pi_2(\sigma_{n_k})(t) - \Pi_2(\sigma_0)(t)|^p$$

$$\leq C_p N(q)\left(\int_0^t (t-s)^{2(q-1)}\left(\mathbb{E}\|\sigma_{n_k}(s) - \sigma_0(s)\|^p_{\mathscr{L}_2^0}\right)^{\frac{2}{p}} ds\right)^{\frac{p}{2}}$$

$$\leq C_p N(q)\left(\frac{p-2}{a_1}\right)^{\frac{p-2}{2}} b^{\frac{a_1}{2}} \int_0^t \mathbb{E}\|\sigma_{n_k}(s) - \sigma_0(s)\|^p_{\mathscr{L}_2^0} ds$$

$$\to 0, \text{ as } k \to \infty.$$

Hence, the sequence $\{\Pi_2(\sigma_n)(t)\}$ has a convergent subsequence, which implies that $\{\Pi_2(\sigma_n)(t)\} \subset L^p(\Omega, \mathcal{H})$ is relatively compact for every $t \in [0, b]$.

On the other hand, we have

$$\mathbb{E}|\Pi_2(\sigma_n)(t_2) - \Pi_2(\sigma_n)(t_1)|^p$$

$$\leq 4^{p-1}\mathbb{E}\left|\int_{t_1}^{t_2} (t_2 - s)^{q-1}K_q(t_2 - s)\sigma_n(s)dW(s)\right|^p$$

$$+ 4^{p-1}\mathbb{E}\left|\int_0^{t_1-\delta} (t_1 - s)^{q-1}\big(K_q(t_2 - s) - K_q(t_1 - s)\big)\sigma_n(s)dW(s)\right|^p$$

$$+ 4^{p-1}\mathbb{E}\left|\int_{t_1-\delta}^t (t_1 - s)^{q-1}\big(K_q(t_2 - s) - K_q(t_1 - s)\big)\sigma_n(s)dW(s)\right|^p$$

$$+ 4^{p-1}\mathbb{E}\left|\int_0^{t_1} \big((t_2 - s)^{q-1} - (t_1 - s)^{q-1}\big)K_q(t_2 - s)\sigma_n(s)dW(s)\right|^p$$

$$\leq 4^{p-1}C_p N(q)\left(\int_{t_1}^{t_2} (t_2 - s)^{2(q-1)}\Big(\mathbb{E}\|\sigma_n(s)\|_{\mathscr{L}_2^0}^p\Big)^{\frac{2}{p}}ds\right)^{\frac{p}{2}}$$

$$+ 4^{p-1}C_p \sup_{s\in[0,t_1-\delta]} \|K_q(t_2 - s) - K_q(t_1 - s)\|^p$$

$$\times \left(\int_0^{t_1-\delta} (t_1 - s)^{2(q-1)}\Big(\mathbb{E}\|\sigma_n(s)\|_{\mathscr{L}_2^0}^p\Big)^{\frac{2}{p}}ds\right)^{\frac{p}{2}}$$

$$+ 8^{p-1}C_p N(q)\left(\int_{t_1-\delta}^t (t_1 - s)^{2(q-1)}\Big(\mathbb{E}\|\sigma_n(s)\|_{\mathscr{L}_2^0}^p\Big)^{\frac{2}{p}}ds\right)^{\frac{p}{2}}$$

$$+ 4^{p-1}C_p N(q)\left(\int_0^{t_1} \big((t_2 - s)^{q-1}\right.$$

$$\left. - (t_1 - s)^{q-1}\big)^2\Big(\mathbb{E}\|\sigma_n(s)\|_{\mathscr{L}_2^0}^p\Big)^{\frac{2}{p}}ds\right)^{\frac{p}{2}}.$$

According to Lemma 5.5(ii) and $\{\sigma_n(t)\} \subset K$ for a.e. $t \in [0, b]$, the right-hand side of this inequality tends to zero as $t_2 \to t_1$ uniformly with respect to n. So $\{\Pi_2(\sigma_n)\}$ is an equicontinuous set. Thus from Lemma 1.2, we obtain that the sequence $\{\Pi_2(\sigma_n)\} \subset \mathcal{C}$ is relatively compact.

Property (i) ensures that $\Pi_2 : L^p([0, b], \mathscr{L}_2^0) \to \mathcal{C}$ is a bounded linear operator. Then it is continuous with respect to the topology of weak sequential convergence, that is, the weak convergence $\sigma_n \rightharpoonup \sigma$ ensures

$\Pi_2(\sigma_n) \rightharpoonup \Pi_2(\sigma)$. Taking into account that $\{\Pi_2(\sigma_n)\}$ is relatively compact, we arrive at the conclusion that $\Pi_2(\sigma_n) \to \Pi_2(\sigma)$ strongly in $\mathcal{C}$. $\square$

Lemma 5.14. *Let the sequence $\{\sigma_n\} \subset L^p([0,b], \mathscr{L}_2^0)$ be integrably bounded:*

$$\|\sigma_n(t))\|_{\mathscr{L}_2^0} \leq v(t), \quad \text{for a.e. } t \in [0,b],$$

where $v \in L^p([0,b], \mathbb{R}^+)$. Assume that

$$\beta(\{\sigma_n(t)\}) \leq \eta(t)$$

for a.e. $t \in [0,b]$, where $\eta \in L^p([0,b], \mathbb{R}^+)$. Then we have

$$\beta(\{\Pi_2(\sigma_n)(t)\}_{n=1}^\infty) \leq 2\left(c_2 \int_0^t |\eta(s)|^p ds\right)^{\frac{1}{p}}$$

for all $t \in [0,b]$, where $c_2 \geq 0$ is the constant in Lemma 5.13(i).

Proof. For any $\varepsilon > 0$, choose $\delta \in (0, \varepsilon)$ such that for every set $m \subset [0,b]$, $\mathrm{mes}(m) < \delta$, then

$$\int_m |v(s)|^p ds < \varepsilon.$$

Take m_δ and G_δ from Lemma 5.2. Then any sequence $\{g_n\}_{n=1}^\infty$, $g_n \in G_\delta$ is weakly compact by

$$\|g_n(t)\|_{\mathscr{L}_2^0} \leq \left(v(t) + 2\eta(t)\right) + \delta, \quad \text{for a.e. } t \in [0,b].$$

Thus, from Lemma 5.13(ii), it follows that the sequence $\{\Pi_2\sigma_n(t)\}_{n=1}^\infty$ is relatively compact in $\mathcal{C}$. This means that $\Pi_2(G_\delta)$ is relatively compact in $\mathcal{C}$.

By Lemma 5.13(i), we have σ_n and corresponding function g_n satisfying

$$\mathbb{E}|\Pi_2(\sigma_n)(t) - \Pi_2(g_n)(t)|^p$$

$$\leq C_p N(q)\left(\frac{p-2}{a_1}\right)^{\frac{p-2}{2}} b^{\frac{a_1}{2}} \int_0^t \mathbb{E}\|\sigma_n(s) - g_n(s)\|_{\mathscr{L}_2^0}^p ds$$

$$\leq C_p N(q)\left(\frac{p-2}{a_1}\right)^{\frac{p-2}{2}} b^{\frac{a_1}{2}} \left(\int_{[0,t]\backslash m_\delta} \mathbb{E}\|\sigma_n(s) - g_n(s)\|_{\mathscr{L}_2^0}^p ds\right.$$

$$\left. + \int_{[0,t]\cap m_\delta} \mathbb{E}\|\sigma_n(s)\|_{\mathscr{L}_2^0}^p ds\right)$$

$$\leq C_p N(q) \left(\frac{p-2}{a_1}\right)^{\frac{p-2}{2}} b^{\frac{a_1}{2}} \left(\int_{[0,t]\setminus m_\delta} (2\eta(s) + \delta)^p ds + \varepsilon\right)$$

$$\leq c_2 \left(\int_0^t (2\eta(s) + \varepsilon)^p ds + \varepsilon\right).$$

Then the relatively compact set $\Pi_2(G_\delta)$ forms a $[c_2(\int_0^t (2\eta(s)+\varepsilon)^p ds+\varepsilon)]^{\frac{1}{p}}$-net for the set $\{\Pi_2(\sigma_n)(t)\}_{n=1}^\infty$, proving the theorem due to the arbitrary choice of ε.   □

With the similar proof of Lemma 5.13, we can get the following result.

Lemma 5.15. *Let $\{\sigma_n\}$ be a semicompact sequence in $L^p([0,b], \mathscr{L}_2^0)$. Then $\{\sigma_n\}$ is weakly compact in $L^p([0,b], \mathscr{L}_2^0)$, and $\{\Pi_2(\sigma_n)\}$ is relatively compact in $\mathcal{C}$. Moreover, if $\sigma_n \rightharpoonup \sigma$, then $\Pi_2(\sigma_n) \to \Pi_2(\sigma)$.*

Lemma 5.16. *Let the sequence $\{f_n\} \subset L^p([0,b], L^p(\Omega, \mathcal{H}))$ be L^p-integrably bounded:*

$$\|f_n(t)\| \leq v_1(t)$$

for all $n = 1, 2, \ldots$, and a.e. $t \in [0,b]$, where $v_1(t) \subset L^p([0,b], \mathbb{R}^+)$. Assume that

$$\beta(f_n(t)) \leq \eta_1(t)$$

for a.e. $t \in [0,b]$, where $\eta_1 \in L^p([0,b], \mathbb{R}^+)$. Then we have

$$\beta(\{\Pi_1(f_n)(t)\}) \leq 2\left(c_3 \int_0^t |\eta_1(s)|^p ds\right)^{\frac{1}{p}}$$

for any $t \in [0,b]$, where $c_3 = N(q)\left(\frac{p-1}{pq-1}\right)^{p-1} b^{pq}$.

Theorem 5.5. *Let conditions (H_A), $(H_1)'$, (H_2), $(H_3)'$, $(H_4)'$, and (H_5) be satisfied. Then inclusion (5.1) has at least one mild solution for each initial value $x_0 \in \mathcal{H}$.*

Proof. For the same $\mathcal{M}_0$, as the reason for Theorem 5.4, we see that $\mathcal{M}_0$ is a closed and convex subset of $\mathcal{C}$.

Claim 1. The multioperator $\mathcal{F} = S \circ Sel_{\Sigma}$ has closed graph with compact values.

Let $x_n \subset \mathcal{M}_0$ with $x_n \to x$ and $y_n \in \mathcal{F}(x_n)$ with $y_n \to y$. We shall prove that $y \in \mathcal{F}(x)$. By the definition of $\mathcal{F}$, there exist $\sigma_n \in Sel_{\Sigma}(x_n)$ such that

$$y_n(t) = S_q(t)x_0 + \Pi_2(\sigma_n)(t) + \int_0^t (t-s)^{q-1} K_q(t-s) f(s, x_n(s)) ds.$$

We need to prove that there exists $\sigma \in Sel_{\Sigma}(x)$ such that for a.e. $t \in [0, b]$,

$$y(t) = S_q(t)x_0 + \Pi_2(\sigma)(t) + \int_0^t (t-s)^{q-1} K_q(t-s) f(s, x(s)) ds.$$

In view of $(H_4)'$ we have that $\{\sigma_n\}$ is bounded in $L^p([0, b], \mathscr{L}_2^0)$, one obtains $\sigma_n \rightharpoonup \sigma$ in $L^p([0, b], \mathscr{L}_2^0)$ (see Lemma 1.4). Since Sel_{Σ} is weakly u.s.c. with weakly compact and convex values (see Lemma 5.4), in view of Lemma 1.5, we have $\sigma \in Sel_{\Sigma}(x)$.

We see that $\{\sigma_n\}$ is integrably bounded by $(H_4)'$, and the following inequality holds by (H_5):

$$\beta(\{\sigma_n(t)\}) \leq k_1(t)\beta(\{x_n(t)\}).$$

Since the sequence $\{x_n\}$ converges in $\mathcal{C}$, then $\beta(\{\sigma_n(t)\}) = 0$ for a.e. $t \in [0, b]$, thus $\{\sigma_n\}$ is a semicompact sequence. By Lemma 5.15, we may assume, without loss of generality, that there exists $\sigma \in Sel_{\Sigma}(x)$ such that

$$\sigma_n \rightharpoonup \sigma \text{ and } \Pi_2(\sigma_n) \to \Pi_2(\sigma),$$

and by the same methods as Theorem 5.4, we have

$$\int_0^t (t-s)^{q-1} K_q(t-s) f(s, x_n(s)) ds$$
$$\to \int_0^t (t-s)^{q-1} K_q(t-s) f(s, x(s)) ds,$$

which implies

$$y_n(t) = S_q(t)x_0 + \Pi_2(\sigma_n)(t) + \int_0^t (t-s)^{q-1} K_q(t-s) f(s, x_n(s)) ds$$

$$\rightarrow S_q(t)x_0 + \Pi_2(\sigma)(t) + \int_0^t (t-s)^{q-1} K_q(t-s) f(s, x(s))ds$$
$$= y(t).$$

It remains to show that, for $x \in \mathcal{M}_0$ and $\{\sigma_n\}$ chosen in $Sel_\Sigma(x)$, the sequence $\{\Pi_2(\sigma_n)\}$ is relatively compact in $\mathcal{C}$. Hypotheses $(H_4)'$ and (H_5) imply that $\{\sigma_n\}$ is semicompact. Using Lemma 5.15, we obtain that $\{\Pi_2(\sigma_n)\}$ is relatively compact in $\mathcal{C}$. Thus $\mathcal{F}(x)$ is relatively compact in $\mathcal{C}$, together with the closeness of $\mathcal{F}$, then $\mathcal{F}(x)$ has compact values.

Claim 2. The multioperator $\mathcal{F}$ is u.s.c.

In view of Lemma 1.7, it suffices to check that $\mathcal{F}$ is a quasicompact multivalued map. Let U be a compact set. We prove that $\mathcal{F}(U)$ is a relatively compact subset of $\mathcal{C}$.

Assume that $\{y_n\} \subset \mathcal{F}(U)$. Then

$$y_n(t) = S_q(t)x_0 + \Pi_2(\sigma_n)(t) + \int_0^t (t-s)^{q-1} K_q(t-s) f(s, x_n(s))ds,$$

where $\sigma_n \in Sel_\Sigma(x_n)$, for a certain sequence $\{x_n\} \subset U$. Hypotheses $(H_4)'$ and (H_5) yield the fact that $\{\sigma_n\}$ is semicompact and then it is a weakly compact sequence in $L^p([0, b], \mathscr{L}_2^0)$. Similar arguments as the previous proof of closeness imply that $\{\Pi_2(\sigma_n)\}$ is relatively compact in $\mathcal{C}$. Thus, $\{y_n\}$ converges in $\mathcal{C}$, so the multioperator $\mathcal{F}$ is u.s.c.

Claim 3. The multioperator $\mathcal{F}$ is a condensing multioperator.

We first need an MNC constructed suitably for our problem. For a bounded subset $\Lambda \subset \mathcal{M}_0$, let $\mathrm{mod}_C(\Lambda)$ be the modulus of equicontinuity of the set of functions Λ given by

$$\mathrm{mod}_C(\Lambda) = \lim_{\delta \to 0} \sup_{x \in \Lambda} \max_{|t_2 - t_1| < \delta} \|x(t_2) - x(t_1)\|.$$

Let χ be real MNC defined on bounded set $D \subset \mathcal{C}$ by

$$\chi(D) = \sup_{t \in [0,b]} e^{-Lt} \beta(D(t));$$

here, the constant L is chosen such that

$$
l_1 = 2 \sup_{t \in [0,b]} \left(c_2 \int_0^t e^{-Lp(t-s)} k_1^p(s) ds \right)^{\frac{1}{p}}
$$
$$
+ 2 \sup_{t \in [0,b]} \left(c_3 \int_0^t e^{-Lp(t-s)} k_2^p(s) ds \right)^{\frac{1}{p}}
$$
$$
< 1,
$$

where $k_1(t), k_2(t)$ are the functions from condition (H_5).

Consider the function $\nu(\Lambda) = \max_{D \in \Delta'(\Lambda)} \left(\chi(D), \mathrm{mod}_C(D) \right)$ in the space of C, where $\Delta'(\Lambda)$ is the collection of all countable subsets of Λ.

To show that $\mathcal{F}$ is ν-condensing, let $\Lambda \subset \mathcal{M}_0$ be a bounded set in $\mathcal{M}_0$ such that

$$
\nu(\Lambda) \leq \nu(\mathcal{F}(\Lambda)). \tag{5.9}
$$

We will show that Λ is relatively compact. Let $\nu(\mathcal{F}(\Lambda))$ be achieved on a sequence $\{y_n\} \subset \mathcal{F}(\Lambda)$, i.e.,

$$
\nu(\{y_n\}) = \left(\chi(\{y_n\}), \mathrm{mod}_C(\{y_n\}) \right).
$$

Then

$$
y_n(t) = S_q(t) x_0 + \Pi_2(\sigma_n)(t) + \int_0^t (t-s)^{q-1} K_q(t-s) f(s, x_n(s)) ds,
$$

where $\{x_n\} \subset \Lambda$ and $\sigma_n \in Sel_\Sigma(x_n)$. Now inequality (5.9) implies

$$
\chi(\{y_n\}) \geq \chi(\{x_n\}). \tag{5.10}
$$

It follows from (H_5) that

$$
\beta(\{\sigma_n(t)\}) \leq k_1(t) \beta(x_n(t))
$$

and

$$
\beta(\{f(t, x_n(t))\}) \leq k_2(t) \beta(x_n(t))
$$

for $t \in [0, b]$. Then

$$
\beta(\{\sigma_n(t)\}) \leq k_1(t) e^{Lt} \left(\sup_{s \in [0,t]} e^{-Ls} \beta(x_n(t)) \right) \leq k_1(t) e^{Lt} \chi(\{x_n\}),
$$

$$\beta(\{f(t, x_n(t))\}) \leq k_2(t)e^{Lt}\left(\sup_{s \in [0,t]} e^{-Ls}\beta(x_n(t))\right) \leq k_2(t)e^{Lt}\chi(\{x_n\}).$$

Now the application of Lemma 5.14 for Π_2 and Lemma 5.16 for Π_1 yield

$$e^{-Lt}\beta(\{\Pi_2(\sigma_n)(t)\}) \leq 2e^{-Lt}\left(c_2\int_0^t e^{Lps}k_1^p(s)ds\right)^{\frac{1}{p}}\chi(\{x_n\})$$

$$\leq 2\left(c_2\int_0^t e^{-Lp(t-s)}k_1^p(s)ds\right)^{\frac{1}{p}}\chi(\{x_n\}),$$

$$e^{-Lt}\beta(\{\Pi_1(f)(t)\}) \leq 2e^{-Lt}\left(c_3\int_0^t e^{Lps}k_2^p(s)ds\right)^{\frac{1}{p}}\chi(\{x_n\})$$

$$\leq 2\left(c_3\int_0^t e^{-Lp(t-s)}k_2^p(s)ds\right)^{\frac{1}{p}}\chi(\{x_n\}),$$

for any $t \in [0, b]$. Putting this relation together with (5.10), we obtain

$$\chi(\{x_n\}) \leq \chi(\{y_n\})$$
$$= \sup_{t \in [0,b]} e^{-Lt}\beta(y_n(t))$$
$$\leq \sup_{t \in [0,b]} e^{-Lt}\beta(\{\Pi_2(\sigma_n)(t)\}) + \sup_{t \in [0,b]} e^{-Lt}\beta(\{\Pi_1(f)(t)\})$$
$$\leq l_1\chi(\{x_n\}).$$

Therefore, $\chi(\{x_n\}) = 0$. This implies $\beta(x_n(t)) = 0$. Thus for every $t \in [0, b]$, $\{\Pi_1(f)(t)\}$ is relatively compact in $L^p(\Omega, \mathcal{H})$; this together with equicontinuous of $\{\Pi_1(f)(t)\}$ (due to Lemma 5.10) implies that $\{\Pi_1(f)\}$ is relatively compact in $\mathcal{C}$.

Using $(H_4)'$ and (H_5) again, one gets that $\{\sigma_n\}$ is a semicompact sequence. Then, Lemma 5.15 ensures that $\{\Pi_2(\sigma_n)\}$ is relatively compact in $\mathcal{C}$. This yields that $\{y_n\}$ is relatively compact in $\mathcal{C}$. Hence, $\mathrm{mod}_\mathcal{C}(\{y_n\}) = 0$. Finally, $\nu(\{y_n\}) = (0,0)$, and $\nu(\{x_n\}) = (0,0)$, so the map $\mathcal{F}$ is ν-condensing.

From Theorem 1.9, we deduce that the fixed point set $\mathrm{Fix}\mathcal{F}$ is a nonempty compact set. $\qquad\square$

Lemma 5.17. *Let $p > 2$, $pq > 1$, and $\tilde{\sigma}(\cdot, x)$ be L^p-integrable for every $x \in \mathcal{H}$. Assume that $\{T(t)\}_{t \geq 0}$ is equicontinuous and there exists $k(t) \in L^p([c, b], \mathbb{R}^+)$ such that $\beta(\tilde{\sigma}(t, D)) \leq k(t)\beta(D)$ for every bounded set $D \subset$*

$L^p(\Omega, \mathcal{H})$, and $\beta(f(t, D)) \leq k_2(t)\beta(D)$; here, $k_2(t)$ is the function from condition (H_5).

Suppose in addition that

(i) *for any compact subset $K \subset \mathcal{H}$, there exist $\delta > 0$ and $L_K \in L^1([c, b], \mathbb{R}^+)$ such that*

$$\|\tilde{\sigma}(t, x_1) - \tilde{\sigma}(t, x_2)\|^p_{\mathscr{L}^0_2} \leq L_K(t)|x_1 - x_2|^p,$$

for a.e. $t \in [c, b]$ and each $x_1, x_2 \in B_\delta(K)$;
(ii) *there exists $\gamma_1(t) \in L^1([c, b], \mathbb{R}^+)$ such that $\|\tilde{\sigma}(t, x)\|^p_{\mathscr{L}^0_2} \leq \gamma_1(t)(c' + |x|^p)$ for a.e. $t \in [c, b]$ and every $x \in \mathcal{H}$, where c' is arbitrary, but fixed.*

Then integral equation (5.5) admits a unique solution for every $\phi(t) \in C([c, b], L^p(\Omega, \mathcal{H}))$. Moreover, the solution of (5.5) depends continuously on ϕ.

Proof. Let $\phi(t) \in C([c, b], L^p(\Omega, \mathcal{H}))$ be fixed. Write

$$B(\phi, \xi) = \left\{ x \in C([c, \xi], L^p(\Omega, \mathcal{H})) : \max_{t \in [c,\xi]} \|x(t) - \phi(t)\|^p \leq \kappa \right\}$$

with

$$2^{p-1}N(q)\left[\left(\frac{p-1}{pq-1}\right)^{p-1}(\xi - c)^{pq}\left(2^{p-1}c_0\left(\kappa + \max_{t \in [c,\xi]} \|\phi(t)\|^p\right) + c_1\right)\right.$$
$$\left. + C_p\left(\frac{p-2}{a_1}\right)^{\frac{p-2}{2}}(\xi - c)^{\frac{a_1}{2}}\|\gamma_1\|_{L[0,b]}\left(c' + 2^{p-1}\left(\kappa + \max_{t \in [c,\xi]} \|\phi(t)\|^p\right)\right)\right]$$
$$\leq \kappa.$$

Let us define the operator $\mathcal{P}$:

$$\mathcal{P}(x)(t) = \phi(t) + \int_0^t (t - s)^{q-1}K_q(t - s)f(s, x(s))ds$$
$$+ \int_0^t (t - s)^{q-1}K_q(t - s)\tilde{\sigma}(s, x(s))dW(s).$$

Then $\mathcal{P}$ mapping $B(\phi, \xi)$ into itself is continuous due to conditions (i) and (ii). Moreover, from the proof of Theorem 5.5, we see that $\mathcal{P}$ is condensing map. Thus, there is a fixed point of $\mathcal{P}$, denoted by x, which is a local solution to equation (5.5).

In the sequel, the operator $\mathcal{P}$ is a mapping from $C([c, b], L^p(\Omega, \mathcal{H}))$ to $C([c, b], L^p(\Omega, \mathcal{H}))$. Define the operator $\Psi : [c, b] \times C([c, b], L^p(\Omega, \mathcal{H})) \to C([c, b], L^p(\Omega, \mathcal{H}))$ with

$$\Psi(t, y)(s) = \begin{cases} y(s), & s \in [c, t], \\ y(t), & s \in [t, b]. \end{cases}$$

Put

$$J = \{t \in [c, b] : y^t \in C([c, b], L^p(\Omega, \mathcal{H})), y^t = \Psi(t, \mathcal{P}(y^t))\}.$$

Note that $x^\xi = \Psi(\xi, \mathcal{P}(x^\xi))$ with $x^\xi = \Psi(\xi, x)$, which means $\xi \in J$, i.e., $J = \emptyset$. Moreover, it is easy to see that $[c, t] \subset J$ for all $t \in J$.

Let $\{t_n\} \subset J$ be a monotonically increasing sequence such that t_n tends to $t_0 = \sup J$ as $n \to \infty$. Noticing that

$$\Psi(t_m, x^{t_m}) = \Psi(t^m, \mathcal{P}(x^{t_m})) \text{ and } \Psi(t_m, x^{t_n}) = \Psi(t^m, \mathcal{P}(x^{t_n}))$$

on $[c, t_m]$ when $m \leq n$, we obtain that $x^{t_m}(s) = x^{t_n}(s)$ for all $s \in [0, t_m]$. Also, note that

$$\|x^{t_m}(t_0) - x^{t_n}(t_0)\|^p = \|x^{t_m}(t_m) - x^{t_n}(t_n)\|^p.$$

Therefore, by the continuity of ϕ we conclude, using a similar argument with that in Lemma 5.10, that

$$\|x^{t_m}(t_0) - x^{t_n}(t_0)\|^p \to 0, \text{ as } n, m \to \infty.$$

Accordingly, $\lim_{n \to \infty} x^{t_n}(t_0)$ exists. Consider the function

$$x^{t_0}(s) = \begin{cases} x^{t_n}(s), & s \in [c, t_n], \\ \lim_{n \to \infty} x^{t_n}(t_0), & s \in [t_0, b], \end{cases}$$

where $n \to \infty$, which maps $[c, b]$ into $L^p(\Omega, \mathcal{H})$. It follows from the equicontinuity of the family $\{x^{t_n}\}$ that x^{t_0} is continuous. Also, note that $x^{t_0}(t) = \mathcal{P}(x^{t_0})(t)$ for all $t \in [c, t_0)$. Moreover, it is easy to verify, by

Theorem 1.1, that

$$
\begin{aligned}
x^{t_0}(t_0) &= \lim_{n \to \infty} \left(\phi(t_n) + \int_c^{t_n} (t_n - s)^{q-1} K_q(t_n - s) f(s, x^{t_n}(s)) ds \right. \\
&\quad \left. + \int_c^{t_n} (t_n - s)^{q-1} K_q(t_n - s) \tilde{\sigma}(s, x^{t_n}(s)) dW(s) \right) \\
&= \phi(t_0) + \int_c^{t_0} (t_0 - s)^{q-1} K_q(t_0 - s) f(s, x^{t_0}(s)) ds \\
&\quad + \int_c^{t_0} (t_0 - s)^{q-1} K_q(t_0 - s) \tilde{\sigma}(s, x^{t_0}(s)) dW(s).
\end{aligned}
$$

Thus, we find that $x^{t_0} = \Psi(t_0, \mathcal{P}(x^{t_0}))$, which yields that $t_0 \in J$.

Next, we show that $t_0 = b$. If this is not the case, then $t_0 < b$. Put

$$
\begin{aligned}
\tilde{\phi}(t) &= \phi(t) + \int_c^{t_0} (t - s)^{q-1} K_q(t - s) f(s, x^{t_0}(s)) ds \\
&\quad + \int_c^{t_0} (t - s)^{q-1} K_q(t - s) \tilde{\sigma}(s, x^{t_0}(s)) dW(s)
\end{aligned}
$$

with $\tilde{\phi}(t) \in C([c, b], L^p(\Omega, \mathcal{H}))$. As we can see there exists $\xi' > 0$ such that the following integral equation

$$
\begin{aligned}
x(t) &= \tilde{\phi}(t) + \int_{t_0}^t (t - s)^{q-1} K_q(t - s) f(s, x(s)) ds \\
&\quad + \int_{t_0}^t (t - s)^{q-1} K_q(t - s) \tilde{\sigma}(s, x(s)) dW(s)
\end{aligned}
$$

has a solution $z \in C([t_0, t_0 + \xi'], L^p(\Omega, \mathcal{H}))$.

Denote

$$
x^{t_0 + \xi'}(s) = \begin{cases} x^{t_0}(s), & s \in [c, t_0], \\ z(s), & s \in [t_0, t_0 + \xi'], \\ z(t_0 + \xi'), & s \in [t_0 + \xi', b]. \end{cases}
$$

Then it is clear that $x^{t_0+\xi'} \in C([c, b], L^p(\Omega, \mathcal{H}))$. Moreover, one finds that

$$x^{t_0+\xi'}(t) = \phi(t) + \int_c^{t_0} (t-s)^{q-1} K_q(t-s) f(s, x^{t_0+\xi'}(s)) ds$$

$$+ \int_c^{t_0} (t-s)^{q-1} K_q(t-s) \tilde{\sigma}(s, x^{t_0+\xi'}(s)) dW(s),$$

for $t \in [c, t_0 + \xi']$, which implies that $x^{t_0+\xi'} = \Psi(t_0 + \xi', \mathcal{P}(x^{t_0+\xi'}))$. This yields that $t_0 + \xi' \in J$, a contradiction.

Finally, let $\phi_n \to \phi_0$ in $C([c, b], L^p(\Omega, \mathcal{H}))$ as $n \to \infty$ and x_n be the solution of the equation 5.5 with the perturbation ϕ_n, i.e.,

$$x_n(t) = \phi_n(t) + \int_c^t (t-s)^{q-1} K_q(t-s) f(s, x_n(s)) ds$$

$$+ \int_c^t (t-s)^{q-1} K_q(t-s) \tilde{\sigma}(s, x_n(s)) dW(s), \quad t \in [c, b], n \geq 1.$$

$$(5.11)$$

On the one hand, we have

$$\chi(\{x_n\}) = \sup_{t \in [c,b]} e^{-Lt} \beta(\{x_n(t)\})$$

$$\leq \sup_{t \in [c,b]} e^{-Lt} \beta(\{\phi_n(t)\}) + \sup_{t \in [c,b]} e^{-Lt} \beta(\{\Pi_1(f)(t)\})$$

$$+ \sup_{t \in [c,b]} e^{-Lt} \beta(\{\Pi_2(\tilde{\sigma})(t)\})$$

$$\leq l_1 \chi(\{x_n\}).$$

Thus $\chi(\{x^n\}) = 0$, then $\beta(\{x^n(t)\}) = 0$. On the other hand, as the reason for the proof of $\mathrm{mod}_C(\{x^n\})$ in Theorem 5.5, it follows that $\mathrm{mod}_C(\{x^n\}) = 0$. Hence, $\{x^n\}$ is relatively compact in $C([c, b], L^p(\Omega, \mathcal{H}))$. We assume, by passing to a subsequence if necessary, that $x_n \to \tilde{x}$ in $C([c, b], L^p(\Omega, \mathcal{H}))$ as $n \to \infty$. Therefore, taking the limit in (5.11) as $n \to \infty$, one finds, again by Theorem 1.1, that $\tilde{x}$ is the solution of equation (5.5) with the perturbation ϕ_0. The proof is completed. $\qquad \square$

Theorem 5.6. *Under the conditions in Theorem 5.5, $\Theta(x_0)$ is a compact R_δ-set.*

Proof. We also consider inclusion (5.6). Clearly, its solution set $\Theta^n(x_0)$ is nonempty and compact in $\mathcal{C}$.

We show that each sequence $\{y_n\}$ such that $y_n \in \Theta^n(x_0)$ for all $n \geq 1$ has a convergent subsequence $y_{n_k} \to y \in \Theta(x_0)$.

At first we notice that

$$y_n(t) = S_q(t)x_0 + \int_0^t (t-s)^{q-1} K_q(t-s) f(s, y_n(s)) ds$$

$$+ \int_0^t (t-s)^{q-1} K_q(t-s) \tilde{\sigma}_n(s) dW(s)$$

for $t \in [0, b]$, where $\tilde{\sigma}_n \in L^p([0, b], \mathscr{L}_2^0)$ is such that $\tilde{\sigma}_n(s) \in \Sigma_n(s, y_n(s))$ for almost every $s \in [0, b]$.

For any $r \geq 1$ we obtain

$$\beta(\{\tilde{\sigma}_n(s)\}_{n \geq 1}) = \beta(\{\tilde{\sigma}_n(s)\}_{n \geq r})$$
$$\leq \beta\big(\Sigma(s, B(\{y_n(s)\}_{n \geq r}, 3d_r))\big)$$
$$\leq k_1(s)\beta\big(B(\{y_n(s)\}_{n \geq r}, 3d_r)\big)$$
$$\leq k_1(s)\big(\beta(\{y_n(s)\}_{n \geq r}) + 3d_r\big)$$
$$= k_1(s)(\bar{\rho}(s) + 3d_r),$$

where $\bar{\rho}(s) = \beta(\{y_n(s)\}_{n \geq 1})$. Now

$$\beta(\{y_n(t)\}_{n \geq 1}) \leq \beta\left(\int_0^t (t-s)^{q-1} K_q(t-s) f(s, y_n(s)) ds\right)$$

$$+ \beta\left(\int_0^t (t-s)^{q-1} K_q(t-s) \tilde{\sigma}_n(s) dW(s)\right)$$

$$\leq 2\left(c_3 \int_0^t k_2^p(s) \bar{\rho}^p(s) ds\right)^{\frac{1}{p}}$$

$$+ 2\left(c_2 \int_0^t k_1^p(s)(\bar{\rho}(s) + 3d_r)^p ds\right)^{\frac{1}{p}}$$

for every $r \geq 1$. Since $d_r \to 0$ as $r \to \infty$, we obtain

$$\bar{\rho}(s) \leq 2\left(c_3 \int_0^t k_2^p(s) \bar{\rho}^p(s) ds\right)^{\frac{1}{p}} + 2\left(c_2 \int_0^t k_1^p(s) \bar{\rho}^p(s) ds\right)^{\frac{1}{p}}.$$

Thus,

$$\bar{\rho}^p(s) \leq 2^{2p-1}\left(c_3 \int_0^t k_2^p(s) \bar{\rho}^p(s) ds + c_2 \int_0^t k_1^p(s) \bar{\rho}^p(s) ds\right)$$

$$\leq 2^{2p-1} \int_0^t \left(c_3 k_2^p(s) + c_2 k_1^p(s) \right) \bar{\rho}^p(s) ds.$$

By Gronwall inequality we get $\bar{\rho}(s) = 0$ and, as a consequence, $\beta(\{y_n(t)\}_{n \geq 1}) = 0$. This also implies that $\beta(\{\tilde{\sigma}_n(s)\}_{n \geq 1}) = 0$.

Since the equicontinuity of $\{y_n\}$ is proved in Lemma 5.10. This implies the existence of a subsequence $\{y_{n_k}\}$ which is convergent on $[0, b]$. Denote the limit by y.

Since $\beta(\{\tilde{\sigma}_n(s)\}_{n \geq 1}) = 0$, we can assume, up to subsequence, that $\tilde{\sigma}_n(s) \to \tilde{\sigma}_0(s) \in \mathscr{L}_2^0$ for $s \in [0, t]$. Together with the above discussion, we have

$$y(t) = S_q(t)x_0 + \int_0^t (t-s)^{q-1} K_q(t-s) f(s, y(s)) ds$$
$$+ \int_0^t (t-s)^{q-1} K_q(t-s) \tilde{\sigma}_0(s) dW(s).$$

Since $\{\Sigma_n\}$ satisfies conditions (H_2), $(H_3)'$, and $(H_4)'$ for each $n \geq 1$, from Lemma 5.9(ii) and (vi). By using the similar method in Lemma 5.4, we can prove that $\tilde{\sigma}_0(t) \in \Sigma(t, y(t))$ for a.e. $t \in [0, b]$.

It follows that $\sup\{d(y, \Theta(x_0)) : y \in \Theta^n(x_0)\} \to 0$ (an easy proof by contradiction). Therefore, $\sup\{d(y, \Theta(x_0)) : y \in \overline{\Theta^n(x_0)}\} \to 0$, as well. Since $\Theta^n(x_0)$ is compact and $\Theta^{n+1}(x_0) \subset \Theta^n(x_0)$, $\beta(\Theta^n(x_0)) = \beta(\overline{\Theta^n(x_0)}) \to 0$, as $n \to \infty$ and $\Theta(x_0) = \bigcap_{n=1}^{\infty} \overline{\Theta^n(x_0)}$.

By the same methods as Theorem 5.4, in view of Lemma 5.17, we can know that $\Theta^n(x_0)$ is contractible for all $n \geq 1$. Consequently, we conclude that $\Theta(x_0)$ is an R_δ-set. The proof is completed. $\square$

5.3 NOTES AND REMARKS

Random differential inclusions play an important role in characterizing many social, physical, biological, and engineering problems. In recent years, many authors have extensively investigated the existence, uniqueness, stability, invariant measures, and other qualitative behaviors of solutions to differential inclusions. Theory of problems concerning stochastic evolution inclusions may be found in several monographs (see, for example, [102, 137,

205]). A generalization of differential inclusions to "stochastic differential inclusions" called multivalued stochastic differential equations are obtained by replacing the term $\sigma(t)$ in the differential inclusions

$$\begin{cases} x'(t) \in Ax(t) + f(t, x(t)) + \sigma(t), & t \in [0, b], \\ x(0) = x_0, \end{cases}$$

by a matrix Σ times the generalized derivative of Brownian motion. In this case it is convenient to write, analogously to stochastic evolution inclusions as follows:

$$\begin{cases} dx(t) \in Ax(t)dt + f(t, x(t))dt + \Sigma(t)dW(t), & t \in [0, b], \\ x(0) = x_0. \end{cases}$$

However, the question of fractional stochastic evolution inclusions with unbounded operators involving Caputo derivative in Hilbert spaces has not been studied extensively. Recently Toufik [225] has studied the existence of solutions of the following fractional evolution inclusions:

$$\begin{cases} {}^C_0D^q_t x(t) \in Ax(t) + f(t, x(t)) + \Sigma(t, x(t))\dfrac{dW(t)}{dt}, & t \in [0, b], \\ x(0) = x_0. \end{cases}$$

$$(5.12)$$

By using fractional calculation, multivalued map and Bohnenblust-Karlin fixed point theorem, an existence result of mild solutions for stochastic system (5.12) is obtained under the assumption of compact semigroup. Further, controllability problem is discussed for system (5.12).

In Chapter 5, we investigate the existence and the topological structure of solution sets of (5.12). The material is due to Zhou and Peng [306].

REFERENCES

[1] R. P. Agarwal and B. Ahmad, Existence of solutions for impulsive anti-periodic boundary value problems of fractional semilinear evolution equations, Dyn. Contin. Discrete Impuls. Syst. Ser. A Math. Anal., 18(2011), 457-470.

[2] S. Agarwal and D. Bahuguna, Existence of solutions to Sobolev-type partial neutral differential equations, J. Appl. Math. Stoch. Anal., 2006(2006), (Art. ID 16308), 10 pages.

[3] R. P. Agarwal, M. Belmekki and M. Benchohra, A survey on semilinear differential equations and inclusions involving Riemann-Liouville fractional derivative, Adv. Differential Equations, 2009(2009), (Art. ID 981728), 47 pages.

[4] R. P. Agarwal, M. Benchohra and S. Hamani, A survey on existence results for boundary value problems of nonlinear fractional differential equations, Acta Appl. Math., 109(2010), 973-1033.

[5] R. P. Agarwal, Y. Zhou and Y. He, Existence of fractional neutral functional differential equations with bounded delay, Comput. Math. Appl., 59(2010), 1095-1100.

[6] R. P. Agarwal, Y. Zhou, J. R. Wang et al., Fractional functional differential equations with causal operators in Banach spaces, Math. Comput. Modelling, 54(2011), 1440-1452.

[7] O. P. Agrawal, J. A. T. Machado and J. Sabatier, Fractional Derivatives and their Application, Nonlinear Dynam., Springer-Verlag, Berlin, (2004).

[8] N. U. Ahmed, Semigroup Theory with Applications to System and Control, Longman Scientific Technical, New York, (1991).

[9] R. R. Akhmerov, M. I. Kamenskii, A. S. Potapov et al., Measures of Noncompactness and Condensing Operators, Birkhäser, Boston, Basel, Berlin, (1992).

[10] M. R. S. Ammi and D. F. M. Torres, Regularity of solutions to higher-order integrals of the calculus of variations, Internat. J. Systems Sci., 39(2008), 889-895.

[11] J. Andres, G. Gabor and L. Górniewicz, Topological structure of solution sets to multi-valued asymptotic problems, Z. Anal. Anwend., 19(2000), 35-60.

[12] J. Andres and M. Pavlačková, Topological structure of solution sets to asymptotic boundary value problems, J. Differential Equations, 248(2010), 127-150.

[13] A. Anguraj, P. Karthikeyan and G. M. N'Guérékata, Nonlocal Cauchy problem for some fractional abstract differential equations in Banach spaces, Commun. Math. Anal., 6(2009), 31-35.

[14] V. V. Anh and N. N. Leonenko, Spectral analysis of fractional kinetic equations with randomdata, J. Stat. Phys., 104(2001), 1349-1387.

[15] L. Arnold, Stochastic Differential Equations: Theory and Applications, John Wiley & Sons, New York, (1974).

[16] N. Aronszajn, Le correspondant topologique de l'unicité dans la théorie des équations différen-tielles, Ann. of Math., 43(1942), 730-738.

[17] C. Atkinson and A. Osseiran, Rational solutions for the time-fractional diffusion equation, SIAM J. Appl. Math., 71(2011), 92-106.

[18] J. P. Aubin and H. Frankowska, Set-Valued Analysis, Boston, Berkhäuser, (1990).

[19] R. Bader and W. Kryszewski, On the solution sets of differential inclusions and the periodic problem in Banach spaces, Nonlinear Anal., 54(2003), 707-754.

[20] Z. Bai and H. Lu, Positive solutions for boundary value problem of nonlinear fractional differential equation, J. Math. Anal. Appl., 311(2005), 495-505.

[21] A. Bakowska and G. Gabor, Topological structure of solution sets to differential problems in Fréchet spaces, Ann. Polon. Math., 95(2009), 17-36.

[22] K. Balachandran and J. P. Dauer, Controllability of functional differential systems of Sobolev type in Banach spaces, Kybernetika, 34(1998), 349-357.

[23] K. Balachandran, S. Kiruthika and J. J. Trujillo, On fractional impulsive equations of Sobolev type with nonlocal condition in Banach spaces, Comput. Math. Appl., 62(2011), 1157-1165.

[24] K. Balachandran and J. Y. Park, Nonlocal Cauchy problem for abstract fractional semilinear evolution equations, Nonlinear Anal., 71(2009), 4471-4475.

[25] K. Balachandran and J. Y. Park, Controllability of fractional integrodifferential systems in Banach spaces, Nonlinear Anal.: TMA, 3(2009), 363-367.

[26] K. Balachandran, Y. Zhou and J. Kokila, Relative controllability of fractional dynamical systems with delays in control, Commun. Nonlinear Sci. Numer. Simul., 17(2012), 3508-3520.

[27] K. Balachandran, Y. Zhou and J. Kokila, Relative controllability of fractional dynamical systems with distributed delays in control, Comput. Math. Appl., 64(2012), 3201-3209.

[28] E. Balder, Necessary and Sufficient Conditions for L_1-strong-weak Lower Semicontinuity of Integral Functional, Nonlinear Anal., 11(1987), 1399-1404.

[29] D. Baleanu and S. I. Muslih, Lagrangian formulation of classical fields within Riemann-Liouville fractional derivatives, Phys. Scripta, 72(2005), 119-121.

[30] D. Baleanu, J. A. Tenreiro Machado and Z. B. Guvenc (Editors), New Trends in Nanotechnology and Fractional Calculus Applications, Springer, (2010).

[31] D. Baleanu, J. A. Tenreiro Machado and A. C. J. Luo, Fractional Dynamics and Control, Springer, New York, (2012).

[32] D. Baleanu and J. J. Trujillo, On exact solutions of a class of fractional Euler-Lagrange equations, Nonlinear Dynam., 52(2008), 331-335.

[33] J. Banaś and K. Goebel, Measure of Noncompactness in Banach Spaces, Marcel Dekker, Inc., New York, (1980).

[34] J. Banasiak and L. Arlotti, Perturbations of Positive Semigroups with Applications, Springer, New York, (2006).

[35] M. Belmekki and M. Benchohra, Existence results for fractional order semilinear functional differential equations with nondense domain, Nonlinear Anal.: TMA, 72(2010), 925-932.

[36] M. Benchohra and S. Hamani, Nonlinear boundary value problems for differential inclusions with Caputo fractional derivative, Topol. Methods Nonlinear Anal., 32(2008), 115-130.

[37] M. Benchohra, J. Henderson, S. K. Ntouyas et al., Existence results for fractional order functional differential equations with infinite delay, J. Math. Anal. Appl., 338(2008), 1340-1350.

[38] M. Benchohra, J. Henderson, S. K. Ntouyas et al., Existence results for fractional functional differential inclusions with infinite delay and application to control theory, Fract. Calc. Appl. Anal., 11(2008), 35-56.

[39] I. Benedetti, V. Taddei and M. Väth, Evolution problems with nonlinear boundary conditions, J. Dyn. Diff. Equat., 25(2013), 477-503.

[40] P. Benevieri and M. Furi, A simple notion of orientability for Fredholm maps of index zero between Banach manifolds and degree theory, Ann. Sci. Math. Québec, 22(1998), 131-148.

[41] D. A. Benson, S. W. Wheatcraft and M. M. Meerschaert, Application of a fractional advection-dispersion equation, Water Resources Res., 36(2000), 1403-1412.

[42] M. N. Berberan-Santos, Relation between the inverse Laplace transforms of $I(t^\beta)$ and $I(t)$: Application to the Mittag-Leffler and asymptotic inverse power law relaxation functions, J. Math. Chem., 38(2005), 265-270.

[43] S. Bochner and A. E. Taylor, Linear functionals on certain spaces of abstractly valued functions, Ann. of Math., 39(1938), 913-944.

[44] D. Bothe, Multivalued perturbation of m-accretive differential inclusions, Israel J. Math., 108(1998), 109-138.

[45] A. Boucherif and R. Precup, Semilinear evolution equations with nonlocal initial conditions, Dynam. Systems Appl., 16(2007), 507-516.

[46] A. Bressan and Z. P. Wang, Classical solutions to differential inclusions with totally disconnected right-hand side, J. Differential Equations, 246(2009), 629-640.

[47] H. Brezis, Analyse Fonctionelle, Théorie et Applications, Masson Editeur, Paris, (1983).

[48] F. E. Browder and C. P. Gupta, Topological degree and nonlinear mappings of analytic type in Banach spaces, J. Math. Anal. Appl., 26(1969), 390-402.

[49] D. Bugajewska and P. Kasprzak, On the existence, uniqueness and topological structure of solution sets to a certain fractional differential equation, Comput. Math. Appl., 59(2009), 1108-1116.

[50] S. Bushnaq, S. Momani and Y. Zhou, A reproducing kernel Hilbert space method for solving integro-differential equations of fractional order, J. Optim. Theory Appl., 156(2013), 96-105.

[51] V. S. Buslaev and C. Sulem, On asymptotic stability of solitary waves for nonlinear Schrödinger equations, Ann. I. H. Poincaré-AN, 20(2003), 419-475.

[52] L. Byszewski, Theorems about existence and uniqueness of solutions of a semi-linear evolution nonlocal Cauchy problem, J. Math. Anal. Appl., 162(1991), 494-505.

[53] L. Byszewski and V. Lakshmikantham, Theorem about the existence and uniqueness of a solution of a nonlocal abstract Cauchy problem in a Banach space, Appl. Anal., 40(1991), 11-19.

[54] O. Carja, M. Necula and I. I. Vrabie, Viability, Invariance and Applications, Elsevier Science Ltd, (2007).

[55] A. N. Carvalho, T. Dlotko and M. J. D. Nascimento, Nonautonomous semilinear evolution equations with almost sectorial operators, J. Evol. Equ., 8(2008), 631-659.

[56] T. Cazenave, Stable solutions of the logarithmic Schrödinger equation, Nonlinear Anal.: TMA, 7(1983), 1127-1140.

[57] T. Cazenave, Semilinear Schrödinger Equations, Courant Lecture Notes in Mathematics, vol. 10, New York University, American Mathematical Society, Providence, RI, (2003).

[58] T. Cazenave and A. Haraux, An Introduction to Semilinear Evolution Equations, Clarendon Press, Oxford, (1998).

[59] T. Cazenave and P.-L. Lions, Orbital stability of standing waves for some nonlinear Schrödinger equations, Comm. Math. Phys., 85(1982), 549-561.

[60] Y.-K Chang, V. Kavitha and M. Mallika Arjunan, Existence and uniqueness of mild solutions to a semilinear integrodifferential equation of fractional order, Nonlinear Anal., 71(2009), 5551-5559.

[61] Y.-K. Chang and J. J. Nieto, Some new existence results for fractional differential inclusions with boundary conditions, Math. Comput. Modelling, 49(2009), 605-609.

[62] A. Chen, F. Chen and S. Deng, On almost automorphic mild solutions for fractional semilinear initial value problems, Computers and Mathematics with Applications, 59(2010), 1318-1325.

[63] F. Chen, X. Luo and Y. Zhou, Existence results for nonlinear fractional difference equation, Adv. Difference Equ., 2011(2011), (Art. ID 713201), 12 pages.

[64] F. Chen, J. J. Neito and Y. Zhou, Global attractivity for nonlinear fractional differential equations, Nonlinear Anal.: RWA, 13(2012), 287-298.

[65] D. H. Chen, R. N. Wang and Y. Zhou, Nonlinear evolution inclusions: Topological characterizations of solution sets and applications, J. Funct. Anal., 265(2013), 2039-2073.

[66] F. Chen and Y. Zhou, Attractivity of fractional functional differential equation, Comput. Math. Appl., 62(2011), 1359-1369.

[67] G. Conti, V. Obukhovskii and P. Zecca, On the topological structure of the solution set for a semilinear functional-differential inclusion in a Banach space, Topology in Nonlinear Analysis, Banach Center Publications, Inst. Math. Polish Acad. Sci., Warszawa, 35(1996), 159-169.

[68] C. Corduneanu, Functional Equations with Causal Operators, Stability and Control: Theory, Methods and Applications, vol. 16, Taylor & Francis, London, (2002).

[69] S. Cuccagna, Stabilization of solutions to nonlinear Schrödinger equations, Comm. Pure Appl. Math., 54(2001), 1110-1145.

[70] M. A. Darwish, J. Henderson and S. K. Ntouyas, Fractional order semilinear mixed type functional differential equations and inclusions, Nonlinear Stud., 16(2009), 197-219.

[71] J. P. Dauer and N. I. Mahmudov, Approximate controllability of semilinear functional equations in Hilbert spaces, J. Math. Anal. Appl., 273(2002), 310-327.

[72] F. S. De Blasi and J. Myjak, On the structure of the set of solutions of the Darboux problem for hyperbolic equations, Proc. Edinb. Math. Soc., 29(1986), 7-14.

[73] F. S. De Blasi and J. Myjak, On the solutions sets for differential inclusions, Bull. Pol. Acad. Sci. Math., 12(1985), 17-23.

[74] A. Debbouchea and D. Baleanu, Controllability of fractional evolution nonlocal impulsive quasilinear delay integro-differential systems, Comput. Math. Appl., 62(2011), 1442-1450.

[75] A. Debbouche and D. F. M. Torres, Approximate controllability of fractional nonlocal delay semilinear systems in Hilbert spaces, Int. J. Control, 86(2013), 949-963.

[76] K. Deimling, Ordinary differential equations in Banach spaces, Lecture Notes Mathematics, no. 596, Springer, Berlin, (1977).

[77] K. Deimling, Nonlinear Functional Analysis, Springer-Verlag, (1985).

[78] K. Diethelm and A. D. Freed, On the solution of nonlinear fractional order differential equations used in the modeling of viscoplasticity, In: F. Keil, W. Mackens, H. Voss and J. Werther (eds.) Scientific Computing in Chemical Engineering II: Computational Fluid Dynamics, Reaction Engineering and Molecular Properties, 217-224. Springer-Heidelberg, (1999).

[79] R. Delaubenfels, Existence Families, Functional Calculi and Evolution Equations, Springer-Verlag, Berlin, (1994).

[80] O. Diekmann, S. A. van Gils, S. M. Lunel et al., Delay Equations: Functional-, Complex-, and Nonlinear Analysis, Springer-Verlag, New York, (1995).

[81] K. Diethelm, The Analysis of Fractional Differential Equations, Lecture Notes in Math., Springer, New York, (2010).

[82] T. Donchev, E. Farkhi and B. S. Mordukhovich, Discrete approximations, relaxation, and optimization of one-sided Lipschitzian differential inclusions in Hilbert spaces, J. Differential Equations, 243(2007), 301-328.

[83] J. Dugundji, An extension of Tietzeąfs theorem, Pacific. J. Math., 1(1951), 353-367.

[84] N. Dunford and J. T. Schwartz, Linear Operators, Wiley, New York, (1988).

[85] R. Eid, S. I. Muslih, D. Baleanu et al., On fractional Schrödinger equation in α-dimensional fractional space, Nonlinear Anal.: RWA, 10(2009), 1299-1304.

[86] S. D. Eidelman and A. N. Kochubei, Cauchy problem for fractional diffusion equations, J. Differential Equations., 199(2004), 211-255.

[87] M. M. El-Borai, Some probability densities and fundamental solutions of fractional evolution equations, Chaos Solitons Fractals, 14(2002), 433-440.

[88] M. M. El-Borai, The fundamental solutions for fractional evolution equations of parabolic type, J. Appl. Math. Stoch. Anal., 3(2004), 197-211.

[89] M. M. El-Borai, Semigroup and some nonlinear fractional differential equations, Appl. Math. Comput., 149(2004), 823-831.

[90] A. El-Sayed and A. Ibrahim, Multivalued fractional differential equations, Appl. Math. Comput., 68(1995), 15-25.

[91] K. J. Engel and R. Nagel, One-parameter Semigroups for Linear Evolution Equations, GTM 194, Springer, Berlin, (2000).

[92] M. M. Fall and T. Weth, Nonexistence results for a class of fractional elliptic boundary value problems, J. Funct. Anal., 263(2012), 2205-2227.

[93] M. Fečkan, J. R. Wang and Y. Zhou, Controllability of fractional functional evolution equations of Sobolev type via characteristic solution operators, J. Optim. Theory Appl., 156(2013), 79-95.

[94] M. Fečkan, J. R. Wang and Y. Zhou, Periodic solutions for nonlinear evolution equations with non-instantaneous impulses, Nonauton. Dyn. Syst., 1(2014), 93-101.

[95] M. Fečkan, Y. Zhou and J. R. Wang, On the concept and existence of solutions for impulsive fractional differential equations, Commun. Nonlinear Sci. Numer. Simul., 17(2012), 3050-3060.

[96] G. Fibich, Singular solutions of the subcritical nonlinear Schrödinger equation, Phys. D, 240(2011), 1119-1122.

[97] A. Floer and A. Weinstein, Nonspreading wave packets for the cubic Schrödinger equation with a bounded potential, J. Funct. Anal., 69(1986), 397-408.

[98] G. F. Franklin, J. D. Powell and A. Emami-Naeini, Feedback Control of Dynamic Systems, Addison-Wesley, (1986).

[99] G. Gabor and A. Grudzka, Structure of the solution set to impulsive functional differential inclusions on the half-line, Nonlinear Differential Equations Appl., 19(2012), 609-627.

[100] G. Gabor and M. Quincampoix, On existence of solutions to differential equations or inclusions remaining in a prescribed closed subset of a finite-dimensional space, J. Differential Equations, 185(2002), 483-512.

[101] L. Gaul, P. Klein and S. Kempfle, Damping description involving fractional operators, Mech. Syst. Signal Pr., 5(1991), 81-88.

[102] L. Gawarecki and V. Mandrekar, Stochastic Differential Equations in Infinite Dimensions with Applications to Stochastic Partial Differential Equations, Springer Berlin Heidelberg, (2011).

[103] W. G. Glockle and T. F. Nonnenmacher, A fractional calculus approach of self-similar protein dynamics, Biophys. J., 68(1995), 46-53.

[104] R. Gorenflo, Y. Luchko and P. P. Zabrejko, On solvability of linear fractional differential equations in Banach spaces, Fract. Calc. Appl. Anal., 2(1999), 163-176.

[105] L. Górniewicz, Topological Fixed Point Theory of Multivalued Mappings, Topol. Fixed Point Theory Appl., 2nd edn., Springer, Dordrecht, 4(2006).

[106] L. Górniewicz and M. Lassonde, Approximation and fixed points for compositions of RęÄ-maps, Topol. Appl., 55(1994), 239-250.

[107] L. Górniewicz and T. Pruszko, On the set of solutions of the Darboux problem for some hyperbolic equations, Bull. Acad. Polon. Math., 28(1980), 279-286.

[108] D. J. Guo, V. Lakshmikantham and X. Z. Liu, Nonlinear Integral Equations in Abstract Spaces, Kluwer Academic, Dordrecht, (1996).

[109] B. L. Guo and Y. P. Wu, Orbital stability of solitary waves for the nonlinear derivative Schrödinger equation, J. Differential Equations, 123(1995), 35-55.

[110] A. Hanyga, Multidimensional solutions of time-fractional diffusion-wave equations, Proc. R. Soc. Lond. Ser. A Math. Phys. Eng. Sci., 458(2002), 933-957.

[111] N. Hayashi, E. I. Kaikina and P. I. Naumkin, Asymptotics for fractional nonlinear heat equations, J. London Math. Soc., 72(2005), 663-688.

[112] H.-P. Heinz, On the behaviour of measure of noncompactness with respect to differentiation and integration of vector-valued functions, Nonlinear Anal.: TMA, 7(1983), 1351-1371.

[113] J. Henderson and A. Ouahab, Fractional functional differential inclusions with finite delay, Nonlinear Anal., 70 (2009), 2091-2105.

[114] J. Henderson and A. Ouahab, Impulsive differential inclusions with fractional order, Comput. Math. Appl., 59(2010), 1191-1226.

[115] D. Henry, Geometric Theory of Semilinear Parabolic Equations, Springer, Berlin, (1981).

[116] H. Henríquez, M. Pierri and P. Táboas, On S-asymptotically ω-periodic functions on Banach spaces and applications, J. Math. Anal. Appl., 343(2008), 1119-1130.

[117] E. Hernández, D. O'Regan and K. Balachandran, On recent developments in the theory of abstract differential equations with fractional derivatives, Nonlinear Anal.: TMA, 73(2010), 3462-3471.

[118] E. Hernández, D. O'Regan and K. Balachandran, Existence results for abstract fractional differential equations with nonlocal conditions via resolvent operators, Indag. Math. (N.S.), 24(2013), 68-82.

[119] R. Hilfer, Applications of Fractional Calculus in Physics, World Scientific, Singapore, (2000).

[120] S. Hu and N. S. Papageorgiou, Handbook of Multivalued Analysis (Theory), Kluwer Academic Publishers, Dordrecht Boston, London, (1997).

[121] S. Hu and N. S. Papageorgiou, On the topological regularity of the solution set of differential inclusions with constraints, J. Differential Equations, 107(1994), 280-289.

[122] L. Hu, Y. Ren and R. Sakthivel, Existence and uniqueness of mild solutions for semilinear integro-differential equations of fractional order with nonlocal initial conditions and delays, Semigroup Forum, 79(2009), 507-514.

[123] S. Iqbal, J. Pecaric and Y. Zhou, Generalization of an inequality for integral transforms with kernel and related results, J. Inequal. Appl., 2010(2010), (Art. ID 948430), 17 pages.

[124] F. Isaia, On a nonlinear integral equation without compactness, Acta Math. Univ. Comenian., LXXV(2006), 233-240.

[125] O. K. Jaradat, A. Al-Omari and S. Momani, Existence of the mild solution for fractional semilinear initial value problems, Nonlinear Anal., 69(2008), 3153-3159.

[126] W. Jiang, The controllability of fractional control systems with control delay, Comput. Math. Appl., 64(2012), 3153-3159.

[127] H. Jiang, F. Liu, I. Turner et al., Analytical solutions for the multi-term timespace Caputo-Riesz fractional advection-diffusion equations on a finite domain, J. Math. Anal. Appl., 389(2012), 1117-1127.

[128] F. Jiao and Y. Zhou, Existence of solutions for a class of fractional boundary value problem via critical point theory, Comput. Math. Appl., 62(2011), 1181-1199.

[129] F. Jiao and Y. Zhou, Existence results for fractional boundary value problem via critical point theory, Int. J. Bifurcat. Chaos, 22(2012), 1-17.

[130] M. I. Kamenskii, P. Nistri, V. V. Obukhovskii et al., Optimal feedback control for a semilinear evolution equation, J. Optim. Theory Appl., 82(1994), 503-517.

[131] M. Kamenskii, V. Obukhovskii and P. Zecca, Condensing Multivalued Maps and Semilinear Differential Inclusions in Banach Spaces, de Gruyter Ser. Nonlinear Anal. Appl., vol. 7, Walter de Gruyter, Berlin, (2001).

[132] L. V. Kantorovich and G. P. Akilov, Functional Analysis, Pergamon Press, Oxford, (1982).

[133] T. D. Ke, V. Obukhovskii, N. C. Wong et al., On a class of fractional order differential inclusions with infinite delays, Appl. Anal., 92(2013), 115-137.

[134] A. A. Kilbas, H. M. Srivastava and J. J. Trujillo, Theory and Applications of Fractional Differential Equations, In: North-Holland Mathematics Studies, vol. 204, Elsevier, Amsterdam, (2006).

[135] M. Kirane, Y. Laskri and N.-e. Tatar, Critical exponents of Fujita type for certain evolution e-quations and systems with spatiotemporal fractional derivatives, J. Math. Anal. Appl., 312(2005), 488-501.

[136] V. Kiryakova, Generalized Fractional Calculus and Applications, Pitman Research Notes in Mathematics, vol. 301, Longman, (1994).

[137] M. Kisielewicz, Stochastic Differential Inclusions and Applications, New York: Springer, (2013).

[138] J. Klafter, S. C. Lim and R. Metzler (Editors), Fractional Dynamics in Physics, World Scientific, Singapore, (2011).

[139] V. Kolmanovskii and A. Myshkis, Introduction to the Theory and Applications of Functional Differential Equations, Kluwer Academic, Dordrecht, (1999).

[140] M. Krasnoschok and N. Vasylyeva, On a nonclassical fractional boundary-value problem for the Laplace operator, J. Differential Equations, 257(2014), 1814-1839.

[141] M. G. Kreĭn and M. A. Rutman, Linear operators leaving invariant a cone in a Banach space, Amer. Math. Soc. Transl. Ser., 10(1962), 199-325.

[142] S. Kumar and N. Sukavanam, Approximate controllability of fractional order semilinear systems with bounded delay, J. Differential Equations, 252(2012), 6163-6174.

[143] V. Lakshmikantham and S. Leela, Nonlinear Differential Equations in Abstract Spaces, Pergamon Press, New York, (1969).

[144] V. Lakshmikantham, S. Leela and J. V. Devi, Theory of Fractional Dynamic Systems, Cambridge Scientific Publishers, Cambridge, (2009).

[145] V. Lakshmikantham, S. Leela, Z. Drici et al., Theory of Causal Differential Equations, In: Chui, C.K. (ed.) Atlantis Studies in Mathematics for Engineering and Science, vol. 5, World Scientific, Singapore, (2010).

[146] V. Lakshmikantham and A. S. Vatsala, Basic theory of fractional differential equations, Nonlinear Anal.: TMA, 69(2008), 2677-2682.

[147] Y.-H. Lan, H. X. Huang and Y. Zhou, Observer-based robust control of α $(1 < \alpha < 2)$ fractional-order uncertain systems: an LMI approach, IET Control Theory Appl., 6(2012), 229-234.

[148] Y.-H. Lan and Y. Zhou, LMI-based robust control of fractional-order uncertain linear systems, Comput. Math. Appl., 62(2011), 1460-1471.

[149] Y.-H. Lan and Y. Zhou, High-order D^{α}-type iterative learning control for fractional-order nonlinear time-delay systems, J. Optim. Theory Appl., 156(2013), 153-166.

[150] Y.-H. Lan and Y. Zhou, D^{α}-type iterative learning control for fractional-order linear time-delay systems, Asian J. Control, 15(2013), 669-677.

[151] Y. Li, The positive solutions of abstract semilinear evolution equations and their applications, Acta Anal. Funct. Appl., 39(1996), 666-672.

[152] Y. Li, Existence and uniqueness of periodic solution for a class of semilinear evolution equations, J. Math. Anal. Appl., 349(2009), 226-234.

[153] F. Li, J. Liang and H. K. Xu, Existence of mild solutions for fractional integrodifferential equations of Sobolev type with nonlocal conditions, J. Math. Anal. Appl., 391(2012), 510-525.

[154] C. F. Li, X. N. Luo and Y. Zhou, Existence of positive solutions of boundary value problem for fractional differential equations, Comput. Math. Appl., 59(2010), 1363-1375.

[155] K. Li, J. Peng and J. Jia, Cauchy problems for fractional differential equations with Riemann-Liouville fractional derivatives, J. Funct. Anal., 263(2012), 476-510.

[156] X. Li and J. Yong, Optimal Control Theory for Infinite Dimensional Systems, Birkhäuser, Boston, (1995).

[157] J. Liang and T. J. Xiao, Abstract degenerate Cauchy problems in locally convex spaces, J. Math. Anal. Appl., 259(2001), 398-412.

[158] J. H. Lightbourne and S. M. Rankin, A partial functional differential equation of Sobolev type, J. Math. Anal. Appl., 93(1983), 328-337.

[159] W. Lin, Global existence theory and chaos control of fractional differential equations, J. Math. Anal. Appl., 332(2007), 709-726.

[160] L. Liu, F. Guo, C. Wu et al., Existence theorems of global solutions for nonlinear Volterra type integral equations in Banach spaces, J. Math. Anal. Appl., 309(2005), 638-649.

[161] Z. Liu and X. Li, On the controllability of impulsive fractional evolution inclusions in Banach spaces, J. Optim. Theory Appl., 156(2013), 167-182.

[162] C. Lizama and G. M. N'Guérékata, Bounded mild solutions for semilinear integro differential equations in Banach spaces, Integr. Equ. Oper. Theory, 68(2010), 207-227.

[163] C. Lizama and F. Poblete, Regularity of mild solutions for a class of fractional order differential equations, Appl. Math. Comput., 224(2013), 803-816.

[164] C. Lizama and R. Ponce, Bounded solutions to a class of semilinear integro–differential equations in Banach spaces, Nonlinear Anal., (2011), 3397-3406.

[165] Y. Luchko, Boundary value problems for the generalized time-fractional diffusion equation of distributed order, Fract. Calc. Appl. Anal., 12(2009) 409-422.

[166] Y. Luchko, Initial-boundary-value problems for the generalized multi-term time-fractional diffusion equation, J. Math. Anal. Appl., 374(2011), 538-548.

[167] A. C. J. Luo and V. S. Afraimovich (Editors), Long-range Interaction, Stochasticity and Fractional Dynamics, Higher Education Press and Springer, (2010).

[168] J. A. T. Machado, V. Kiryakova and F. Mainardi, Recent history of fractional calculus, Commun. Nonlinear Sci. Numer. Simul., 16(2011), 1140-1153.

[169] R. L. Magin, Fractional calculus models of complex dynamics in biological tissues, Comput. Math. Appl., 59(2010), 1586-1593.

[170] N. I. Mahmudov, Approximate controllability of semilinear deterministic and stochastic evolution equations in abstract spaces, SIAM J. Control Optim., 42(2003), 1604-1622.

[171] N. I. Mahmudov, Approximate controllability of fractional Sobolev-type evolution equations in Banach spaces, Abst. Appl. Anal., 2013(2013), (Art ID 502839), 9 pages.

[172] N. I. Mahmudov and A. Denker, On controllability of linear stochastic systems, Internat. J. Control, 73(2000), 144-151.

[173] N. I. Mahmudov and S. Zorlu, Approximate controllability of fractional integro-differential equations involving nonlocal initial conditions, Bound. Value Probl., 50(2013), 1-13.

[174] N. I. Mahmudov and S. Zorlu, On the approximate controllability of fractional evolution equations with compact analytic semigroup, J. Comput. Appl. Math., 259(2014), 194-204.

[175] F. Mainardi, P. Paraddisi and R. Gorenflo, Probability Distributions Generated by Fractional Diffusion Equations, J. Kertesz, I. Kondor(Eds.), Econophysics: An Emerging Science, Kluwer, Dordrecht, (2000).

[176] A. B. Malinowska and D. F. M. Torres, Introduction to the Fractional Calculus of Variations, Imp. Coll. Press Adv. Phys. Texts, London, (2012).

[177] H. Markus, The Functional Calculus for Sectorial Operators, Oper. Theory Adv. Appl., Birkhauser-Verlag, Basel, 169(2006), 19-60.

[178] M. A. Mckibben, Discovering Evolution Equations with Applications, vol. I, Chapman & Hall Mathematics, Boca Raton, (2011).

[179] M. M. Meerschaert, D. A. Benson, H. Scheffler et al., Stochastic solution of space-time fractional diffusion equations, Phys. Rev. E, 65(2002), 1103-1106.

[180] M. M. Meerschaert, E. Nane and H. P. Scheffler, Stochastic model for ultraslow diffusion, Stochastic Process. Appl., 116(2006), 1215-1235.

[181] A. I. Mees, Dynamics of feedback systems, Wiley, New York, (1981).

[182] R. Metzler and J. Klafter, The Random Walk's Guide to Anomalous Diffusion: A Fractional Dynamics Approach, Phys. Rep., 339(2000), 1-77.

[183] F. Metzler, W. Schick, H. G. Kilian et al., Relaxation in filled polymers: a fractional calculus approach, J. Chem. Phys., 103(1995), 7180-7186.

[184] K. S. Miller and B. Ross, An Introduction to the Fractional Calculus and Differential Equations, John Wiley, New York, (1993).

[185] H. Mönch, Boundary value problems for nonlinear ordinary differential equations of second order in Banach spaces, Nonlinear Anal.: TMA, 4(1980), 985-999.

[186] G. M. Mophou and G. M. N'Guérékata, Mild solutions for semilinear fractional differential equations, Electron. J. Differential Equations, 21(2009), 1-9.

[187] G. M. Mophou and G. M. N'Guérékata, Existence of mild solution for some fractional differential equations with nonlocal conditions, Semigroup Forum, 79(2009), 315-322.

[188] G. M. Mophou and G. M. N'Guérékata, Existence of mild solutions of some semilinear neutral fractional functional evolution equations with infinite delay, Appl. Math. Comput., 216(2010), 61-69.

[189] J. Mu and H. Fan, Positive mild solutions of periodic boundary value problems for fractional evolution equations, J. Appl. Math., 2012 (2012), Article ID 691651, 13 pages.

[190] J. Mu, Y. Zhou and S.B. Huang, Bounded solutions of fractional evolution equations in Banach spaces, to appear.

[191] K. Naito, On controllability for a nonlinear Volterra equation, Nonlinear Anal., 18(1992), 99-108.

[192] K. Naito and J.Y. Park, Approximate controllability for trajectories of a delay Volterra control system, J. Optim. Theory Appl., 61(1989), 271-279.

[193] G. M. N'Guérékata, A Cauchy problem for some fractional abstract differential equation with nonlocal conditions, Nonlinear Anal.: TMA, 70(2009), 1873-1876.

[194] N. Nyamoradi and Y. Zhou, Infinitely many solutions for a perturbed nonlinear fractional boundary value problems depending on two parameters, Eur. Phys. J. Special Topics, 222(2013), 1997-2011.

[195] V. Obukhovskii and P. Rubbioni, On a controllability problem for systems governed by semilinear functional differential inclusions in Banach spaces, Topol. Methods Nonlinear Anal., 15(2000), 141-151.

[196] D. O'Regan, Fixed point theorems for weakly sequentially closed maps, Arch. Math., 36(2000), 61-70.

[197] N. Özdemir, D. Karadeniz and B. B. İskender, Fractional optimal control problem of a distributed system in cylindrical coordinates, Phys. Lett. A, 373(2009), 221-226.

[198] D. N. Pandey, A. Ujlayan and D. Bahuguna, On a solution to fractional order integrodifferential equations with analytic semigroups, Nonlinear Anal.: TMA, 71(2009), 3690-3698.

[199] A. Pazy, Semigroups of Linear Operators and Applications to Partial Differential Equations, Appl. Math. Sci., vol. 44, Springer, New York, (1983).

[200] F. Periago and B. Straub, A functional calculus for almost sectorial operators and applications to abstract evolution equations, J. Evol. Equ., 2(2002), 41-68.

[201] B. J. Pettis, On the integration in vector spaces, Trans. Amer. Math. Soc., 44(1938), 277-304.

[202] I. Podlubny, Fractional Differential Equations, Academic Press, San Diego, (1999).

[203] R. Ponce, Bounded mild solutions to fractional integro-differential equations in Banach spaces, Semigroup Forum, 87(2013), 377-392.

[204] R. Ponce, Hölder continuous solutions for fractional differential equations and maximal regularity, J. Differential Equations, 255(2013), 3284-3304.

[205] G. D. Prato and J. Zabczyk, Stochastic Equations in Infinite Dimensions, Cambridge: Cambridge University Press, (1992).

[206] M. H. M. Rashid and Y. El-Qaderi, Semilinear fractional integro-differential equations with compact semigroup, Nonlinear Anal.: TMA, 71(2009), 6276-6282.

[207] K. Rykaczewski, Approximate controllability of differential inclusions in Hilbert spaces, Nonlinear Anal., 75 (2012), 2701-2712.

[208] J. Sabatier, O. P. Agrawal and J. A. T. Machado, Advances in Fractional Calculus, Theoretical Developments and Applications in Physics and Engineering, Springer-Verlag, Berlin, (2007).

[209] R. Sakthivel and Y. Ren, Approximate controllability of fractional differential equations with state-dependent delay, Result Math., 63(2013), 949-963.

[210] R. Sakthivel, Y. Ren and N. I. Mahmudov, On the approximate controllability of semilinear fractional differential systems, Comput. Math. Appl., 62(2011), 1451-1459.

[211] R. Sakthivela, N. I. Mahmudov and J. J. Nieto, Controllability for a class of fractional-order neutral evolution control systems, Appl. Math. Comput., 218(2012), 10334-10340.

[212] R. Sakthivela, S. Suganya and S. M. Anthoni, Approximate controllability of fractional stochastic evolution equations, Comput. Math. Appl., 63(2012), 660-668.

[213] S. G. Samko, A. A. Kilbas and O. I. Marichev, Fractional Integral and Derivatives: Theory and Applications, Gordon and Breach Science Publishers, Longhorne, PA, (1993).

[214] R. S. Schatten, Norm Ideals of Continuous Operators, Springer, New York, (1970).

[215] W. R. Schneider and W. Wayes, Fractional diffusion and wave equation, J. Math. phys., 30(1989), 134-144.

[216] T. I. Seidman, Invariance of the reachable set under nonlinear perturbations, SIAM J. Control Optim., 25(1987), 1173-1191.

[217] X. B. Shu, Y. Z. Lai and Y. M. Chen, The existence of mild solutions for impulsive fractional partial differential equations, Nonlinear Anal.: TMA, 74(2011), 2003-2011.

[218] V. Staicu, On the solution sets to nonconvex differential inclusions of evolution type, Discrete Contin. Dyn. Syst., 2(1998), 244-252.

[219] V. Staicu, On the solution sets to differential inclusions on unbounded interval, Proc. Edinb. Math. Soc., 43(2000), 475-484.

[220] P.-L. Sulem and C. Sulem, The Nonlinear Schrödinger Equation: Self-focusing and Wave Collapse, Hardback, (1999).

[221] V. E. Tarasov, Fractional Dynamics: Application of Fractional Calculus to Dynamics of Particles, Fields and Media, Springer, (2010).

[222] A. Z. -A. M. Tazali, Local existence theorems for ordinary differential equations of fractional order, Ordinary and Partial Differential Equations, Lecture Notes in Math., Springer, Dundee, 964(1982), 652-665.

[223] H. R. Thieme, Semiflows generated by Lipschitz perturbations of non-densely defined operators, Differential Integral Equations, 3(1990), 1001-1229.

[224] Y. Tian and Y. Zhou, Positive solutions for multipoint boundary value problem of fractional differential equations, J. Appl. Math. Comput., 38(2012), 417-427.

[225] G. Toufik, Existence and controllability results for fractional stochastic semilinear differential inclusions differential equations and dynamical systems, Differ. Equ. Dynam. Syst., 23(2015), 225-240.

[226] R. Triggiani, On the stabilizability problem in Banach space, J. Math. Anal. Appl., 52(1975), 383-403.

[227] R. Triggiani, Addendum: a note on the lack of exact controllability for mild solutions in Banach spaces, SIAM J. Control Optim., 18(1980), 98-99.

[228] T.-P. Tsai, Asymptotic dynamics of nonlinear Schrödinger equations with many bound states, J. Differential Equations, 192(1995), 225-282.

[229] M. Vath, Topological Analysis: from the Basics to the Triple Degree for Nonlinear Fredholm Inclusions, de Gruyter, Berlin, (2012).

[230] I. I. Vrabie, Compactness Methods for Nonlinear Evolutions, Pitman Monographs and Surveys in Pure and Applied Mathematics, second edition, Longman and John Wiley & Sons, 75(1995).

[231] Y. Wang, Global existence and blow up of solutions for the inhomogeneous nonlinear Schrödinger equation in $\mathbb{R}^2$, J. Math. Anal. Appl., 338(2008), 1008-1019.

[232] J. R. Wang, J. H. Deng and Y. Zhou, A new method in the study of impulsive fractional differential equation in Banach spaces, Fixed Point Theory, 15(2014), 285-296.

[233] J. R. Wang, X. W. Dong and Y. Zhou, Analysis of nonlinear integral equations with Erdelyi-Kober fractional operator, Commun. Nonlinear. Sci. Numer. Simul., 17(2012), 3129-3139.

[234] J. R. Wang, X. W. Dong and Y. Zhou, Existence, attractive and stability of solutions for quadratic urysohon fractional integral equations, Commun. Nonlinear Sci. Numer. Simul., 17(2012), 545-554.

[235] J. R. Wang, Z. Fan and Y. Zhou, Nonlocal controllability of semilinear dynamic systems with fractional derivative in Banach spaces, J. Optim. Theory Appl., 154(2012), 292-302.

[236] J. R. Wang, M. Fečkan and Y. Zhou, On the new concept of solutions and existence results for impulsive fractional evolution equations, Dyn. Partial Differ. Equ., 8(2011), 345-361.

[237] J. R. Wang, M. Fečkan and Y. Zhou, Ulam's type stability of impulsive ordinary differential equations, J. Math. Anal. Appl., 395(2012), 258-264.

[238] J. R. Wang, M. Fečkan and Y. Zhou, Nonexistence of periodic solutions and asymptotically periodic solutions for fractional differential equations, Commun. Nonlinear Sci. Numer. Simul., 18(2013), 246-256.

[239] J. R. Wang, M. Fečkan and Y. Zhou, Relaxed controls for nonlinear fractional impulsive evolution equations, J. Optim. Theory Appl., 156(2013), 13-32.

[240] J. R. Wang, M. Fečkan and Y. Zhou, Fractional order iterative functional differential equations with parameter, Appl. Math. Modelling, 37(2013), 6055-6067.

[241] J. R. Wang, M. Fečkan and Y. Zhou, Presentation of solutions of impulsive fractional Langevin equations and existence results, Eur. Phys. J. Special Topics, 222(2013), 1855-1872.

[242] J. R. Wang, M. Fečkan and Y. Zhou, Controllability of Sobolev type fractional evolution systems, Dyn. Partial Differ. Equ., 11(2014), 71-87.

[243] J. R. Wang, M. Fečkan and Y. Zhou, Weakly Picard operators method for modified fractional iterative functional differential equations, Fixed Point Theory, 15(2014), 297-310.

[244] J. R. Wang, M. Fečkan and Y. Zhou, On the stability of first order impulsive evolution equations, Opuscula Math., 34(2014), 639-657.

[245] J. R. Wang, Michal Fečkan and Y. Zhou, Approximate controllability of Sobolev type fractional evolution systems with nonlocal conditions, J. Optim. Theory Appl., to appear.

[246] J. R. Wang, L. Lv and Y. Zhou, Boundary value problems for fractional differential equations involving Caputo derivative in Banach spaces, J. Appl. Math. Comput., 38(2012), 209-224.

[247] J. R. Wang, L. Lv and Y. Zhou, New Concepts and results in stability of fractional differential equations, Commun. Nonlinear Sci. Numer. Simul., 17(2012), 2530-2538.

[248] J. R. Wang, W. Wei and Y. Yang, Fractional nonlocal integrodifferential equations of mixed type with time-varying generating operators and optimal control, Opuscula Math., 30(2010), 217-234.

[249] J. R. Wang, W. Wei and Y. Zhou, Fractional finite time delay evolution systems and optimal controls in infinite dimensional spaces, J. Dyn. Control Syst., 17(2011), 515-535.

[250] J. R. Wang and Y. Zhou, A class of fractional evolution equations and optimal controls, Nonlinear Anal.: RWA, 12(2011), 262-272.

[251] J. R. Wang and Y. Zhou, Analysis of nonlinear fractional control systems in Banach spaces, Nonlinear Anal.: TMA, 74(2011), 5929-5942.

[252] J. R. Wang and Y. Zhou, Time optimal controls problem of a class of fractional distributed system, Int. J. Dyn. Syst. Differ. Equs., 3(2011), 363-382.

[253] J. R. Wang and Y. Zhou, Existence of mild solutions for fractional delay evolution systems, Appl. Math. Comput., 218(2011), 357-367.

[254] J. R. Wang and Y. Zhou, Existence and controllability results for fractional semilinear differential inclusions, Nonlinear Anal.: RWA, 12(2011), 3642-3653.

[255] J. R. Wang and Y. Zhou, Mittag-Leffler-Ulam stabilities of fractional evolution equations, Appl. Math. Lett., 25(2012), 723-728.

[256] J. R. Wang and Y. Zhou, Complete controllability of fractional evolution systems, Commun. Nonlinear Sci. Numer. Simul., 17(2012), 4346-4355.

[257] J. R. Wang, Y. Zhou and M. Fečkan, On recent developments in the theory of boundary value problems for impulsive fractional differential equations, Comput. Math. Appl., 64(2012), 3008-3020.

[258] J. R. Wang, Y. Zhou and M. Fečkan, Nonlinear impulsive problems for fractional differential equations and Ulam stability, Comput. Math. Appl., 64(2012), 3389-3405.

[259] J. R. Wang, Y. Zhou and M. Fečkan, Abstract Cauchy problem for fractional differential equations, Nonlinear Dynam., 71(2013), 685-700.

[260] J. R. Wang, Y. Zhou and M. Fečkan, On the nonlocal Cauchy problem for semilinear fractional order evolution equations, Cent. Eur. J. Math., 12(2014), 911-922.

[261] J. R. Wang, Y. Zhou and Z. Lin, On a new class of impulsive fractional differential equations, Appl. Math. Comput., 242(2014), 649-657.

[262] J. R. Wang, Y. Zhou and M. Medved, On the solvability and optimal controls of fractional integrodifferential evolution systems with infinite delay, J. Optim. Theory Appl., 152(2012), 31-50.

[263] J. R. Wang, Y. Zhou and M. Medved, Picard and weakly Picard operators technique for nonlinear differential equations in Banach spaces, J. Math. Anal. Appl., 389(2012), 261-274.

[264] J. R. Wang, Y. Zhou and M. Medved, Qualitative analysis for nonlinear fractional differential equations via topological degree method, Topol. Methods Nonlinear Anal., 40(2012), 245-271.

[265] J. R. Wang, Y. Zhou and M. Medved, Existence and stability of fractional differential equations with Hadamard derivative, Topol. Methods Nonlinear Anal., 41(2013), 113-133.

[266] J. R. Wang, Y. Yang and W. Wei, Nonlocal impulsive problems for fractional differential equations with time-varying generating operators in Banach spaces, Opuscula Math., 30(2010), 361-381.

[267] J. R. Wang, Y. Zhou and W. Wei, A class of fractional delay nonlinear integrodifferential controlled systems in Banach spaces, Commun. Nonlinear Sci. Numer. Simul., 16(2011), 4049-4059.

[268] J. R. Wang, Y. Zhou and W. Wei, Impulsive problems fractional evolution equations and optimal controls in infinite dimensional spaces, Topol. Methods Nonlinear Anal., 38(2011), 17-43.

[269] J. R. Wang, Y. Zhou and W. Wei, Study in fractional differential equations by means of topological degree method, Numer. Funct. Anal. Optim., 33(2012), 216-238.

[270] J. R. Wang, Y. Zhou and W. Wei, Optimal feedback control for semilinear fractional evolution equations in Banach spaces, Systems Control Lett., 61(2012), 472-476.

[271] J. R. Wang, Y. Zhou and W. Wei, Stabilization of solutions to nonlinear impulsive evolution equations, Kybernetika, 48(2012), 1211-1228.

[272] J. R. Wang, Y. Zhou and W. Wei, Fractional Schrödinger equations with potential and optimal controls, Nonlinear Anal.: RWA, 13(2012), 2755-2766.

[273] J. R. Wang, Y. Zhou and W. Wei, Cauchy problems for fractional differential equations via Picard and weakly Picard operators technique, Fixed Point Theory, 14(2013), 219-234.

[274] J. R. Wang, Y. Zhou, W. Wei et al., Nonlocal problems for fractional integrodifferential equations via fractional operators and optimal controls, Comput. Math. Appl., 62(2011), 1427-1441.

[275] R. N. Wang, D. Chen and T. J. Xiao, Abstract fractional Cauchy problems with almost sectorial operators, J. Differential Equations, 252(2012), 202-235.

[276] R. N. Wang, Q. H. Ma and Y. Zhou, Topological Theory of Non-autonomous Parabolic Evolution Inclusions on a Noncompact Interval and Applications, Math. Ann., 362(2015), 173-203.

[277] R. N. Wang, Q. M. Xiang and Y. Zhou, Fractional delay control problems: topological structure of solution sets and its applications, Optimization, 63(2014), 1249-1266.

[278] R. N. Wang and Y. H. Yang, On the Cauchy problems of fractional evolution equations with nonlocal initial conditions, Results Math., 63(2013), 15-30.

[279] R. N. Wang and P. X. Zhu, Non-autonomous evolution inclusions with nonlocal history conditions: global integral solutions, Nonlinear Anal., 85(2013), 180-191.

[280] R. N. Wang, P. X. Zhu and Q. H. Ma, Multi-valued nonlinear perturbations of time fractional evolution equations in Banach spaces, Nonlinear Dynam., 80(2015), 1745-1759.

[281] W. Wei and X. Xiang, Optimal feedback control for a class of nonlinear impulsive evolution equations, Chinese J. Engineering Math., 23(2006), 333-342.

[282] X. Xiang and H. Kuang, Delay systems and optimal controls, Acta Math. Appl. Sin., 16(2000), 27-35.

[283] H. Ye, J. Gao and Y. Ding, A generalized Gronwall inequality and its application to a fractional differential equation, J. Math. Anal. Appl., 328(2007), 1075-1081.

[284] G. Zaslavsky, Fractional kinetic equation for hamiltonian chaos, chaotic advection, tracer dynamics and turbulent dispersion, Phys. D, 76(1994), 110-122.

[285] G. Zaslavsky, Chaos, Fractional Kinetics, and Anomalous Transport, Phys. Rep., 371(2002), 461-580.

[286] E. Zeidler, Nonlinear Functional Analysis and Its Application II/A, New York, Springer-Verlag, (1990).

[287] S. Zhang, Existence of solution for a boundary value problem of fractional order, Acta Math. Sci., 26(2006), 220-228.

[288] S. Zhang, Positive solutions to singular boundary value problem for nonlinear fractional differential equation, Comput. Math. Appl., 59(2010), 1300-1309.

[289] L. Zhang and Y. Zhou, Fractional Cauchy problems with almost sectorial operators, Appl. Math. Comput., 257(2015), 145-157.

[290] Y. Zhou, Existence and uniqueness of fractional functional differential equations with unbounded delay, Int. J. Dyn. Syst. Diff. Equs., 1(2008), 239-244.

[291] Y. Zhou, Existence and uniqueness of solutions for a system of fractional differential equations, Fract. Calc. Appl. Anal., 12(2009), 195-204.

[292] Y. Zhou, Basic Theory of Fractional Differential Equations, World Scientific, Singapore, (2014).

[293] Y. Zhou (Editor), Advances in Fractional Differential Equations, Comput. Math. Appl., 59(2010), 1047-1376.

[294] Y. Zhou (Editor), Advances in Fractional Differential Equations (II), Comput. Math. Appl., 62(2011), 821-1618.

[295] Y. Zhou (Editor), Advances in Fractional Differential Equations (III), Comput. Math. Appl., 64(2012), 2965-3484.

[296] Y. Zhou (Editor), Control and Optimization of Fractional Systems, J. Optim. Theory Appl., 156(2013), 1-182.

[297] Y. Zhou (Editor), Recent Advances in Fractional Differential Equations, Appl. Math. Comput., 257(2015), 1-602.

[298] Y. Zhou and E. Casaa (Editors), Optimal Control of Fractional Order Systems, Optimization, 63(2014), 1153-1279.

[299] Y. Zhou, H. B. Gu, L. Peng et al., Topological structure for fractional evolution inclusions, to appear.

[300] Y. Zhou and F. Jiao, Existence of extremal solutions for discontinuous fractional functional differential equations, Int. J. Dyn. Syst. Diff. Equs., 2(2009), 237-252.

[301] Y. Zhou and F. Jiao, Nonlocal Cauchy problem for fractional evolution equations, Nonlinear Anal.: RWA, 11(2010), 4465-4475.

[302] Y. Zhou and F. Jiao, Existence of mild solutions for fractional neutral evolution equations, Comput. Math. Appl., 59(2010), 1063-1077.

[303] Y. Zhou, F. Jiao and J. Li, Existence and uniqueness for p-type fractional neutral differential equations, Nonlinear Anal.: TMA, 71(2009), 2724-2733.

[304] Y. Zhou, F. Jiao and J. Li, Existence and uniqueness for fractional neutral differential equations with infinite delay, Nonlinear Anal.: TMA, 71(2009), 3249-3256.

[305] Y. Zhou, F. Jiao and J. Pecaric, On the Cauchy problem for fractional functional differential equations in Banach spaces, Topol. Methods Nonlinear Anal., 42(2013), 119-136.

[306] Y. Zhou and L. Peng, Topological structure for fractional stochastic evolution inclusions, to appear.

[307] Y. Zhou, X. H. Shen and L. Zhang, Cauchy problem for fractional evolution equations with Caputo derivative, Eur. Phys. J. Special Topics, 222(2013), 1747-1764.

[308] Y. Zhou, C. Ionescu, J. A. T. Machado (Editors), Fractional Dynamics and Its Applications, Nonlinear Dynam., 80(2015), 1661-1897.

[309] Y. Zhou, V. E. Tarasov, J. J. Trujillo et al. (Editors), Dynamics of Fractional Partial Differential Equations, Eur. Phys. J. Special Topics, 222(2013), 1743-2011.

[310] Y. Zhou, L. Zhang and X. H. Shen, Existence of mild solutions for fractional evolutions, J. Integral Equ. Appl., 25(2013), 557-586.

[311] Q. J. Zhu, On the solution set of differential inclusions in Banach space, J. Differential Equations, 93(1991), 213-237.

[312] L. P. Zhu and G. Li, Nonlocal differential equations with multivalued perturbations in Banach spaces, Nonlinear Anal., 69(2008), 2843-2850.

Printed in the United States
By Bookmasters